G.S. Isserson and the
War of the Future

G.S. Isserson and the War of the Future

Key Writings of a Soviet Military Theorist

G.S. Isserson

Translated and edited by Richard W. Harrison

Foreword by Lieutenant General Paul K. Van Riper, USMC (Ret.)

McFarland & Company, Inc., Publishers
Jefferson, North Carolina

LIBRARY OF CONGRESS CATALOGUING-IN-PUBLICATION DATA

Names: Isserson, G. S., author. | Harrison, Richard W., 1952– translator.
Title: G.S. Isserson and the war of the future : key writings of a Soviet military theorist / G.S. Isserson ; translated and edited by Richard W. Harrison ; foreword by Lieutenant General Paul K. Van Riper, USMC (Ret.).
Description: Jefferson, North Carolina : McFarland & Company, Inc., Publishers, 2016 | Includes bibliographical references and index.
Identifiers: LCCN 2016009382 | ISBN 9781476662367 (softcover : acid free paper) ∞
Subjects: LCSH: Operational art (Military science) | Isserson, G. S. | Defensive (Military science) | Military art and science—Soviet Union.
Classification: LCC U162 .I843 2016 | DDC 355.0201—dc23
LC record available at http://lccn.loc.gov/2016009382

BRITISH LIBRARY CATALOGUING DATA ARE AVAILABLE

ISBN (print) 978-1-4766-6236-7
ISBN (ebook) 978-1-4766-2390-0

© 2016 Richard W. Harrison. All rights reserved

No part of this book may be reproduced or transmitted in any form or by any means, electronic or mechanical, including photocopying or recording, or by any information storage and retrieval system, without permission in writing from the publisher.

On the cover: Brigade Commander Isserson, ca. 1936–39. This picture may have been taken while he was teaching at the General Staff Academy (courtesy Russian State Military Archives).

Printed in the United States of America

McFarland & Company, Inc., Publishers
Box 611, Jefferson, North Carolina 28640
www.mcfarlandpub.com

Table of Contents

Foreword by Lieutenant General Paul K. Van Riper, USMC (Ret.) 1
Preface by Richard W. Harrison 6
About the Translation 9
A Short Biography of G.S. Isserson 10

One—*The Evolution of Operational Art* (1932) 15

Two—"The Fundamentals of the Deep Operation" (1933) 73

Three—"The Fundamentals of the Defensive Operation" (1938) 155

Four—"The Fundamentals of Conducting Operations" (1939) 211

Five—*The New Forms of Struggle* (1940) 235

Six—"The Development of the Theory of Soviet Operational Art in the 1930s" (1965) 285

Chapter Notes 309
Index 321

Foreword by Lieutenant General Paul K. Van Riper, USMC (Ret.)

It is my sincere hope that Richard Harrison's *G.S. Isserson and the War of the Future* will help motivate members of the American defense community to end the prolonged cycle of interest followed by disinterest in Russian military affairs. This unwelcome cycle has hindered the U.S. military's ability to learn from past Russian and Soviet efforts to develop a theory and practice of operational art. It has also hindered the potential of U.S. military leaders to anticipate undesirable Russian military actions. Scores of defense professionals will need to read and study Harrison's important work if I am to realize this hope. May the following words encourage them to do just that.

Over the past 240 years America's military interest in Russia has tended to wax and wane with its overall state of attentiveness to Russian affairs. American patriots grew keenly aware of the Russian Empire during the American Revolutionary War, as Catherine the Great chose to carry on trade with the colonists while officially remaining neutral. After the Revolution, American public interest in Russia diminished, only to intensify during the Civil War when Russia offered moral support to the Union and with the purchase of Alaska. Thereafter it declined but rose once more in 1900 as the two nations joined the Eight-Nation Alliance fighting to stem the spread of the Boxer Rebellion in China, and remained elevated as the two nations joined the intense competition for trade in Manchuria. It heightened again in 1905 when the United States mediated the Treaty of Portsmouth that ended the Russo-Japanese War. Over the following years Americans grew less conscious of Russia—that is, until the nation sent its soldiers and Marines to assist the White Army of the anti–Bolshevik forces during the Russian Civil War. After the Bolshevik victory and establishment of the Union of Soviet Socialist Republics in 1922, American military concern with Russia—the heart of the matter—dimmed again for nearly two decades. Then in 1941 the United States joined with the USSR and the other Allies to fight the Axis powers in World War II. For the next half-century American concern with the military forces of the USSR remained intense. After the demise of the USSR in 1991, the American military gave little thought to the new Russian Federation's military except to monitor its activities in Chechnya and its small contribution to the Bosnian peacekeeping force.

Thus, not surprisingly, when Harrison's *The Russian Way of War: Operational Art, 1904–1940* appeared in 2001, establishing his reputation as an authority on the imperial Russian army and the successor Soviet army, hardly any U.S. military professionals considered Russia a threat to international security. These professionals had shifted their focus from the Soviet menace and a parallel interest in operational art to what the 2001 *National Security Strategy of the United States of America* identified as "contemporary threats such as the proliferation of nuclear, chemical and biological weapons, terrorism, and international crime."[1] Professional discussions revolved around "global issues" and "operations other than war." This pivoting of attention was unfortunate, for American military officers had only begun to understand and appreciate the potential of operational art in modern warfare. Moreover, it reflected their general unawareness of the gaps they had in their knowledge of the Russian military, gaps produced by the cycles of attention and inattention to Russia and its armed forces.

Harrison's next book, *Architect of Soviet Victory in World War II: The Life and Theories of G.S. Isserson*, appeared in 2010. Though the Russo-Georgian War in August 2008 drew the U.S. government's notice along with its ire, it was the war against a global Islamist insurgency and especially the campaigns in Afghanistan and Iraq that most concerned the American defense community. Wars of insurgency—wrongfully called irregular wars—caught the interest of practitioners and theorists alike. Few thought wars involving the fire and maneuver of large, generally similar forces—what some inappropriately identified as regular wars—were likely in the near term. Then as now, they needed to recognize the relevance of this important book, which revealed the profound conceptual thought of a remarkably talented Soviet officer. More important, they overlooked the reality that Isserson's intellectual approach and the historical analyses he applied to solving the strategic and operational problems of his age provided a model for answering their own contemporary problems.

Harrison's translated and edited compendium of Georgii S. Isserson's major theoretical works, *G.S. Isserson and the War of the Future*, arrives when the U.S. defense community once more finds itself troubled by Russia's recent military activities. Though Russia's behavior is disquieting in one way, it is fortuitous in another, for these latest events might encourage defense professionals to study this essential book, as well as its two predecessors, all with great care. Harrison identifies his two Isserson books as companion volumes; thus, we need to read them in tandem.

In the present work, Harrison includes, in chronological order, six of Isserson's most important publications on operational art, the first five written between 1932 and 1940, and the last, actually two journal articles, written in 1965. Each of these documents offers members of today's U.S. defense community exceptional insights on the study and exercise of operational art, insights that remain useful despite changed circumstances.

Unless the reader is familiar with personalities and events in the Soviet military during the 1920s and 1930s, I suggest he or she first read Chapter Six, "The Development of the Theory of Soviet Operational Art in the 1930s." Written many years after the German army severely tested the theory during the latter stages of World War II, it provides historical context for the other Isserson writings in chapters one through five. In addition, as a primary source for our understanding of the evolution of Soviet military thought, it offers insights on how the U.S. military might advance its own conceptual ideas today.

Although a reader may approach the other five chapters in any order, there is logic to reading them sequentially. To provide a flavor of what they contain as well as to identify some of the elements that have applicability today, I offer the following observations.

In *The Evolution of Operational Art*, Isserson surveys the progression of operational theory and practice from the Napoleonic era to the Russian Civil War. His historical analyses and examination of the enduring factors that affect the progression of operational thought—minus some proselytizing of Marxist-Leninist ideas—rival in clarity and focus many recent books claiming to discern the development and principal attributes of operational art.[2] For this reason alone it is an excellent place for new students of operational art to begin, and for more advanced students to ponder why so many of Isserson's early ideas survive yet today.

In "The Fundamentals of the Deep Operation" Isserson explores the then emerging concept of deep operations. He establishes at the outset that he offers no direct guide for action; he means simply to suggest what to do, not how to do it. Isserson also cautions the reader that he only intends to treat offensive operations. Furthermore, he states that he will limit details to the approach march, meeting engagement, and breakthrough. Present-day concept developers would do well to constrain their exploration of a problem as neatly as Isserson does in this paper, for this is the only way to frame properly one's study efforts.

Following a discussion of the novel conditions that the belligerents experienced in World War I, when a solid front allowed no room for maneuver, Isserson proposes deep operations as the way to carry an attack into an enemy's operational rear. He argues that such operations would begin with a deep engagement (what later we came to know as deep battle) to break through the enemy's front line defenses. These operations would open up a path for forces echeloned to continue the attack into the operational rear, thus leading to the enemy's collapse. Central to his concept was the creation of diverse units able to operate in depth and in three dimensions. The U.S. military's approach to air-ground operations today mirrors many of Isserson's concepts. Although contemporary readers responsible for the development of modern forces will find little direct application of the numerous calculations Isserson offers on the type, size, and composition of units or the tasks the Soviet army might have assigned them, they will gain valuable knowledge by studying how Isserson thought through a unique operational problem.

In a note, Isserson tells the reader that "The Fundamentals of the Defensive Operation" merely summarizes his lectures; therefore, one should regard it as the result of research that serves to shape the question of how to organize and conduct a defense. American concept developers would profit if they heeded Isserson's advice to frame or structure the question (or problem) before they set out to solve it. Isserson hopes that further study (and, by implication, readers' comments) will enable him to produce a document of "a more developed, systematized and finished character." Those writing contemporary U.S. operating concepts would do well to emulate his reticent attitude, rather than to assert, as they often do, that their writing provides complete and lasting answers to the problems they address.

Isserson contends in the first page of his Introduction to this summary of his lectures that militaries have traditionally focused their intellectual efforts on offensive operations, leaving defensive operations as an afterthought. He is particularly critical of the sole focus on tactical defense at the expense of operational defense. The latter, he argues, requires a grasp of the defense as a holistic system rather than an assembly of isolated components. He was clearly ahead of his time with this observation; for it is only in the last few years that U.S. military officers have begun to think of operational problems in system terms, not systems analysis, but considering a problem as a whole rather than the sum of its parts.

A survey of the U.S. military's current research efforts, doctrine, and professional educational curricula would show it has minimal interest in the operational defense. Study and assessment of Isserson's findings and judgments can rectify this problem as it provides the basis for advancing the U.S. military's knowledge beyond the tactical defense to operational defense.

An editor's note reveals that "The Fundamentals of Conducting Operations" "is actually a chapter from a projected 'operational manual.'" This note also discloses that other parts of the manual did not survive the purges of 1937 and 1938. If the quality of this one chapter is indicative of those destroyed, we indeed lost extraordinary material. The document certainly meets its stated purpose to provide a "theoretical base" for "organizing and conducting operations." Few U.S. military manuals achieve this crucial goal. Early on, Isserson establishes that

> the determination of the tasks of military operations corresponding to the political goals of the war; the employment of the armed forces for achieving these goals; the waging of war as a whole—on land, air and sea, and the organization of the country's resources for feeding war—comprise the field of **military strategy as the continuation of politics.**
>
> The employment of the armed forces for the resolution of tasks assigned by strategy comprises the **field of operational art.**

U.S. military manuals did not incorporate such a Clausewitzian understanding of war until well after the Vietnam War. Isserson defines terms, outlines his operational ideas, describes formations, their roles, and organization, and the general concept of offensive operations in clear and concise prose. One wishes that U.S. joint doctrinal publications met this same standard. Concept and doctrine writers will learn much from examining the construction and content of this unique document.

The New Forms of Struggle is Isserson's 1940 historical analyses of two different types of then modern war. These were the Spanish Civil War (1936–1939) and Germany's invasion of Poland in 1939. Isserson acknowledges that Germany's invasion of France in May 1940 is certain to offer additional insights on the emerging character of war.

Isserson's straightforward narrative demonstrates the power of this form of analysis when compared to the prevailing American tendency towards "lessons learned," which produce reports too often filled with diagrams, tables, charts, and graphs and a ponderous and obtuse style of writing. Those who intend to study recent and past wars would reap much from reviewing the manner in which Isserson approached the task. They would also benefit if they recognized Clausewitz's admonition, as Isserson obviously does, that "no two wars are ever the same." Finally, they would do well to heed Isserson's caution:

> To pass indifferently past the events of this war, only in order to not upset one's established conception of the old "classical" forms of struggle; to reduce everything to the fact that this was only an isolated case and that nothing new has happened; to dispassionately describe events, while only formally fixing the facts—means to not understand anything in the new manifestations of historical development and to imitate an ostrich, having chosen his tactics so convenient for military conservatism.

For the many reasons I have touched on above, Isserson's six publications have much to offer to those involved in force development, that is, writing military concepts and doctrine and designing future organizations. They also contain information of value to those responsible for professional military education. This translation of Isserson's principal works is of invaluable importance to the U.S. defense establishment because it can help advance our armed forces knowledge and proficiency. This book certainly deserves a place in any military professional's library.

Lieutenant General Paul K. Van Riper served more than 41 years in the U.S. Marine Corps, where he commanded at every level from infantry platoon to division. He saw combat during five tours, earning a Purple Heart as well as two Silver Star medals for valor. He is a graduate of the Army's Airborne and Ranger schools, the Marine Corps Amphibious Warfare School, the Navy's College of Command and Staff, and the Army War College. He holds the Kim T. Adamson Chair at Marine Corps University.

Preface by Richard W. Harrison

I first came across G.S. Isserson several years ago, while conducting research for my dissertation on the development of Russian-Soviet operational art, and was immediately struck by the volume and profundity of his writings, which contributed so much to the Red Army's military theory. Even more amazing was the fact that a man who had been instrumental in advancing his country's military thinking had been relegated to a position of near total obscurity in the official accounts of the time.

Later, I decided that Isserson's life and work would be the ideal vehicle for a more in-depth study of Soviet operational art during the interwar period. Such a work would not only be a valuable contribution to the existing literature in the field, but also an act of historical justice in bringing the man and his thinking to light.

The resulting biography was well received, and I resolved to take the matter further and produce a translation of Isserson's major works to serve as a companion volume to the previous book. It is my hope that these two works, taken together, will accomplish the goal of according Isserson in his proper place among the great military theoreticians of the twentieth century.

The present work is organized into six parts. The first five consist of complete translations of Isserson's most important works between 1932 and 1940. The sixth entry is a translation of his serialized memoirs, which were published following his return from labor camp and internal exile and which provide a historical context for the preceding works. Taken together, these form a thematic whole, revealing in depth Isserson's thinking about the conduct of operations in a future war.

The Evolution of Operational Art was published in 1932, while Isserson was an instructor at the Frunze Academy's newly founded operational department. This work marked a radical shift from his previous exercises in military history and serves as the ideological foundation for much of his later theoretical efforts. It was here that Isserson first broached many of the ideas that came to constitute the theory of the deep operation.

The first part of this work involves a Marxist interpretation of the development of warfare from Napoleonic times to the present, with particular emphasis on the role of technology. The second lays out for the first time Isserson's *idée fixe*—the decisive role of depth in modern war and the necessity of switching to a "strategy of depth," as opposed to the "linear strategy" of World War I.

"The Fundamentals of the Deep Operation" is a compilation of lectures Isserson read to his students in the Frunze Academy's operational department during the 1932–33 academic year. This was an internal academy document, which has never been released and remains classified. I came across it while doing dissertation research in Moscow at the Military History Institute.

In this work, Isserson moves from the realm of abstraction to offer a detailed guide for conducting modern offensive operations in their various forms. These are the meeting operation, generally conducted when the situation is still quite fluid, and the breakthrough operation against an entrenched defender. The author's vehicle is the combined-arms "shock army," consisting of large numbers of tanks, aircraft, motorized infantry, and airborne forces.

"The Fundamentals of the Defensive Operation" was published in 1938 as an internal General Staff Academy document, while Isserson was serving there as a professor. This is a rarity among Soviet military works of the period, in that it examines the defense at anything above the tactical level. As such, it offers a different insight into an army that is often viewed as being devoted exclusively to the offensive.

In this work, Isserson skillfully grafts his ideas on the importance of depth in modern warfare to defensive preparations. The conclusion Isserson drew is that only a deeply echeloned defense can halt the deep offensive operation, as was demonstrated more than once during the Great Patriotic War.

"The Fundamentals of Conducting Operations" appeared as an internal General Staff Academy document in 1939. This work is an excerpt from an earlier attempt by the General Staff to compile an operational manual. However, due to the military purges of 1937–38, most of the original manuscript was lost. Isserson, as the author, was able to either retrieve some of the text, or reconstruct from his notes only part of the larger work.

In contrast to Isserson's previous works, which focused on conducting operations at the army level, this chapter breaks new ground by examining operations at the *front* (army group) level, as well. He also sought to integrate such new organizational developments as the cavalry-mechanized group into his theory.

Isserson's last major work was *The New Forms of Struggle*, which appeared in 1940, when his star was already setting. During the military purge, many of the deep operation's proponents were arrested and executed and their views declared anathema. This work, which was undoubtedly sanctioned from above, marks the deep operation's return to respectability in the Red Army.

In this work, Isserson examines the experience of the Spanish Civil War, which he dismisses as not being applicable to modern conditions. By contrast, he sees the German-Polish War of 1939 as something fundamentally new that the Soviet Union's military-political leadership would be wise to heed.

Isserson wrote a follow-up volume to this work, which included an examination of the German army's campaign in the West in 1940 and in the Balkans a year later. However, he was arrested after submitting the manuscript to the publisher and returned to Moscow only fourteen years later. This work may have been published later under another name, but in any event has never been found.

Isserson's two-part memoir, "The Development of the Theory of Soviet Operational Art in the 1930s," appeared in the monthly *Military-Historical Journal* in early 1965. Here, Isserson recounts in a highly subjective fashion the origins of Soviet operational art in the 1920s and the theory's full flowering during the next decade, while at the same time offering insights into the various personalities behind these events.

This article also has a poignant significance as one of the last manifestations of the Khrushchev-era "thaw," when the tragic events of the recent past could be addressed with some honesty. The onset of Leonid Brezhnev's neo–Stalinist policies soon made such efforts impossible for another twenty years.

This is by no means an exhaustive account of Isserson's works. He also wrote extensively on military history and tactics and later even dabbled in philosophy. However, for all his erudition in other fields, it is Isserson's decisive contribution to the development of the Red Army's operational art for which he should be most remembered. If the works presented here serve to advance that understanding, then this will have been time well spent.

Acknowledgments

I should like to thank Col. David M. Glantz (USA, Ret.), whose encouragement during this and other projects have bolstered me greatly.

Thanks also to Lt. Gen. Paul K. Van Riper (USMC, Ret.), an early and enthusiastic supporter of this project, who kindly consented to write the foreword.

Finally, as always, I would like to thank my wife Yelena, who has throughout exhibited a superhuman patience for my work.

About the Translation

Any attempt to translate the works of G.S. Isserson presents a number of interesting problems. While his prose style is infinitely superior to that of most Soviet authors writing on military topics during his time, and afterwards, there are nevertheless a number of difficulties to be overcome. The chief among these is his highly elliptical and eccentric way of expressing his thoughts, which probably had its roots in the author's early years, when he knew German as well as his native Russian.

Another is Isserson's often strained attempt to fit his military theories to the Procrustean bed of Marxist-Leninist ideology. Unlike many hack writers of a later era, he seems to have been quite sincere in his beliefs, although they often detract from a clearer understanding of the matter at hand. This explains the appearance of an excessive number of "developments," "transformations," "stages" and other jargon in his works.

Although some amount of editing was involved in preparing this work, nothing has been removed and, in fact, some things have been added. These chiefly have to do with people, places and events that may be obscure for the average reader. These I have sought to explain in numerous editor's notes (identified as such) that follow the text, and among the notes are a handful that Isserson himself included. The final piece also includes a few notes from the editors of the journal in which it was published. Also, Isserson subsequently made a number of corrections to the texts in his own distinct and crabbed handwriting. In some cases he simply inserted new material, while in others he excised entire passages and replaced them with something new. I have decided to retain the original text, while at the same time offering the reader the opportunity to view the author's changes. Those passages which Isserson excised are enclosed by brackets [ ], while passages enclosed within braces { } denote parts he later added.

Additionally, Isserson sought in several places to emphasize this or that point by either noting this in italics, by underlining a particular passage, or otherwise distinguishing one section of text from another. All of these distinctions have been retained and appear here as they did in the original.—R.W.H.

A Short Biography of G. S. Isserson

Readers of this book may already be familiar with my biography of Georgii Samoilovich Isserson, perhaps the least well known of the previous century's great military theoreticians. The favorable reception accorded that work has encouraged me to embark on this latest effort to bring his theoretical heritage to a wider audience. In that sense, this volume of his most important works should be viewed as a documentary companion to the biography. While the events of Isserson's life may already be known to the reader, a brief sketch of the man's life is nonetheless called for.

Isserson was born on June 16, 1898, in Kaunas, now in Lithuania, a small town in one of the westernmost provinces of the Russian Empire. His father was a Russian Jew who had managed to overcome the multiple discriminations of the time to become a medical doctor. His mother was a German Jew from Königsberg (now Kaliningrad), then the capital of East Prussia. She passed on to the younger Isserson her love of music and knowledge of German, the latter of which would prove to be useful in his professional life. He also had an older sister, Lyusi, who was to die in the Holocaust during World War II.

Fittingly, for a military theoretician, war was to play an important role in Isserson's life. His father was called up during the Russo-Japanese War and World War I broke out while Isserson was still a student at the local high school. In 1916 Isserson enrolled in the University of Petrograd, with the evident intention of becoming a lawyer, a fitting profession for someone of his quarrelsome nature. However, he was drafted into the dying tsarist army in early 1917. His educational background and obvious intellectual gifts qualified him for an officer's rank and he underwent basic training just outside the capital.

The Bolshevik coup of November 1917 marked not only the end of the unstable Provisional Government, which had ruled Russia since the overthrow of the Romanov dynasty that same March; it also signaled the end of the old Russian army, which began to fall apart at an accelerated pace. Isserson was one of the many hundreds of thousands who simply walked away.

However, military life had evidently made an impression on the young man, and in June 1918 he joined the Worker's and Peasant's Red Army (RKKA), the new regime's response to the many anti–Bolshevik movements springing up around the country. Isserson's first duties were chiefly administrative ones with the local military commissariat in the Vologda area of northern Russia, to where his family had relocated. He later joined the Communist Party and served as a commissar to several units. He took part in a number of actions against White Guard units and other anti–Soviet forces during the Civil War in the north and was wounded in August 1919. He recovered in time to take part in the Red Army's final campaign there, which ended with the capture of Archangel in January 1920.

Although the Civil War in Russia was winding down, the Polish offensive against the Soviet Republic in the spring of 1920 opened a new front in the struggle. Isserson's 18th Rifle Division was hurriedly transferred to Belorussia, where it took part in the first, albeit unsuccessful, counteroffensive against the Poles as part of the Western Front. The second effort was more successful, and in a little over a month the Western Front's armies pushed the Poles back nearly to the Vistula River. However, the Soviets had greatly overextended themselves during the headlong advance and a Polish counteroffensive took them completely by surprise. Isserson's division, which was far to the west, was cut off and forced to cross the border into Germany, where it was interned by the German authorities. Isserson spent several months in confinement before being repatriated home later that year.

After his return, Isserson elected to continue his military career and he was quickly accepted into the RKKA General Staff Academy. His stay at the academy was interrupted by frequent postings outside of Moscow, as a result of which he completed the academy course only in the summer of 1924. It was during this time that his first written works appeared.

Following his graduation from the academy, Isserson spent several years in various command and staff postings in the Moscow and Leningrad military districts. He also briefly traveled to Germany as a member of a military delegation. Following this assignment, he was transferred back to Moscow to serve in the RKKA staff's operational directorate. It was during this time that he courted and married Yekaterina Ivanovna Fedulova, the daughter of a prosperous merchant from the Altai region of Russia. However, Isserson proved to be a tyrannical and unfaithful husband, and Yekaterina Ivanovna divorced him in 1951. Their only child, Irena Georgievna, was born in 1928.

In addition to his regular duties, Isserson also continued to regularly publish various works during these years, dealing with the German army and questions of recent military history. Among the latter was a study of the Russian army's ill-fated East Prussian operation of 1914 and the German army's March 1918 offensive along the Western Front.

Isserson's career reached a turning point in the autumn of 1929, when he was transferred back to Moscow as a junior instructor at the Frunze Military Academy, the country's premier military-educational institution. This move coincided with the beginning of Stalin's radical transformation of Soviet society, embodied in the first of his five-year plans. This and succeeding economic plans quickly transformed the Soviet Union into a modern industrial state, with an army to match.

Isserson's sojourn at the academy marked the beginning of his maturation as a military writer, and the academy was at the forefront of the new ideas then agitating the Red Army's leading minds.

Among the most important of these was how to employ the new generation of long-range weapons (aircraft, tanks, airborne forces, and mechanized troops) to overcome the enemy's multi-layered defense in order to restore maneuver to warfare and avoid a repetition of the trench deadlock of 1914–18. These ideas soon became known under the rubric of the "deep engagement" (*glubokii boi*) and the "deep operation" (*glubokaya operatsiya*).

The theory of the deep engagement, which was divided into offensive and defensive components, as well as the meeting engagement, sought to combine the new weapons in such a way as to strike and defeat the enemy throughout the entire depth of his tactical formation. The same was true of the deep operation, which was also divided into offensive and defensive components and the meeting operation. Here, the focus was overwhelmingly on the offensive operation, which would take advantage of the deep battle's success in piercing the enemy's tactical defense by exploiting the breakthrough with mechanized and cavalry formations and driving into the operational depth.

Isserson's first major contribution to this debate was *The Evolution of Operational Art*, which appeared in 1932. In this work he traced the evolution of warfare from Napoleonic times, through the era of Moltke, World War I, and to the present, highlighting the forms predominating in each.

This was followed by "The Fundamentals of the Deep Operation," which he completed in 1933. As opposed to the previous work, which was highly theoretical in its approach, this was a deeply practical primer on conducting the deep operation, and contained two of its major components: the deep meeting operation and the breakthrough operation.

While at the academy he also published two other important works. One was a series of lectures on the deep engagement, which was published in 1933 as *Lectures on Deep Tactics*. Another was a lengthy study entitled *Military Art in the Age of the National Wars of the Latter Half of the XIX Century*, in which he examined the Austro-Prussian War of 1866 and the Franco-Prussian War of 1870–71 in the light of the evolution of military art. The latter work is also a prime example of the Marxist method of historical analysis as applied to military affairs.

This academic idyll came to an end in 1933, when Isserson was posted to the Belorussian Military District as commander and commissar of the 4th Rifle Division. At the beginning of 1936 he was transferred back to Moscow as deputy chief of the RKKA General Staff's operational directorate. He was shortly afterwards assigned to head the department of army operations in the newly created General Staff Academy. This offered Isserson the opportunity to propagate his ideas on the conduct of the deep operation to the upcoming generation of Soviet military leaders, many of whom came to occupy high command and staff positions during World War II.

However, what would later be viewed as the zenith of the Red Army's power and influence came to an abrupt end in June 1937 with the beginning of Stalin's purge of the military. The wave of repressions that swept over the armed forces eventually led to the arrest and/or execution of a significant part of its higher command echelon. Isserson survived the bloodletting, even though many of those arrested had been associated to one degree or another with the notion of the deep operation, which now became, by extension, "subversive."

Nevertheless, Isserson struggled manfully on and continued his prolific writing, although he was forced to be circumspect. One work from this period is his "Fundamentals of the Defensive Operation," which appeared in 1938.

Another work from this time was Isserson's "Fundamentals of Conducting Operations," which was published in 1939. Of particular interest here are Isserson's comments on conducting operations at the *front* (army group) level, in which the *front* would carry out strategic tasks in a theater of military activities (TVD).

Later that same year Isserson was appointed chief of staff of the 7th Army, which was to spearhead the attack on Finland. However, this posting was a disaster and the

Soviets failed to penetrate the Finnish defenses, while suffering heavy casualties. He was quickly relieved and relegated to a rear-area post, after which he was demoted to the rank of colonel and posted back to the academy.

Despite these setbacks, Isserson threw himself into a new work, *The New Problems of Struggle*. This work, which appeared in 1940, was an operational-strategic examination of the recent civil war in Spain and the German-Polish war of 1939. It was also an attempt to resurrect the tenets of the deep operation from the anathema to which they had been consigned during the purges.

However, Isserson's days as a free man were numbered and he was arrested on June 7, 1941, two weeks before the German attack on the Soviet Union.

Isserson was interrogated throughout the summer and fall and forced to refute the most fantastic accusations imaginable, even as the Germans struck deep into Soviet territory. In January 1942 he was sentenced to death for various "anti–Soviet" crimes, although this was inexplicably commuted to ten years of confinement in a labor camp, followed by five years of internal exile. He spent the next several years in camps in northern Kazakhstan and Siberia. Although he survived, the ordeal ruined his health forever.

Isserson's conviction was overturned in June 1955, as a result of the post–Stalin "thaw." He was allowed to return to Moscow and secure an apartment and pension. He even married a younger woman, but the pair divorced after only a year. He remarried a few years later, though this union did not prove any happier. Nor were his relations with his daughter and grandson any better.

Following his return from exile, Isserson was offered the opportunity to resume his work on military theory at a time when the Soviet army was attempting to come to terms with the introduction of nuclear weapons. However, he turned down the offer in a fit of pique, thus sidelining himself from subsequent developments. Thereafter he limited himself primarily to occasional articles and angry letters to the editors of various journals.

Isserson's health continued to decline and he began to suffer more from heart troubles contracted in camp. He died on April 27, 1976, and his ashes were interred in one of the walls of Moscow's prestigious Novodevichy Cemetery.—R.W.H.

One

The Evolution of Operational Art (1932)

Moscow. 1932 Gosudarstvennoe Voennoe Izdatel'stvo

G. Isserson—The Evolution of Operational Art. While examining the basic problems of operational art in their historical and theoretical context, particularly while critically analyzing the operational heritage of the past (the Napoleonic era, the era of the Franco-Prussian War of 1870–1871, and the era of the Imperialist War), the author lays out the contours for resolving operational problems in conditions of a future revolutionary-class war, when the chief operational form will be the deep destruction operation. The work contains several controversial passages and should serve as the impetus for the further study of a number of the theses put forward by the author. The work is chiefly directed at the RKKA's[1] senior and high-ranking command personnel and academy students.

—G.S. Isserson

TABLE OF CONTENTS

Author's Introduction — 16

Part I: The Operational Heritage of the Past

1. The Paths of Development of Our Operational Art — 16
2. The Evolution of Operational Art Before the World War — 22
3. The Evolution of Operational Art During the World War — 29

Part II: The Justification for a Deep Strategy

1. The Fundamentals of Our Operational Art — 41
2. The Evolution of the Nature of Operations in a Future War — 44
3. The Present Correlation of Offensive and Defensive Means — 48
4. The Offensive's Deep Operational Formation — 52
5. The Deep Entry into the Modern Operation — 56
6. The Deep Breakthrough and the Smashing of the Front — 61
7. The Art of Managing the Deep Operation — 65
8. From the Theory to the Practice of the Deep Operation — 69

Author's Introduction

At major turning points in history, when in a grandiose struggle old social foundations are overthrown and a new society is being constructed—the phenomena of armed struggle, as a continuation of politics, are subjected to radical and fundamental changes.

A revolution is taking place in the evolution of military art. It forces us to put forward and resolve in a different and new way all of the fundamental problems of organizing and waging the armed struggle of the proletariat.

In reevaluating the bases of the old military art, and in resolving a number of new and contemporary problems, our Marxist, military-scientific study finds an enormous field for work.

This work, *The Evolution of Operational Art*, represents an attempt at such a study of the nature of operations in a future war.

This new subject, which has of yet been little studied, is presented in an historical and general-theoretical context, in order to resolve on this basis the following task—to construct an applied theory of modern operational art.

The work thus presupposes a further specific and calculated working out of the theses put forth here.

It stands to reason that this is a research work and cannot aspire to a full and final resolution of the problem.

The opposite is intended; that a broad discussion of this matter will call forth a further development of our military-theoretical thinking in the field of operational art.

This would justify to a significant degree the work's purpose.

—G. Isserson. Moscow, 16 October 1932

Part I: The Operational Heritage of the Past

1. The Paths of Development of Our Operational Art

Modern operational art, as the study of the conduct of the operation, faces a number of new problems. There is still much that has not been studied and remains unresolved in this field.

The colossal changes in equipment, arms and combat organization, which have been reflected in the evolution of tactics, are still far from having been sufficiently considered by theory on the scale of the armed front as a whole.

The modern operation is developing in completely different political conditions and on a completely different material-technical base, and there is not yet a sufficiently specific prospect for the organization of military activities and the development of their operational forms.

The entire experience of recent wars, so rich in the tactical realm, still conceals the true nature of the future operation under a blanket of obscurity.

This tenet is aggravated by the fact that the World War basically did not yield a single operation that could truly be considered an operational solution to the problem of achieving victory.

Individual local operations that achieved their goal—the true defeat of the enemy; for example, the defeat of Samsonov's[2] army—failed to play any kind of significant role in the overall context of the war.

Even 1918, with its grandiose mortal clashes, failed to resolve the problem of overcoming the front at the operational level and was the apotheosis of the dead end which the military art in the imperialist age had reached.

The World War concluded under the banner of unresolved difficulties in the organization and conduct of the offensive operation.

These difficulties were determined by the great defensive strength of the positional front, the absence among the soldiers of political stimuli for overcoming it, the superiority of the means of defense over the means of attack, the necessity of concentrating enormous weapons of suppression, the complexity of organizing and conducting offensive actions, etc.—in other words; from the military point of view they were completely localized in the sphere of *tactics*.

This exerted an enormous influence on the conduct of all operations in 1918.

"It was necessary to place tactics above strategy," wrote Ludendorff.[3] And truly they did not attack where it was required by considerations of operational utility, but where it was possible according to tactical conditions. They developed the attack not along the axis where it promised an operational result, but where the front was most easily tactically broken. The German offensive in March 1918[4] is the most striking example of this. The positional being of the war masterfully defined its operational consciousness. It proved impossible to overcome the new conditions of struggle. The main problem was that there could not be found a soldier who possessed sufficiently convincing political convictions to attempt to resolve this task with his own blood.

In order for the new soldier to find within himself sufficient strength to overcome the enemy's resistance in an open attack, it was necessary for class volition to be awoken in the masses; that the class contradictions unfold in an open armed class struggle, and; that the imperialist war be transformed into a civil war.

Our civil war of 1918–1921, with its destructive *deep* blows, up to the final rout of the enemy, undoubtedly marked the beginning of a new era in the history of military art and radically changed the entire nature of armed struggle.

"The revolutionary wars," wrote Clausewitz[5] about the wars of the French Revolution, "turned all the old ways upside down swept away everything from Chalons to Moscow." One does not have to be a Clausewitz to understand the entire mobile nature of our revolutionary-class war by changing the direction of the last words from east to west.

However, the operational content of the new era, in the sense of commanding major combat forces and richly endowed with modern equipment, is far from having been revealed for the future.

The changes that have taken place during the time separating us from the end of the civil war are immeasurable in their significance.

They force us to set forth in a different way the problem of the correlation of the qualitative power of defensive and offensive weapons, revealing an obvious tendency toward the predominance of the latter.

In these conditions, the problem of overcoming the front of fire and the opportunities for breaking through it throughout the entire depth acquires new significance for us.

In essence, all of our theoretical thinking in the field of military affairs is directed toward resolving this task.

As in the capitalist countries after the World War, and here following the civil war, the evolution of military art is proceeding on a different class basis, under the banner of seeking out new tactical forms of attack and employing the new technical means of struggle.

During the short period following the World War, which is an entire age in the field of military art, tactics is undergoing greater changes than in the half-century preceding the World War. During this time all the manuals are being reworked and reissued; a new tactics has been created in a few years.

It is not without interest to note that tactics is changing this rapidly for the first time in the history of the development of military affairs.

Prussia entered the wars of 1866 and 1870 with a field manual issued in 1847 and then only changed it in 1888. Germany entered the World War with the 1888 field manual. Thus the Germans changed their field manual only once over a long period of 70 years.

During the period of rapid socialist construction, we have issued the Provisional Field Manual of 1925, then replaced it with a permanent field manual in 1929, and in 1932, we are once again on the eve of revising it. Thus we are already drawing up our third field manual in seven years.

This rapid pace of field manual creativity, natural in conditions of colossally progressing technology, is a general phenomenon in the development of military affairs following the World War.

However, this reflected and predetermined the evolution of military affairs chiefly in the area of tactics.

The problems of struggle along an armed front, as a whole, and the conduct of military activities at the operational level, were placed on the back burner and have attracted the attention of military-scientific investigation to a much lesser degree.

To be sure, general questions of waging war within the framework of politics, strategy and the economy have generated a great deal of interest in the literature. However, the practical problems of waging military activities along an armed front and the problems of the techniques of waging the modern operation are reflected in a weak and pale manner in the literature.

The Germans, in the conditions of their military system, limited by the conditions of Versailles,[6] go no further in their press than examining the World War's operations. Having created a rich military theory following the war of 1870–1871, they are still digesting the teachings of Schlieffen.[7] Their military writer Groener[8] offers much that is interesting in this regard, although it's doubtful that much here is new.

In France, the country possessing the most concentrated military system of imperialism, Culmann's[9] *Strategy* has appeared which is considered the final word in studying the operation in the capitalist countries. At the same time, Culmann does not elaborate an integral operational system, while highlighting only a number of individual problems connected with it, and, what is most important, his outlook on the future does by no means fully envisage the examination of everything new at the operational level.

As concerns an entire series of bourgeois military writers (Ludendorff, Immanuel,[10] Metzsch,[11] Requin,[12] and Fuller[13]), who are substituting any hint of a scientific theory for conducting the operation with fantasizing, which is far from justified, as to the prospects of a future war; their writings, which reflect the class nature of the contradictions of modern capitalism, show best of all just how little the problems of modern operational art have yet to be studied by scientific theory.

Our literature has an obvious advantage in this regard.

Comrade Triandafillov's[14] work, *The Nature of Operations of Modern Armies*, doubtlessly occupies an undoubtedly high place in the literature of the modern operation. According to the volume of questions raised and the nature of their presentation, this work puts forward an entire operational system that resolves a number of important problems in their practical context.

However, it's important to keep in mind that before his tragic death, comrade Triandafillov had radically changed his views on a series of fundamental questions. His inquisitive mind was already building a different and far-seeing prospect on the basis of our new achievements.[15] Cruel happenstance denied him the opportunity of developing his new system of operational views. In the meantime, life has moved far ahead.

As a result, the study of the modern operation is today at a far from sufficient level and remains the least elaborated part of military art. The fact that this is not taking place for the first time in history can hardly serve as consolation.

Under the conditions of capitalism, the theory of military affairs systematically lagged behind practice, and this was reflected most of all in operational questions. Tactics is, to a significant degree, practice, tested in exercises and maneuvers. The conduct of the operation is more of a theory in peacetime and is not subject to experimental testing. Naturally, it is much easier to employ a new weapon in more narrow confines than to organize its mass employment. Thus tactics has more than once left operational art behind. However, at present such a situation is even less tolerable. The completely altered conditions of struggle along the armed front, the new human material and new weapons forcefully demand new ways and forms for employing them on a mass, operational scale in which quantity is transformed into a completely different quality.

Before the age of imperialism, given the comparatively limited size of armed forces (the Prussian army had 500,000 men in 1870), problems of conducting the operation had not acquired the status of an independent theoretical discipline, as they were completely resolved within the confines of a specific war plan. All the questions that arose before Moltke[16] in preparing the war of 1870 were resolved in the practical elaboration of his deployment against France.

Now, with mass armies, extremely complicated equipment, the enormous depth of columns and the difficulty of deploying them in a combat formation, the complexity of the rear services, and an entire host of other complicating factors, the conduct of the operation puts forth problems, the resolution of which does not fit into a specific deployment plan and requires the erection of an overall theoretical base.

The operational worker is now in need of a definite theory of conducting operations in his practical work.

Operational art, as the study of the operation, thus acquires the significance of a very important discipline for the practice of operational work and the command of large troop formations.

The topicality of the problems of operational art is also conditioned by other prerequisites. It is quite obvious that radical changes in the field of technology and tactics are bringing about no less radical changes in the field of conducting operations.

"Changes in the nature of tactics," Clausewitz said, "must influence strategy as well. In the given case, if tactical phenomena possess a different nature than in another, then strategic phenomena must change as well; otherwise, they will not be consistent and logical."

This obvious rule has not always been understood. During the age of Moltke, with its new firepower and altered tactics, they nevertheless approached the waging of the battle from the point of view of Napoleonic military art. In this sense, Moltke was a great reformer, having understood the new conditions and requirements of his time.

However, in 1914 the conduct of operations had left the age of Moltke far behind in its forms and methods.

All the factors of armed struggle had grown quantitatively and therefore qualitatively. However, the operational control of these factors had not undergone any kind of qualitative changes. Even now, if one thinks about the organization of the operation, as it is usually presented, it is difficult to perceive any kind of major changes.

Corps form up in a single line, attack axes are assigned, and tasks are assigned according to lines.... But, this is how it was done in 1914 and, if you dig further back, how it was in Moltke's era!

Operational art is proving to be intolerably conservative. At the same time, the conditions of our time and 1914, much less the age of Moltke, are completely incommensurate. The entire arsenal of the basic factors of armed struggle has changed. New weaponry, new tactics and a new solider are all inevitably bringing about enormous, radical changes in the field of conducting operations.

After all, it's quite obvious that a change in factory equipment and the exploitation of new machines will radically change the entire productive process and the organization of production as a whole. In the military field, this obviously predetermines a different organizational structure for troop organisms.

According to this point of view, the conduct of the modern operation requires the most serious review. However, this should be done not only from the point of view of the material factors of struggle and the new soldier. This would be completely inadequate. The operation is the tool of strategy, and strategy is the tool of policy. Thus the operation is still not the highest degree of armed struggle. It itself is an element, subordinated to war as a whole.

"Of the new phenomena in the field of military art," as comrade Lenin excerpted from the work of Clausewitz, "one should ascribe only the most insignificant part to new inventions and new ideas; the majority should be ascribed to new social relations and new social conditions" (*Leninskii Sbornik*, vol. XII, p. 421).

Completely altered social conditions, a different socio-political milieu, a different economy and the different nature of our future war as a revolutionary-class one, are naturally bringing forth a different nature of the operation itself. We are now in the most favorable circumstances for defining this nature. Marxist-Leninist teaching on war introduces complete clarity into questions of the nature of armed struggle. Also, a number of party documents and Comintern[17] resolutions have brilliantly encapsulated this teaching applicable to individual problems of a future war.

"A future world war," as we read in the resolutions of the VI Comintern Congress, "will not only be a mechanized war, during which enormous amounts of material resources will be employed, but also a war that will embrace multi-million masses and the majority of the population of the belligerent countries."

This is how the Comintern congress resolves one of the fundamental problems regarding the specific weight of equipment and the masses in a future war, and thus in this war's operations.

Only on the basis of Marxist-Leninist teaching on war can the theory of our operational art be built.

Thus an entire sum of factors with a completely new qualitative content—different socio-political conditions, a different arsenal of the technical means of struggle, new tactical forms of the engagement, and, finally, the enormous practical importance of the theory of conducting the operation and the insistent need for it—determine the bases for the development of our operational art.

At the same time, it should be borne in mind that operational art, as the study of conducting the operation, is an extremely young discipline. It essentially traces its origins to the period following the World War, when it first occupied an independent place in the hierarchy of military disciplines.

Before the World War military art included two basic parts: strategy, as the study of war, and tactics as the study of the engagement. What is more, this two-tiered system revealed the great lagging behind of military theory from practice.

As early as the second half of the 19th century, the evolution of the forms of armed struggle did not fit into the concepts of strategy and tactics alone; it went beyond their boundaries. Armed struggle gave rise to an entire chain of combat events, spread out along the front and broken up in depth, which outgrew the framework of the engagement and thus could not be embraced by the content of tactics; nor did they embrace the phenomena of war as a whole, because they were not considered by strategy as the study of war.

Thus a major gap developed in theory between strategy and tactics, which however had already long ago been filled in the practice of armed struggle by real phenomena of enormous scope and content.

These phenomena required a new concept, which under the name of operational art as the study of the operation, which only following the World War occupied its independent position in a new three-tiered system of dividing military art into strategy as the study of war, operational art as the study of the operation, and tactics as the study of the engagement.

However, having come into being recently as an independent discipline, operational art now faces at the current stage of its development the task of radically reconsidering the entire study of conducting the operation.

As often happens in the evolution of military affairs, the youngest, barely born phenomenon turns out to be already outdated.

Our operational thinking cannot halt on the experience of the World War. The exhausting system of attrition battles, which failed to resolve the problem of the operational breakthrough of the front; the "crawling" pace of the attack, which required four months for the Allies to push back the Germans a mere 100 kilometers—cannot serve as the point of departure for the construction of our theory of conducting the operation.

Proceeding from the revolutionary-class content of our future war as the decisive collision of two mutually-exclusive worlds, we must go much further in our theory and demand significantly more.

The new era of proletarian revolutions and the construction of socialism and revolutionary-class wars are undoubtedly predetermining the arrival of a new era of military art.

"The true liberation of the proletariat, the complete elimination of all class distinctions and the complete socialization of all means of production..., presupposes the creation of a new method of waging war," says Engels[18] (*Sobranie Sochinenii K. Marksa i F. Engel'sa*, vol. VIII, pp. 491–493).

Our operational doctrine is faced with monumental tasks, which could not be and were not resolved by the imperialist war: these are the overcoming of the solid front,[19] the conduct of a deep offensive, that breaks down and breaks through the front of fire throughout its entire operational depth and, finally, the infliction of fatal, smashing attacks for the purpose of finally and completely routing the enemy.

In these conditions the chief task of our operational art is the *substantiation and construction of a theory of the deep destruction operation.*

2. The Evolution of Operational Art Before the World War

The elaboration of a theory of operational art is extremely complex along those paths that must be chosen for it.

"A new method of strategy has never been born, like Minerva, from the head of Jupiter," says Schlichting[20]; "it arises from the conditions of the age and from the combat means proffered by the latter."

All the factors of our age, in their socio-political, economic and military-technical significance, thus offer the seeds for determining the nature of operations in a future war.

These factors should not be viewed, however, only in the stasis of the new age. The tendency of their development, which is important for determining the nature of armed struggle, may be discerned and understood only through the dynamics of the historical process.

In order to comprehend the specific nature of the modern operation, it is necessary to establish the prerequisites and conditions that called forth its birth and determined its evolution up to our time. This historical approach will bring out the prerequisites that determine the further evolution of the operational forms of armed struggle at the present stage of their development.

"In military affairs it is best to substantiate your views of the future on the study of the immediate past" (Schlichting).

In the historical context the phenomenon, known now under the name of the operation, is clearly revealing its chief features, which determine the evolution of its nature.

The conduct of war in the Napoleonic era consisted schematically of two main stages, far from being equal in their spatial and temporal dimensions. These stages were: a large, long march, which gave rise to a long operational line, and a short battle concluding it in a single place. Clausewitz vividly expressed with the following words: "The field of battle is no more than a point to strategy; in the same way the length of the battle comes down to a single moment." And truly, in relation to the extended operational line, the battle during the Napoleonic era was a single point in space and a single moment in time.

One can justifiably call this age of military art the age of *the strategy of a single point*, because the entire task of the commander came down to concentrating all of his forces simultaneously at one point and throwing them into the battle, which represented a single-act tactical phenomenon.

Naturally, there were materiel prerequisites for such an outline of the military art of the Napoleonic era.

Fire was not very effective at the time and its specific weight was insignificant. The main means of acting upon the enemy was the direct attack by troops. This required forming up the entire mass of troops on the battlefield in a single, solid mass of deep attack columns, to which the French Revolution, with its new soldier, burning with the enthusiasm of battle, had given birth.

It's perfectly understandable that from such a concentrated position only a massive attack along internal lines was possible. This attack routed the Frederickian[21] linear combat formation.

At the same time, this concentration in front of the battlefield was due to the condition of the materiel means of struggle.

A characteristic distinction of the combat conditions of the Napoleonic era were that the troops could see (normally 3–4 kilometers) further than they could shoot (200 meters), while a cannon could fire 1,200 meters. The range of fire was considerably less than the range of vision. In these conditions the sides could come together on the battlefield and view each other, but could not act against each other through fire.

This circumstance explains why the meeting engagement, which develops directly "from the march," could not arise during the Napoleonic era; for a meeting engagement, it is necessary that the sides have the opportunity to act through fire as soon as they espy each other on the march.

Thus the pause between the march to the battlefield and the engagement during the Napoleonic era was conditioned by the limited range of fire and enabled the sides to first form up into their combat formation before the start of the engagement and before the battlefield.

This determined an extremely important and characteristic feature of the Napoleonic era's military art. It consisted of the fact that the battle, as a final point crowning a long operational line, was by no means predetermined by the latter, did not organically issue from it and played out like a separate tactical episode. This is most clearly seen in the example of the Italian campaign, which concluded in the distinctive battle of Marengo,[22] and the campaign of 1812, with its crowning battle of Borodino.[23]

Thus the battle during the Napoleonic era was a single-act tactical phenomenon; it had no dimension in space because its scale was only a point; it did not have a temporal dimension because its scale was a moment; it had no depth, because it was waged in place, and; finally, it played out as an independent tactical episode that did not organically flow from the entire march. In these conditions, Napoleon's military art did not yet know the operation in the modern understanding of the word.

The basic indications of the operation were undoubtedly missing at the time.

The battle was the competence of tactics alone, as the independent study of the engagement.

To be sure, each historical age is pregnant with the new and reveals new tendencies and new forms in their embryonic form.

Thus in the Napoleonic era one can already discern the first indications of new forms of armed struggle, which were growing out of the confines of the single battle.

This is evident from the examples of Ulm,[24] Regensburg,[25] Leipzig,[26] and 1814.[27]

However, it was not these new manifestations that were characteristic of the Napoleonic era.

Characteristic was the long operational line crowned by a point as an independent tactical episode. In these conditions strategy had as its chief task the simultaneous concentration of all forces to a single battlefield and yielded its place to tactics once the battle had begun.

Clausewitz described this situation in the following words: "Once the enemy had approached so close as to give a decisive battle, the time of strategy passed and it could rest."

This tenet, which had spread its influence for a long time, played a major conservative role in completely altered conditions and arose in sharp contradiction to the soon-to-be-born phenomenon of the operation.

In the second half of the 19th century all of the conditions which conditioned Napoleon's military art were radically altered.

The flowering of industrial capitalism, the introduction of universal military conscription on the basis of new productive relationships of in bourgeois society, and technical progress on the basis of developed industry—created new prerequisites for the evolution of the military system.

The introduction of the rapid-firing rifled weapons played an enormous role.

Armed with Dreyse[28] rifles, a Prussian battalion during the Moltke era could already fire 4,000 bullets per minute. The distance, to be sure, was still limited (300–400 meters), but soon rose to 1,000 and 1,300 meters (the French Chassepot[29]).

The introduction of the rifled Krupp[30] cannon immediately increased the firing range to 3.5 kilometers.

In these conditions the specific weight of fire rose significantly and fire became the chief *factor of acting* on the enemy and marked the beginning of the *age of fire annihilation.*

But the tactics of fire entered into a major contradiction with the Napoleonic deep column, which made it impossible to employ all fire means while also presenting an easy target for fire. If fire was becoming the main factor, then this required that the greatest possible number of firing units be placed in a single line, so that they could all be committed into the fighting.

All tactical evolution during the second half of the 19th century progressed along the path of deploying the deep column into a broad line of fire, and later into the skirmish line.

The concentration of masses before the battlefield in deep and close attack columns was supposed to disperse into a broad linear deployment, but now on a new qualitative basis of the growing power of fire.

However, for a long time conservative tactics strove to closely heap masses along a narrow space. However, "the soldier proved to be more sensible than the general," says Engels, and common sense led to the broad line of fire.

This tactically significant circumstance had an immediate effect on the nature of armed struggle as a whole, by giving rise to broader combat formations.

Moltke was already teaching that "We lose more in depth than we gain by narrowing the front, and two divisions, advancing 7–10 kilometers from each other, can more easily and better render mutual support than if one division advances directly behind the other."

At the same time, another new factor of great significance during the 19th century led to the spreading of actions in breadth. This factor was the railroad, which sped up the concentration of the army in the theater of military activities, but at the same time, in accordance with the outline of its network, inevitably gathered it in different areas along a lengthy front.

Those 300,000 men whom Napoleon easily led and formed into one concentrated mass, Moltke deployed in 1866 against Austria in three separate armies along a 400-kilometer front. This was, to be sure, conditioned by the outline of the rail network and the configuration of the Bohemian border. However, in 1870 the deployment of the Prussian armies against France initially occupied a front of up to 100 kilometers which, with their movement forward, rose to 150 kilometers.

This widening of the front line seemed wildly improbable at the time and was subjected to harsh condemnation by Moltke's opponents. Conservative military theory had elevated the fundamentals of Napoleonic military art into an unshakeable canon and eternal principle and was unable to explain the conditions and demands of the new era.

Moltke's opponents—Benedek[31] and the French marshals Bazaine[32] and MacMahon[33]—nevertheless still strived to squeeze their armies along narrow spaces and in dense masses, and each time were overwhelmed the Prussians' broader front of fire.

In essence, two eras of military art and two military schools were contending in the national wars of the second half of the 19th century. The advantage, of course, was in favor of that which took into account the new conditions of its time.

This, however, could play its role only because the wars waged by Prussia in the second half of the 19th century had a historically progressive nature and that "the 1870–1871 war was a continuation of the bourgeois-progressive policy of liberating and unifying Germany, which had been going on for decades. The defeat of Napoleon III[34] and his overthrow accelerated this liberation" (Lenin, "On the Peace Program...," 25 March 1916).

Since then military art has gone over to the broad deployment of forces in a single line and armies began to enter the theater of military activities along a single linear front.

This marked the arrival of a new era in the evolution of military art—*the era of the linear strategy.*

At the same time, it was not the quantitative growth of the armed forces that led directly to this broad deployment, for the Prussian army of 1866–1870 still did not outnumber the armies of Napoleon, but rather new materiel factors—weapons and railroads.

New firearms were the initial factor which called forth the broad line, which turned into the broad deployment and the linear strategy. This is the best confirmation of Engels's tenet that "nothing depends to such a degree on economic conditions as the army and navy," and that "armaments, composition, organization, tactics, and strategy depend, most of all, on the level of production achieved at the given moment and on communications routes."

With the onset of the era of linear strategy a number of new phenomena were introduced into the sketch of events being played out in the theater of war, which no longer fit into the framework of a single battlefield as a single point and which, consequently, outgrew the framework of tactics.

Insofar as the armies had begun to enter the war in a broad line, their combat efforts ended up being dispersed along the front and battles began to arise not at a single point, but at different point along the expanded front.

The basic distinguishing feature of armed struggle in the second half of the 19th century was that the single point of the Napoleonic era broke up into a series of individual points, dispersed in space.

To be sure, this was not yet a solid front; this was a broken front of individual points for the application of military efforts.

The war of 1866 began with three separate engagements (Gitschin, Trautenau and Nachod), spread out over a front of 100 kilometers.

The war of 1870 began with two major battles (Spicheren and Worth), which played out simultaneously at a distance of 60 kilometers from each other.

Moltke the strategist was faced with the completely new problem of unifying and directing dispersed combat efforts, which were tactically unconnected in space, for achieving the overall aim of defeating the enemy.

This circumstance was the first characteristic indication of that phenomenon, which in modern terminology is known by the name of the operation.

As is known, Moltke was barely able to cope with this phenomenon. "The greatest strategist," says Schlichting, "lacked a full understanding of how to unify the actions of separate armies in the theater of war."

At the same time, a new factor in the military phenomena in the second half of the 19th century was not only their spread along the front.

Along with the dimension in breadth along the front, there appeared the first, to be sure, still in their embryonic stage, indications of a dimension in depth, and, it follows, of a dimension in time, which the Napoleonic era had absolutely no knowledge of, because battles unfolded literally on the spot, in the course of only a few hours.

The appearance of the second dimension of military activities—the dimension in depth—had its own definite objective prerequisites.

During the second half of the 19th century, given the significantly increased distance of fire effects, the range of fire became equal to the range of vision. It proved possible to strike the enemy with fire as soon as he became visible, for visibility was generally 3–4 kilometers, given average terrain, while the new rifled cannon had the same distance (3.5 kilometers).

This circumstance created completely different conditions for the start of the engagement.

Because it had become possible to strike the enemy on his approach to the battlefield, the pause between the march and the engagement fell by the wayside and the preliminary concentration for the engagement, as in Napoleonic times, proved unattainable.

Each engagement began to break out directly from the march. This explains the appearance of the meeting engagement, which became possible in its modern sense namely during the second half of the 19th century, when the increased range of fire drew even with visibility.

To be sure, this was not understood for a long time; the conservative Prussian generals in 1866 solicitously moved their artillery in the tail of the column, like transport, and wanted to form up prior to the engagement, according to Napoleon's commandments. However, the objective course of events, which was conditioned by the new weaponry, proved stronger and the initiative for beginning the engagement had already moved from the general to the lead columns' security for the march. At the same time, in arising from the march, the engagement had already begun to unfold not in place, but in a certain movement, having acquired the first indications of depth, although still insignificant.

An even more important circumstance proved to be the fact that this tactical depth immediately outgrew the framework of the engagement and became indicative of operational depth.

In the second half of the 19th century the brief bayonet engagement was transformed into a drawn-out firefight and acquired a significantly greater dimension in time.

Battles during the Moltke era were already lasting 10–12 hours. At the same time, they no longer yielded a decisive outcome, as in Napoleonic times. Fire proved to be incapable of resolving its tasks in a single combat act and along a single line. Following the outcome of one battle, the enemy was still not completely defeated; he fell back successively, put himself together along a new line and accepted a new battle.

Thus the chain of combat efforts dispersed into the depth.

In the 1870 war three battles, Colombey—Neuilly, Mars-La-Tour and Gravelotte—St. Privat, were already unfolding consecutively in time. The course of events occupied only six days and in this time the Prussian Second Army carried out a wheel with its left flank, advancing a total of 90 kilometers. This interesting system of battles, broken up in depth, already contained all the indications of the modern operation.

It consisted of individual combat efforts, which Moltke had to unify in space and time for achieving his overall goal.

The same was true of the march to Sedan—a maneuver lasting ten days and demanding an advance in depth of up to 150 kilometers.

Thus the new depth dimension of combat phenomena was unveiled in the second half of the 19th century, although still in an embryonic stage.

The war of 1870 (until the fall of the Second Empire) numbers a total of four links in depth, representing independent battles (Spicheren—Worth, Metz, Sedan, and Paris). This was still a chain of individual combat efforts, unconnected among themselves. Their finale, for the most part, was yet another battle, which in its scale was quite closely related to the battle of Napoleonic times. A number of points spread out in space often led to a single overall point (Königgrätz[35] and Sedan[36]). Strategy could still count as its main task the simultaneous concentration of all available forces at a single place. The distinction consisted, however, in that this concentration was accomplished from a broad deployment, meaning from different directions, leading to the concentric envelopment of the enemy.

The concentric maneuver along exterior converging lines, which gave birth to a "Cannae"[37] on an operational scale, became characteristic of the era of the linear strategy.

This maneuver from various directions still led, to be sure, to a single battle. However, the combat finale of the Moltke era already contained within itself a significant distinction from the preceding age.

It no longer unfolded as an individual tactical episode, independent of a lengthy operational line.

When the engagement began to unfold from the march and the pause between them disappeared, the battle began to organically flow from the maneuver-march and be predetermined by its organization. The march lead directly to the engagement and the maneuver-march naturally grew into the battle. The plan for the latter was already hidden in the organization of the former.

The same Prussian corps, which while deploying in 1866 occupied an extreme flanking position at a distance of 400 kilometers from each other, closed the envelopment at Königgrätz and drew to within 4–5 kilometers of each other. In these conditions, the deployment plan already contained the plan for the forthcoming activities. And because the possibilities for changing the initial disposition of forces were very limited, then the line of deployed corps could not be significantly changed during the course of the offensive.

At a time when Napoleon could organize his march independently of the forthcoming battle, since he had the opportunity to adopt the corresponding combat formation before beginning it, Moltke had to put forward a definite plan for defeating the enemy as the basis for his deployment and maneuver-march.

In his era the organization of military activities was already requiring that the battle be foreseen; that is, it imparted to these actions that nature which is inherent in the concept of the modern operation.

Moltke already had to plan his deployment up to the battle, inclusively.

For the art of command of his age the boundary between the march and the engagement, between the maneuver-march and the battle, and between strategy as the tactics of the theater of military activities and tactics as the conduct of the engagement, had been erased.

The command of armies in the theater of military activities had to immediately grow into the conduct of the battle; that is, to embrace that sphere of competence which includes modern operational art.

At the same time, the distinctive trademark of the strategy of the Napoleonic age to rest when the battle breaks out fell by the wayside and arose in sharp contradiction with the new conditions of commanding troops.

People could not understand this for the longest time. The fundamentals of Napoleonic military art had become too solidly embedded and raised to an immutable principle. On the eve of the battle of Sedan, Moltke lost control of his armies, and only the initiative of subordinate commanders crowned the maneuver-march with the battle's decisive outcome.

For a long time in the second half of the 19th century new phenomena and new conditions could not break through the conservatism of military theory into people's consciousness. And in the beginning of the 20th century, Leer[38] was still building up his own dogmatic strategic system on the bases of Napoleonic military art. At the same time, military activities revealed their new nature as early as the wars of 1866 and 1870: they spread along the front, dispersed in depth and organically grew out of their deployment; that is, they acquired precisely those basic indicators which determine the concept of the operation.

The wars of the second half of the 19th century marked the initial historical boundary along which the operation was born and which marked the beginning of the evolution of its nature.

3. The Evolution of Operational Art During the World War

The age of imperialism offered a broad vista for the further growth of the operation's main indicators—its spread along the front and its dispersion in depth.

The economy of imperialism, with its struggle for markets, sources of raw materials and the investment of capital made a war for the division of the world an inevitable result of the policy of the ruling classes, having brought about a colossal growth in armaments and the size of armies.

This process of broadening the entire military system brought about the further evolution of military art at the turn of the 20th century, but it was itself determined, in turn, by its new demands.

Prussian military doctrine drew from the experience of the war of 1870 the conclusion that given the increased power of fire one should not strive to achieve a result by means of a frontal attack. Schlichting concluded from the experience of 1870 that "Attempts at a purely tactical breakthrough in the future are almost unrealizable."

The results of the battle of Gravelotte—St. Privat—the first example of an attack against a fortified front of fire, an attack in its wild and unrestrained form,—led to this conclusion. Then it had already become clear that the means of fire annihilation were incomparably stronger in the defensive than in the offensive.

The attacker's unprecedented losses, his unsuccessful attacks and, at the same time, the collapse of the defensive line at the very appearance of an insignificant group of troops on its flank, forced people to immediately renounce the frontal attack.

It was recognized as impossible, because it was generally considered unnecessary.

There still remained a lot of free room for maneuver and any position could be outflanked. This was still little understood during the age of Moltke. "And only late in the evening," writes Schlichting, appeared a division along the flank, more likely led by chance than by design, and in the rear of the enemy, and unconsciously taught the gathered commanders how, since the time of Leonidas,[39] one should overcome strong positions."

The entire evolution of military art following the war of 1870 developed under the aegis of shifting the decision from the front to the flanks. This tenet constitutes the entire motif of Schlieffen's teaching.

In this way, the linear strategy strove even more to broaden the front.

"The broad front basically decides," writes Schlieffen, "making envelopments possible and naturally presupposing a stronger and more numerous army ...

"The modern battle will come down to a battle for the flanks.... And in this battle he whose reserves will be not behind the center, but along the extreme flank, will win."

The entire evolution of military art at the turn of the 20th century moved along this path. The striving to lengthen one's flank and extend the front required an increase in the size of armies and better than anything guaranteed their growth in the age of imperialism.

In 1914 the Germans entered the war with a 2,000,000-man army, surpassing by four times their army of 1870.

The competition of the capitalist countries' military systems before the 1914 war essentially consisted of extending one's flank as much as possible, in order to achieve an enveloping position.

The linear strategy had achieved its greatest flowering. Alongside this process there continued the technical evolution of the means of struggle. The new quality of firearms imparted an incomparably greater intensity to combat events.

As early as the age of Moltke firearms could strike an enemy as soon as he became visible.

During 1870–1914 the range of fire essentially underwent an insignificant evolution. It grew from 1,200 meters to 2,000–2,500 meters for infantry weapons, remaining, however, in practice at its previous level. It rose from 3.5 kilometers to 5–6 kilometers for light field artillery, which did not engender any significant changes. As concerns heavy field artillery, to be sure, it increased its range to 11 kilometers, although quantitatively not very much, and was unable to exert a significant influence on an increase in the combat range of weapons.

The evolution of fire means of struggle at the turn of the 20th century was mainly in the way of increasing their rate of fire.

Enormous results were achieved in this field. The number of shots fired in one minute shows the following growth:

	1870	1914	
Rifle	5	12/10	(The numerator indicates the theoretical rate of fire and the denominator the practical rate)
Machine gun	0	500/250	
Cannon	2	20/12	

At the same time, the continuous linear front was transformed into a front of solid fire of enormous power. This was the full flowering the age of fire annihilation, which had arrived as early as the second half of the 19th century, with the introduction of rifled weapons.

It was becoming evident that events of the greatest scope and intensity were ahead, changing all of the conditions of armed struggle and entering into even greater contradiction with the old conservative military theory, the roots of which went back as far as the age of Napoleon.

"A complete revolution in the entire military system," wrote Engels at the time, "brought about by the inclusion of everyone capable of military service into million-man armies and the introduction of firearms of previously unheard of power, have decisively have brought an end to the Bonapartist period of war, having rendered impossible any other kind of war, except a world war of unprecedented cruelty, with an outcome completely impossible to calculate."

The recent events of the imperialist age immediately confirmed this and revealed the increased scope of armed struggle.

In the Russo-Japanese War the battle of Mukden[40] unfolded along a front already 150 kilometers in breadth and lasted three weeks.

The operation's main indicators—the spreading of combat efforts in space and time—had increased enormously according to their quantitative indices.

In 1914 the German armies deployed against France along a 340-kilometer front. They went over to the offensive along just such a broad line of drawn-up corps and fought the Battle of the Marne[41] on a front of 250 kilometers. At the same time, the nature of the front was significantly different from the broken chain of individual points, spread out in space, of the Moltke era. In 1914 this was a continuous front, which had merged into a single line of points.

The nature of the spreading of operations along the front had thus completed its evolution: the point of the Napoleonic era had broken up into a series of individual points in the Moltke era, and in the 20th century the series of individual points had merged into a single unbroken line.

Now the problem consisted only of what limit would the spread of the entire line of the front reach.

At the same time, the concentric maneuver from different directions, which presupposed a certain freedom of maneuver in space, became hardly fit for the sluggish front of the 20th century, which occupied the entire space in its area of activity. The turning maneuver along external lines had replaced it with the entire length of the front, which became characteristic for the new age.

Concentric maneuvers from different directions were now feasible only in independent, local theaters of war, which still retained sufficient freedom for maneuver. They found a place in East Prussia during the beginning period of the 1914 war.

A continuous front in space also called forth the further evolution of the second indicator of the operation—its dispersal in depth.

The matter consisted not in the increased length of the operational line. In 1914, during the offensive to the Marne, this line was 400 kilometers long, although marches in earlier eras had been this long.

However, the new qualitative distinction was the fact that in 1914 a single chain of combat events, unified by an overall operational plan, unfolded throughout the 400-kilometer depth. This was a series of stages in a single operation, or a series of connected consecutive operations, from which each flowed from the preceding one and gave birth to the subsequent one.

Thus the depth of the operation in 1914 achieved a new qualitative character as a single chain of a series of connected combat events.

To be sure, this chain was still not continuous in depth. By no means did battles occupy the entire depth, but rather unfolded along its individual lines.

Combat events occupied only 23% of the time of the entire Marne maneuver-march. In the eastern theater of war this index was even less. Here, in August 1914 fighting occupied 20.7% of the time, and 5.5% in September.

The operational saturation of the depth with combat events was thus still limited, possessing the nature of a broken series of battles. However, what was qualitatively new in this phenomenon was the fact that this broken series constituted a single operational chain.

Thus the operation in the beginning of the 20th century formed as a *chain of combat efforts, continuous along the front, united according to depth, and unified by an overall plan for defeating the enemy or resisting him.*

The main task of operational art, as the study of conducting the operation, became the unification of individual tactical, not directly connected, combat efforts in space—along the front, in time—in the depth for achieving the overall assigned task; in other words—to bring the entire line of combat events into such an active and well-oiled system along the front and in depth, which vigorously and sequentially leads to the defeat of the enemy.

The resolution of this task has placed operational art before a new and complex problem of commanding armies formed up in a solid front in a single line.

One could have foreseen earlier that in conditions of the imperialist war, when both belligerent coalitions equally pursue aggressive and expansionist goals, that armed struggle would take on a cruel and exhausting nature and, given a certain economic equality of the sides, would freeze up in stagnant forms of an attrition struggle.

An entire series of objective prerequisites—the enormous size of the armies, the colossal power of fire, the imperialist goals of the war, which were foreign and hostile to the fighting masses—led by various lines to just this view of the course of events.

As early as 1887 Engels wrote about a future war:

> This would be a worldwide war of unprecedented size and unprecedented power. From eight to nine million soldiers will stifle each other and at the same time will pick Europe so clean as she has never been devoured by swarms of locusts. The destruction caused by the Thirty Years' War, compressed into 3–4 years and spread out over the entire continent; famine, epidemics and the overall barbarization of both the soldiers and the broad masses, caused by deep need; the hopeless confusion of our artificial mechanism in trade, industry and credit will all conclude in general bankruptcy; the collapse of the old states and their traditional state policy; such a collapse that crowns will lie on the pavement by the dozens and no one will be found to raise up these crowns; the absolute impossibility of foreseeing how all of this will end and how the victor will emerge from the struggle—only one result is absolutely sure: a general exhaustion and the creation of conditions for the final victory of the working class.

These brilliant words of Engels foresaw the entire character of imperialism's armed struggle 30 years ahead of time.

The final resolution of the problem, by one means or another, of conducting operations—could in no way be achieved within the framework of the imperialist war.

However, when the operational art of the imperialist age, conducted by the representatives of an old and dying class, proved to be incapable of rising to the level of the new demands of the time and relied on conservative military theory, which traces its roots back to the Napoleonic era—the course of the armed struggle revealed its contradictions much quicker and more vividly, having demonstrated its complete helplessness to achieve at least an operational resolution, even within those confines in which, objectively speaking, this was possible.

By no means were all the new factors in the evolution of the operation's nature taken into account. In order to construct a system of modern operational art, it is vitally important that all of this be revealed in its entirety.

In the Moltke era, when armies did not yet occupy a solid front, their deviation to the right or left, their gathering in one spot, or, just the opposite, their dispersal along different axes, and, finally, the turning of the entire mass in another direction—were still backed up by sufficient freedom of maneuver.

In these conditions, Moltke the strategist possessed a broad field of operational activity, which required active operational control during the very course of events.

However, when the deployed armies formed a continuous line and occupied the entire zone of their deployment—the maneuver of the front in space by means of a change of direction acquired a new qualitative distinction.

Great management skill was required, so that by means of an active operational leadership one could take advantage of each specific situation in order to achieve the actual defeat of the enemy. This could only be achieved by keeping the armies in tight harness, by holding back some and pushing others forward for the purpose of their gaining the flank and rear.

This is what Schlieffen had in mind, when he said that it is necessary to control modern armies like battalions.

The age of the linear strategy of the continuous front by no means yet excluded operational maneuver within its confines. The broken front line, which arose during the course of combat activities, still offered sufficient opportunities for this.

However, the operational art of commanding troops proved incapable of rising to the level of the new demands. It continued to live off the fundamentals of the Napoleonic school and in an age of the continuous deployment fronts it continued to suppose that the conduct of the battle does not fall into the sphere of its competence and that with the start of the battle it can rest. Given this approach, there was nothing for operational art to do but to limit its activity to the grouping of forces and aiming them in specific directions.

The entire evolution of the nature of the operation, which has shown that as early as 1870 the maneuver-march was organically growing into the battle and that each battle in the new conditions of the continuous front of combat activities already contains within itself the prerequisites for the succeeding operation—remained completely misunderstood. Thus it was not understood that operational art in the new conditions requires effective and unbroken control over the course of the entire operation, including the battle.

In this lack of understanding one could feel the deep-rooted conservatism of military theory, which had become frozen on the extremely important question of troop control at the level of the beginning of the 19th century.

The entire conduct of the operation in 1914 came down to defining and directing groups of forces. From the very beginning, armies received their distant reference points and headed for them along specific axes.

This was suggested even before the World War by Bernhardi,[42] who said that modern armies should be like an unleashed arrow. But, as is known, an unleashed arrow is no longer subject to control, which is what happened to the German armies in 1914.

The army formations ended up being fixed to specific axes and were aimed at their distant reference points, without taking into account that possible situation which, upon arising along the path of the offensive, might require a completely different decision.

An operational art that did not correspond to the new conditions of conducting the operation gave birth during the World War to the strategy of long-range aiming, the chief distinction of which became the complete disregard for the facts of the immediate situation.

Taking account of the entire specific situation at the beginning of the battle or employing it upon the battle's conclusion did not constitute the subject of operational concern at each given stage of the development of events.

Moltke taught:

"It's as if each battle is a staging point for new strategic decisions ...

"The materiel and moral consequences of the battle are so huge that a new situation is always created depending on its outcome. Much of what was planned earlier becomes unrealizable and, just the opposite—much of what could not earlier be calculated becomes possible."

This indisputable tenet was forgotten.

The armies advanced along their specific axes, independent of the outcome of the battle and even independent of where it was to take place.

Events played out in their objective development without any kind of influence on the high command, which found out about the outcome of the battle after the event, which had already led to further consequences.

Operational art had removed itself from the effective control of the course of events, leaving them to flow along their assigned axes.

The operation had become uncontrollable summed up operational art's enormous contradiction, which in 1914 could not find its place in the system of conduct of military activities. Operational art, having excluded, according to Napoleonic tradition, the battle from the sphere of its competence, had nothing to do and really did rest during the entire time of the Marne maneuver-march.

Not finding for itself either place or application, German general headquarters hid back in the deep rear and even if it had not existed in 1914 there is little that would have changed in the course of historic events.

As a result, an entire series of extremely favorable operational circumstances, which literally placed success in the Germans' hands, was missed. For example, the French Fifth Army, which had already been jammed between the Sambre and Meuse,[43] got away from certain destruction in the Battle of the Frontiers.[44]

The German general headquarters did not even exert itself with an operational examination of the Battle of the Frontiers and, completely indifferent to the new situation, noted on 27 August 1914 in its war diary: "The German armies are being ordered to advance in the direction of Paris."

This movement changed into an offensive in general, into a long-range offensive, ignoring the given specific situation, jumping over it and thus unfolding indiscriminately and to no purpose in regards to the enemy group of forces.

The situation actually came down to the simple, mechanistic shift of an unchanging grouping of forces in the depth in the space from the Rhine to the Marne.

It was thought sufficient to move one's operational efforts forward, as if this constituted the entire meaning of the operation. It's real goal—the defeat of the enemy's army—was lost in the operational perspective.

Operational art was least of all occupied with the concrete problem of where and how to defeat the enemy; it replaced this with the problem of where and when to arrive.

The attacker began to simply push back with his entire front that enemy who should have been seized, in order to defeat and crush him.

Naturally, the defeat of the enemy's army in these conditions became impossible.

The offensive operation had turned into the push-back operation.

Thus from the linear strategy the idea of crushing and destruction, which had constituted its main meaning, when it arose during the second half of the 19th century, had been emasculated. And this became the first indicator of its degeneration, having uncovered operational art's complete helplessness in the age of imperialism to rise to the new demands of controlling armies in the 20th century.

The conservative influence of false methods of operational leadership later proved to be so deeply rooted that in completely different conditions of revolutionary-class war and with a completely different army—the offensive of the unleashed arrow was repeated in full.

Our 1920 campaign to the Vistula,[45] which once again broadly reproduced the phenomena of the linear strategy, was analogous in its methods of operational leadership to the German march to the Marne.

Again the armies were unchangeably tied in their grouping to definite axes to an enormous depth of 600 kilometers; again long-range goals were assigned and the immediate situation was completely disregarded; once again an indiscriminate straight-line offensive was conducted without regard to the given situation that had arisen, and once again operational leadership was absent during the course of the battle, "resting" deep in the rear.

This resulted, once again, a series of brilliant missed opportunities. The 3rd Cavalry Corps and the 4th Army occupied a position well forward along the Neman, Narew and the Wkra rivers. However, instead of taking advantage of the immediate situation and turning against the flank and rear of the enemy, where a bountiful operational harvest awaited, they each time, like unleashed arrows, aimed straight ahead toward far-away reference points, bypassing the Poles' exposed flank.

The result was the mechanistic shift of an unchanging group of forces from the Dvina to the Vistula and a vividly expressed indiscriminate offensive "in general," about which comrade Stalin spoke: "An indiscriminate forward movement is death for the offensive."[46]

The death sentence was pronounced over the linear strategy at that moment when it degenerated into a simple pushing by a continuous wall, before the front of which the retreating enemy could freely regroup for launching a counterblow; and then it came to pass that the advance of this wall is possible as far forward as far as its movement backwards is inevitable, or its collapse from a simple pin prick in the flank.

In this we see the entire enormous contradiction of operational efforts, frozen in their grouping along a single unchanging direction. When this happened, the operational art that was not cognizant of the ongoing events became hopelessly confused.

Now, some Hentsch[47] or another, as the sole embodiment of the entire system of operational control, had to resolve that which had essentially become insoluble.

After all, a change in the grouping of the armies that had become embroiled in the battle along the entire front of their activities could only be achieved by changing the correlation of forces along individual axes.

But for this it was necessary to feed the front from the rear and to have deep reserves. However, the linear strategy was linear namely because it did not have and did not recognize any operational reserves.

In the spirit of Napoleonic times, the battle was still viewed as a single-act effort, requiring the simultaneous commitment of all available forces into the fray.

They continued to cite the authority of Clausewitz, who wrote:

"All forces which were designated and are available for carrying out a strategic task should be employed simultaneously, as their employment will be all the more absolute and complete the more everything can be compressed into a single moment."

However, that which the great thinker properly sought to establish according to the experience of the Napoleonic wars proved to be completely incorrect in the 20th century, when the operation became a multi-act phenomenon and grew into a series of consecutive operational efforts, dispersed in depth.

And when along the Marne and Vistula, essentially insignificant reserves were required to parry the enemy's attack; the operational leadership did not dispose of a single division.

At this highest point, the development of events according to the linear strategy was essentially finished: for if operational art had put itself in a position in which it was completely helpless to do anything, this means that death had ensued.

And then, naturally, they remembered the old teachings and called upon the shade of Schlieffen for help. The resolution must be sought along the flanks; he will have success whose flank is longer; and in this they still sought to find salvation and began a feverish race to the sea.[48]

However, at the same time they failed to consider that the enemy could do the same and that the extended flanks would essentially only lengthen the already stagnant front.

At this stage operational in the World War ran into even more insoluble contradictions.

As early as the dawn of the 20th century, given the colossal increase in the armed forces, one could foresee that the growth in the size of armies would overtake the length of the fronts containing them, which have a limit set by either nature or the presence of neighboring countries. The striving toward the unlimited spread of one's flank in space, in a competition between two sides, would inevitably lead to its natural limit.

This happened along the Western Front in 1914 as early as the second month of the war. Having spread a distance of 700 kilometers in space, the Western Front's flanks found their limits along the seashore in the north and in neutral Switzerland in the south.

There was no place left to expand. The spreading of combat efforts along the front—that first indicator of the operation—completed its evolution in the World War. The broadened front had reached its natural geographical limit, beyond which it could obviously grow no further, assuming the sea was not preparing to dry up.

At the same time, the linear strategy had arrived at its antithesis. After all, its entire meaning consisted of broadening the front to achieve an envelopment in order to avoid a frontal attack. This possibility had now disappeared; freedom of maneuver along the front had been lost and, it follows, the linear strategy had lost the fundamental idea that had given birth to it.

It carried in its evolution all the factors that inevitably led to its self-negation. Its ideologist, Schlieffen, did not foresee this. His teaching on the all-round strengthening and spread of the enveloping flank, which was expressed in the "Cannae" idea, as the highest expression of the linear strategy, appeared just when this strategy's days were essentially numbered and when there were already lurking all the indicators of its antithesis in the objective course of events. Schlieffen's "Cannae" was undoubtedly *written too late* and its outstanding author should have lived earlier.

When front faced front, it was essentially all over for the linear strategy. There remained nothing else to do but to resort to the *breakthrough*.

That which following the war of 1870 was seen as impossible became necessary during the World War. A stray division appearing on the flank could no longer teach, as had been the case from the time of Leonidas, how to overcome strong positions, because there was no longer any flank.

And then we had to return to the battle of 1870 at Gravelotte—St. Privat and transform the wild and unrestrained attack into a planned breakthrough of the defensive zone. Thus the evolutionary circle had closed. It had led to the great frontal battles of 1918 and created a new stage in the development of armed struggle. It had become evident that the age of the linear strategy had ended and that the solution for the problem of the breakthrough should be sought for along new paths in the evolution of operational art.

At this time the imperialist war had already fully revealed its extended and ennervating nature.

The task of overcoming the front of fire seemed overwhelming and self-sufficient. In its tactical content it had been elevated to an end in itself, and operational art, which obligates us to organize and support the frontal attack, was assigned to serve it. The task undoubtedly required that the superiority of attack means over defensive ones be achieved.

This was an enormous technical problem, which had become topical as early as the onset of the age of fire annihilation. The superiority of defensive weapons over offensive weapons could not be doubted before the World War. This prerequisite was the starting point for the shift in the center of effort to the flanks. Nevertheless, Schlieffen was deeply occupied by the need to supply the German army with powerful attack weapons. Specifically, this was expressed in the plan to form heavy artillery for the field army. The problem seemed completely new and unusual. It's of interest to note that the following perplexing note was attached to Schlieffen's memorandum: "Does the chief of the General Staff really want to turn the heavy artillery into towed artillery with the troops?" Schlieffen laconically replied to this: "By all means."

The German army was the first to introduce heavy artillery for its field troops. But even with these weapons it proved to be completely impossible to resolve the problem of the correlation of the means of defense and attack in favor of the latter.

The entire offensive to the Marne was nowhere able to overcome the front of fire and was capable only of pushing it back. The destruction operation was transformed into the push-back operation and this was one of the factors behind the degeneration of the linear strategy.

When the continuous fronts reached their limit the competition between the means of defense and offensive was the chief axis around which—all the way up to our time—the evolution of the technical means of struggle began to revolve.

From the military-technical point of view, the entire meaning of the World War from the end of 1914 boiled down to the struggle between the means of attack and the means of defense.

At first, the competition was undoubtedly resolved in favor of the latter.

The basic means of fire annihilation—the machine gun—was much easier and cheaper to produce in mass quantities than artillery weaponry—the chief suppression means of this means of fire annihilation. If the number of machine guns in a division rose an average of 20 times in four years of the World War, then the amount of divisional artillery barely increased two times. The superiority of fire remained with the defense. This required an enormous concentration of artillery means of suppression.

The average norm was established at 60 guns per kilometer of front. Actually, this figure was exceeded significantly, reaching concentrations of 100 and more guns per kilometer of front.

However, even such a massing of artillery weapons of suppression could not, in the final analysis, resolve the problem of overcoming the front of fire. Only the forward field of the defensive was actually suppressed; its entire depth, for the most part, remained untouched. The offensive's artillery was incapable of shifting its firepower throughout the entire tactical depth of the defense, because it could not keep up with the advance of the attacking infantry.

The problem was not in the power of fire suppression, but in its mobility, which encountered insuperable obstacles in the battlefield's terrain, which became inaccessible for horse transport and wheeled vehicles.

The entire tragedy of the attacking infantry was that following 3–4 hours of a successful attack, of the 100 guns that had initially supported it, only a paltry number continued to fire. And then the attack would play itself out and expire.

It became obvious that the problem had to be resolved not only by means of quantitatively increasing certain suppression weapons, but by seeking out new ones. It was necessary to create such a suppression weapon for suppressing firepower that would, first of all, be protected against them; that is, armor against the machine gun, and, secondly, would be mobile over any terrain in order to penetrate into the depth of the defense and immediately suppress and fire point-blank on the fire means of annihilation.

This idea, which was summoned up by the demands of the situation, was realized in the construction of the tank, as the combination of the internal combustion engine, tracked movement, armor, and firepower. The very fact of the tank's appearance had enormous significance for resolving the problem of the superiority of attack weapons over defensive ones.

The requirement of suppressing the entire tactical depth of the defense also called forth other means of struggle: there appeared the combat airplane as a transporter of fire in the air, as well as poisonous substances, the employment of which did not have a trajectory and immediately achieved spatial envelopment.

The colossal technical progress during the years of the World War, which were called forth by the new conditions of positional warfare and supported by a high level of industrial development, directed the resolution of the problem of defensive and offensive weapons along the path of the domination of the latter.

However, this happened, at first, theoretically. Practically speaking, attacks with tanks were not successful at first. The reason for this was in the lack of skill in their tactical employment them and in the low level of their combat usage. The tank did not immediately resolve the problem of overcoming the front of fire. The first tactical resolution was achieved by the Germans in 1918, even without the tank. It was only at the end of the war that the new offensive weapons showed the tactical possibility of "breaking into" the front of fire.

However, this took place in conditions when, on the German side, the imperialistic war was growing into a civil war and the masses were turning their weapons against their ruling classes. The question of the strength of the defense should then have received a different political evaluation altogether.

Nevertheless, all the prerequisites for resolving the problem of overcoming the front of fire were on hand during the final period of the World War.

However, it was precisely at this point that operational art proved to be completely impotent.

Having occupied itself completely with the tactical organization and materiel support of the breakthrough, it essentially eliminated itself as the art of conducting the operation. The very tenet that it was necessary to place tactics above strategy speaks of operational art's loss of its purpose.

Operational art, having set tactics against itself, permitted a completely ridiculous contradiction with it to appear. After all, tactics and operational art are phenomena of the same order and differ only in their scale and dynamism. They not only abide together in the process of military activities, but organically grow from one into the other.

If a tactical effort does not grow into an operational achievement, it essentially becomes pointless. A tactical effort is only a step toward the achievement of the goal and can never become an end in itself.

But this is what happened in the events of 1918 and is how the problem is still understood in our time.

For example, Col. Duffour[49] still speaks about the experience of 1918.

"The continuous fortified front ceased to be a simple wall, behind which strategic maneuver can play out to the fullest; *it has became the main goal of this maneuver.*"[50]

The entire problem of the breakthrough was reduced to the tactical breaking in of the front. The entire problem was resolved only at the tactical level. The crowning of tactical efforts with an operational achievement fell completely by the wayside.

"A characteristic feature of the breakthroughs in 1918," writes Gen. Dubeney,[51] "was the fact that only the first phase—the breakthrough of the front—was planned; the development of the operation was lost from view."[52]

Operational art failed to secure the growth of tactical efforts in breaking through the front into a complete operational breakthrough and rout—and this constitutes its bankruptcy during the imperialist war.

The unification of tactical efforts along the front—the fundamental indicator of conducting the operation—remained unresolved.

Combat efforts outside of any system and link, dissipated in space, without any kind of prospect of uniting them for achieving the overall goal.

The attacker in this mortal combat would throw himself first against one, then another, sector of the front and, at best, was able to drive a nail into each. This type of activity, which was obviously doomed and incapable of leading to the achievement of decisive aims, was even elevated into a system of the battle of attrition with limited aims and is even now put forth as an historical necessity of our era and as the most advanced theory of breaking through the fortified front.

There was naturally an entire series of political and economic prerequisites for such a system in the conditions of the imperialist war.

In the conditions of 1918 this or that possible conduct of the breakthrough operation, in the final analysis, could no longer decide the outcome of the war, although it undoubtedly influenced the sides' political, economic and military situation. The solution was maturing along other lines.

However, the political futility of military activities does not yet presuppose their operational absurdity. Operational art consists not only taking into account objective conditions, but also in the fact that, having taken them into account, of overcoming them within the framework of objective possibility.

The system of attrition battles could in no way operationally resolve the problem of breaking through the positional front and was thus senseless. As concerns exhausting the enemy, this system more likely exhausted the attacker than the defender: this was a senseless system of mutual exhaustion, which is vividly clear from a comparison of the men and material expended by the offense and defense in all of the breakthroughs of 1916–1918.

The application of this system revealed the entire helplessness of operational art at a dead end and essentially turned into a senseless system of driving in nails.

But no wall will be destroyed because they start hammering in nails. In order to bring the wall down, it's necessary to undermine its foundations; that is, to get underneath it through the cracks made in it.

Here, however, operational art proved to be even more helpless. The prospects of an operational development of tactical efforts in depth were not foreseen at all.

There were no operational echelons for developing the breakthrough, and in this was vividly revealed the stagnant influence of the already outdated linear strategy.

When a crack in the positional front really did appear, as in the Germans' March breakthrough of 1918, the attacker lacked anyone who could extend the attack into the depth through the newly-formed breach, in order to transform the tactical breakthrough of the front into an operational one and a rout. All of the grandiose efforts for the tactical organization of the breakthrough, all of the technical progress in weaponry, and all of the enormous concentration of men and weapons of suppression—essentially proved to be in vain if the tactical success was powerless to grow into an operational achievement.

It made no sense to try and break down the door, if there was still no one to go in. But this is exactly what happened with all the breakthroughs in 1918.

The imperialist war did not resolve the problem of the breakthrough; it ended, not having proved the possibility of realizing this at the operational level.

If, in the final analysis the German front fell, then this event went beyond the bounds of a solution of the problem by purely military means.

The German front collapsed in 1918, not so much from without as within—under the influence of the powerful process of the revolutionizing of the masses, which led not only to the collapse of the front, but to the overthrow of the monarchy.

Of course, the Entente's[53] colossal economic superiority in men and materiel exerted a huge influence on this process. But there was no place for operational art. And even after the German front fell, because it had lost defensive steadiness, it still took the Allies four months to push back the demoralized German forces a total of only 100 kilometers.

Bear in mind that Foch[54] was not even preparing to decide the outcome of the war in 1918 and was preparing a general offensive for the following year. But before the decisive attack began, the Germans threw away their weapons on the battlefield and a resolution was achieved.

And after this Culmann has the audacity to declare: "During the last four months of the war the French command showed how a breakthrough should be conducted and to what kinds of results it leads."[55]

This stupid braggadocio sounds like cruel irony over the prostration which seized the general staffs of the imperialist countries in the final period of the World War.

The operational art of this era proved helpless to resolve the new problems put forth by the new nature of armed struggle. It hardened at the level of the linear strategy and became powerless when this strategy arrived at its antithesis. The problem of the operational overcoming of the front of fire remained unresolved.

We thus arrive at the present with such an operational approach.

With a completely different political nature of war, with a new army and on a new material-technical base, our operational art must resolve the problem which was not and could not be resolved in the conditions of the imperialist war.

The linear strategy began with brilliant solutions during the era of national wars in the second half of the 19th century.

During the world imperialist war of 1914–1918 it arrived at its self-negation.

And now in the new age of revolutionary-class wars, a new solution must be found. In this lies the enormous task of our operational art.

Part II: The Justification for a Deep Strategy

1. *The Fundamentals of Our Operational Art*

"Any war must be viewed, first of all, according to its probable nature and its general outlines, based upon its political magnitude and relations. Only politics may occupy the leading position, from which extend the main lines directing the war" (Clausewitz).

It's perfectly obvious that along the new paths in the evolution of our operational art it is necessary, first of all, to proceed from the nature of our future war as a revolutionary-class war.

Being the highest manifestation of the class contradictions of two mutually-exclusive social systems, this war will have the nature of a decisive collision of world-historical significance.

In the history of the struggle between peoples and classes, the intensity of this war, according to the radical nature of his goals, will reach its highest pitch.

History vividly shows how the intensity of war has increased with a change in its political character.

The French revolutionary wars immediately threw enormous masses into the struggle, achieving an unprecedented scale. The intensity of the national wars of the second half of the 19th century seemed unusual to those living at the time. In supporting the achievement of policy's goals, this intensity was, nevertheless, historically insignificant. The struggle that was waged for national unification did not arouse the reactionary side to wage war to the last, to the death. The matter was not pushed to the point of the losing side losing its national independence, because such an outcome was not called for by the requirements of politics—agreement was usually achieved easily at the last moment.[56]

The 1914–1918 World War called forth an entirely different intensity. The reactionary imperialist character of the struggle for the division of the world and world hegemony, being the continuation of the exacerbated economic competition between the capitalist countries at the final stage of their development, put forth the goal of the complete economic enslavement of the enemy side and brought the intensity of the war to lengths unheard of in history.

However, the internal contradictions of imperialism led to a situation in which this intensity, which brought about enormous bitterness, at a definite level awoke the proletariat's class self-consciousness and was transformed on the Eastern Front in 1917 into its antithesis and led to revolution, to fraternization between the troops and to the strengthening of the international solidarity of the toiling masses. This was the realization of Lenin's magnificent slogan about transforming the imperialist war into a civil war.

In national-liberation wars the intensity of the struggle acquired a new revolutionary content. This called forth a new power of intensity, flowing from the great arousal of the enslaved popular masses, arising to struggle for the destruction of exploitation.

In the new age of socialist revolutions and revolutionary-class wars, the complex system of socio-political mutual relations predetermines the inevitability of three types of wars—imperialist, national-liberation and revolutionary-class, characterized accordingly by varying their intensity.

The struggle undoubtedly reaches its highest pitch in a civil, revolutionary-class war. This war, being a manifestation of the class struggle at the highest stage of its development, according to the antagonism of the opposing classes, according to the antithesis of the economic systems of socialism and capitalism, according to the decisiveness of its aims, aimed at the overthrow of one of the sides and excluding the other, raises the intensity of struggle to its highest historical limit.

In completing with themselves in history the final stage of wars as social phenomena, while destroying the very institution of war, revolutionary-class wars represent the concentrated expression of struggle in its highest maximum intensity.

In resolving the greatest problem of historical progress—the transition to a new communist society of free labor—these wars must occupy an entire period and unfold in an enormous part of the globe. Our civil war of 1918–1921 was only the first act of these wars and even greater events undoubtedly await us.

"...A long series of wars," said comrade Lenin at the VIII Congress of Soviets, "has up to now decided the fate of all revolutions and of all the greatest revolutions. Our revolution is just such a magnificent revolution. We have completed one stage of war and we must prepare for the second" (vol. XXVI, p. 35).

At this stage of wars we are speaking not about the resolution of some sort of local contradictions between just two given sides in a struggle, but about the resolution of a dispute between two ages, two mutually exclusive systems at the world level.

These wars are resolving the world-historical problem of liberating millions of the exploited masses, and this historical significance of a future war predetermines its decisive nature and enormous intensity.

If in 1887 Engels wrote about a future imperialist war that "...this would be a worldwide war of hitherto unknown scope and unprecedented force...," then it is difficult to find the words to describe the enormous and unprecedented scope and intensity of revolutionary-class wars.

Enormous multi-million masses will be drawn into these wars. If bourgeois military writers (Soldan,[57] Seeckt[58] and Fuller) are preaching the theory of small professional armies, then this reveals best of all the insoluble contradictions among the capitalist countries in the development of their military system, which in no way corresponds to its actual requirements at the present stage.

These wars will be waged on a high material-technical base, which at the present level of industrial development enriches the arsenal of weaponry with means of struggle of unprecedented destructive power.

These wars will require colossal material resources and enormous economic exertion.

Only one outcome is possible in these wars—that is the death of capitalism and the triumph of a new world—the world of socialism. Undoubtedly, never in history has a struggle been waged for the realization of such great goals and never has a single army been called upon to resolve such supreme historical missions.

Such is precisely the calling of our Red Army as the first class army of the proletarian dictatorship.

The nature of struggle along the armed front and the character of operations in our future war are determined by this basis starting factor of the historical significance of revolutionary-class war and "its general outlines, based on political magnitudes and relations."

M.V. Frunze[59] spoke about this struggle:

"...in a class war, in a civil war, the outcome can only be the complete defeat of one of the sides; once the war has started, halfway decisions are impossible" (*Sobranie Sochinenii*, vol. I, p. 400).

"...From the depth of those contradictions which exist between two mutually exclusive worlds, it's obvious that such a collision, when it arrives, will be a decisive collision. This will be a struggle to the death. This will be a struggle to the end, to the victory of one or side or another" (*Sobranie Sochinenii*, vol. III, p. 112).

The decisive nature of the collisions predetermines the decisive character of military operations. These will not be languid and extended attrition operations with a limited aim, but predominantly active and crushing blows with a decisive goal. The modern technical means of war, fast and mobile and highly active in their combat application, also result in a similar nature of operations.

However, the offensive power of the belligerent sides in these decisive operations is far from equal.

For our part, a war against the imperialist predators is historically progressive and thus just. In it we are defending the goals of worldwide-historical significance. As early as the civil war of 1918–1921, we acted as a progressive factor of world significance. Lenin wrote the following about the Polish-Soviet war:

"In the summer of 1920 Soviet Russia acted not only as a force defending itself against violence and the pressure of the Polish White Guards,[60] but actually acted as a worldwide force capable of destroying the Treaty of Versailles and liberating hundreds of millions of people in the majority of the world's countries" (vol. XXV, p. 419).

This worldwide-historic role imparts to us in a future war the greatest strength for overthrowing the class enemy—a strength proceeding from the political conditions of the struggle of the young, historically progressive class against the old and decaying world of capitalism, which has been seized by a cruel economic crisis. Progressive and just war aims have always imparted an enormous offensive power to revolutionary armies; the actions of the armies of the great French Revolution revealed this in a particularly vivid way. As early as the civil war of 1918–1921, the Red Army, being the army of the great socialist revolution, showed its great offensive power. History shows that great historical tasks are resolved on the basis of the offensive and that the revolutionary army must be educated to be ready for the most decisive offensive actions.

Lenin wrote as early as 1905:

"In the final analysis, only force decides the great questions of political freedom and the class struggle; and we must concerns ourselves with the preparation and organization of this force and its active, not only defensive, but offensive as well, employment" (vol. VIII, p. 42).

This commandment of the leader remains the fundamental directive for the development of our entire military system to the present day.

The fundamentals of our military training and our operational art—are the fundamentals of the offensive.

Here there is no contradiction with our peace policy. We are struggling and will struggle with all our might against war. Our peace policy is unchanged. Comrade Stalin's famous words declare: "We don't want an inch of anyone else's land. But as for our land, we will not surrender an inch of our own land to anybody." And in this determination to defend the first socialist country in the world there rests an enormous active force, ready to crush and destroy any attacking class enemy.

The entire meaning of the class struggle transforms for us a progressive war into an historic offensive against any enemy attacking us with shattering with thunderous destructive attacks. Our future war, which is a continuation of the civil war of 1918–1921 at a new level of its development, may thus proceed only from the fundamentals of the *strategy of the attack and destruction.*

"We must organize things in such a way," says comrade Voroshilov,[61] "in order to achieve victory in a future war with few losses and wage this war on the territory of the country that first raises its sword against us" (From a speech to the XI All-Union Congress of the VLKSM[62]). This determines the fundamentals of our operational doctrine of the decisive offensive—the doctrine about which the 1929 field manual says the following: "The Red Army, which is defending the interests of the toilers, must be ready for bold and decisive actions, directed at destroying the armed forces of the class enemies."[63]

This guiding article in the field manual offers us a basis for constructing a theory of our operational art as the art of conducting destructive offensive operations with the most decisive goal of completely overthrowing the enemy.

The tasks of our operational art consist of creating new and brilliant models of military art in new historical conditions, with a new army, on a new material-technical basis, with a new content and in completely new forms.

The great goals of war cannot but call forth equally great operational acts. And in these historical conditions, the strategy of destruction has never yet had such a completely historical justification and such favorable prerequisites for its realization.

2. The Evolution of the Nature of Operations in a Future War

The basis for the construction of the theory of our operational art is the concept of the most decisive offensive operation. The entire character of a future war speaks to the grandiose scope of this operation, thus defining also the further evolution of its main indicators.

The historical development of the operation's character shows that its evolution ran according to two main indicators—the operation's spreading along the front and its dispersal into the depth.

The first of these indicators—the spreading along the front—finally concluded in the 1914–1918 World War. By then the armed struggle already occupied a continuous of front combat activities that had fused into a single line of points of application and spread to the furthest possible geographical limit, further than which it could no longer spread along the front.

We have no reason to suppose that in a future war that history will turn back in the evolution of this indicator of the operation. We do not share and cannot share the point of view espoused by the completely contradictory bourgeois theory of small professional armies, which might, in the opinion of their apologists, change this indication in the opposite direction and bring us once again to a broken front of individual and dispersed in space points of applying combat efforts. History does not move backwards and we must more likely presuppose the opposite—that the spread of operations along the front, if it is supported by geographical conditions, will have a tendency to grow even further.

Our western frontier alone is 3,000 kilometers long and all of it—from the shores of the Arctic Ocean to the Black Sea—is threatened in the case of an intervention.

However, the problem is not confined to just our western frontier, and demands scrupulous strategic attention to an entire series of other frontiers, particularly in the Far East.[64] Strategy has undoubtedly never faced such an enormous special scope of a possible continuous front of struggle. Of course, in these strategic conditions, there can be no talk about the degradation of the indicator of the spreading of operations along the front.

In taking up the evaluation of this question on the operational scale, it is necessary to keep in mind that along the 800-kilometer Soviet-Polish border we may be faced with an average of one division per 10–12 kilometers of front. And because this sector of the strategic front contracts as it moves from east to west, then at the line of the Vistula and San rivers we may already be faced with one division per every 6–8 kilometers of front. At the same time, it is necessary to take into account that the unending growth of the intensity of mobilization has a tendency to further increase this operational density.

Of course, the tactical density may be incomparably higher. The uneven occupation of the front, while pursuing the goal of creating dense shock groups of forces, conditions at the same time the presence of thinly or even unoccupied sectors of the front. This circumstance, in conjunction with the enormous length of our western frontier and the presence along it of several small states,[65] even the coalition attack of which cannot guarantee their close coordination and truly unbroken front, forces us to assume that along our western frontier of military activities there may undoubtedly appear individual operational windows against the overall background of a continuous front. Operational flanks may still be found here; and modern rapidly-moving and fast-acting weapons (motor-mechanized formations and aviation) may, given their skillful employment, create them by their active operations. We should basically see this as their main operational designation in the beginning period of a war.

Our strategic conditions have some general features in common with the conditions of the Franco-German front in the beginning of the 1914 World War, when the German armies could still freely wheel their right flank.

This means that for us the prerequisites of the linear strategy are far from having disappeared. As concerns our individual eastern theaters of military activities, then this strategy will still be fully employed.

The historical process knows no sharp and precisely delineated boundaries between two eras. Having created the prerequisites for new conditions, by the same token it does not yet fully eliminate the possibilities for the old ones and moves from one to the other in a dynamic of dialectical development.

Enveloping maneuvers along external lines in the beginning period of the war are by no means excluded for us. Maintaining that from the first day of the war that front will be opposed by front is undoubtedly a mechanical transference to our front of military activities the conditions of the Franco-German front, where the prerequisites for a linear strategy of enveloping maneuvers really should be considered to have already been excluded.

However, in viewing the entire prospect of the development of military activities, we should foresee the inevitability, or at least the likelihood, of front opposing front, with incomparable greater foundation and much more quickly than happened with the Franco-German front in the beginning of 1914.

To approach with clear foresight this phenomenon, which has been predetermined by the entire historical evolution of the operation's nature; to approach it fully armed with all measures of counteraction, so that the dialectical development of the enveloping linear maneuver growing into the deep frontal attack takes place simultaneously with the immediate shift from some operational methods of action to others, undoubtedly remains the chief and central task of our operational art.

Extremely weighty objective prerequisites force us to make our operational forecast in just this way.

The broadly developing opportunities for operational maneuver now enable us to incomparably more rapidly lengthen our flank with new forces and face the outflanking enemy with a new group of forces than was the case in the World War. Then, during the German offensive to the Marne, the French managed to transfer only 11 infantry and six cavalry divisions to their threatened left flank around Paris. Now, given the modern conditions of rail transport and motorization, we could expect the transfer opposite one of the central sectors of our western front, from one flank to another, of up to one-third of all the armed forces that could be arrayed against us in a comparatively brief time. Moreover, air transport, which is reducing the time for new concentrations to a minimum, is at the service of operational maneuver.

This circumstance, despite the feasibility of the rapid development of flanking maneuvers with modern motor-mechanized and aviation equipment, conditions the undoubted opportunity for opposing a continuous front to the same kind of front of attack. Given this, one should take into account that the means of countering a flanking maneuver now dispose of greater opportunities in both aviation and obstacles.

"Front against front" should not become an unforeseen surprise for our operational art in a future war, as happened to the Germans in 1914. This will most likely be a normal phenomenon in the dynamic of the development of the decisive flanking maneuver into the frontal attack, which must become just as decisive and be calculated to the entire depth of the enemy's position.

This problem brings us to one of the chief problems of modern operational art, to the problem of the evolution of the second operational indication—its breaking up into the depth.

We've seen[66] that during the maneuver period of the World War not everything was achieved in this sense: there was, to be sure, a chain of connected battles in a broken series, which were by no means filled with combat activities throughout the entire depth of the offensive. In a future war the operation's character will undergo its main evolution namely according to this indicator of depth. We undoubtedly must take cognizance of the enormous combat thickening of the entire depth of activities. Essentially, as early as 1918, when the Germans, in their March offensive in Picardy, penetrated to a depth of 60 kilometers; when the combined troops of the Entente in the last months of the war penetrated 100 kilometers into the depth of the German front, they conducted a single unbroken battle throughout the entire depth of this offensive. Combat activities already filled the entire depth of the advance.

In a future war we will collide with this combat depth as a normal phenomenon. This proceeds, first of all, from the depth of the operational organization of the modern combat formation. This includes not only the organized defensive zone, but the depth of the operational organization in any situation.

The immediate fighting line of divisions occupies a tactical depth of 6–8 kilometers. At 8–10 kilometers behind it one must deal with the immediate troop reserves, which form a second line. Still further in the rear, at a distance of 20–25 kilometers from these reserves, one must presuppose the presence of individual groups of army reserves, which form a third line. Finally, all this depth of the operational formation relies on railroads even further in the rear (25–50 kilometers from the third line, depending on conditions), along which new reserves can arrive at any time.

Thus the modern operational organization of the combat formation reaches 60–100 kilometers in depth. If this organization goes over to the defensive, then, while maintaining the same depth, it will take on the appearance of consecutively echeloned fortified zones. At the same time, it is necessary to keep in mind that this depth may be rapidly fed by new reserves and, to a certain degree, be continually supported, if its forward edge were to be broken and fall back; it can grow again by reinforcements transferred from the rear and other sectors of the front. This is conditioned by the possibilities of modern permanent mobilization.

It is becoming obvious that we will have to try and overcome the entire modern operational depth and pass through it in an uninterrupted series of *combat efforts*. Each kilometer will have to be taken in fighting.

If combat activities occupied only 23% of the time during the Marne maneuver-march, then now this percentage of "combat content" throughout the entire course of combat events is nearing 100%. In the beginning of the World War the troops still spent a greater amount of time marching, and only a smaller part in engagements. Now this correlation is radically changing: the troops will spend the greater part of their time in deployed combat formations, and only a smaller part in marches.

This, of course, does not exclude the possibility that the enemy might voluntarily give up part of his territory. Then the operation may develop in leaps, retaining its combat depth only along definite lines. However, even these possibilities are now significantly more limited. In modern conditions, the widely developed prospects for rearguard engagements, obstacles, chemical warfare, and aviation—will force us to pass through even these operational dead zones with a great deal of tactical intensity. At the same time, the shallower the territory of the country, as is the case with our smaller western neighbors, the more limited are their capabilities to give it up.

Thus, as a general tendency, the indicator of the operation breaking up in depth is also acquiring in a future war the same full development to its final limit, which the indicator of the spreading of operations along the front achieved in the World War.

We can presuppose that the breaking up of the operation in depth will achieve greater development in the Western European theater of military activities, and to a lesser degree in ours.

Nevertheless, even for us *a future operation will, according to its depth, no longer be a single chain in a series of intermittent battles, but a continuous chain of merged combat efforts throughout the entire depth.*

This will be a continuous sea of fire and struggle, which spread along the entire front as early as the World War and which will spread throughout the depth in a future war.

Undoubtedly, never in history has armed struggle achieved such a high degree of combat intensity. This intensity is actually historically limited, for after armed struggle has occupied all of the possible fronts of activities on the ground and in the air, there will be no place left for it to spread.

Thus the main factor in the evolution in the nature of the modern operation, which determines its new and enormous intensity, is *its depth.*

It is impossible to imagine the modern operation as a single-act operational effort along a single line in space. The modern deep operational formation requires a continuous series of unbroken operational efforts merging into a single whole.

In operational terminology, this is known as a series of consecutive operations. But this is essentially no longer a true definition. *A series of consecutive operations is the modern operation*, which without the indicator of depth is now losing its chief meaning and is becoming a historically conservative concept, no longer corresponding to its character's new conditions.

We are faced with the growth of the operation in a new dimension, the dimension of depth, in which a series of consecutive operational efforts is merging into a single overall concept of the modern deep operation.[67]

In current conditions it is already necessary to speak not about a series of consecutive operations, but rather about *a series of consecutive strategic efforts, about a series of separate campaigns in a single war.*

This circumstance is of enormous historical significance in the evolution of the operation's nature and is radically changing the forms and methods of its conduct, and speaks to the fact that *we are at the dawn of a new era of military art and must move from a linear strategy to a strategy of depth.*

3. The Present Correlation of Offensive and Defensive Means

The nature of the modern operation, which faces every offensive with the necessity of overcoming the enormous depth of fire resistance, demands, first of all, the offensive's materiel provisioning with the corresponding weaponry.

The chief factor in the resolution of this problem is the correlation of the specific weight of the means of offense and defense achieved at the current stage of industrial development.

Theoretically, this question was resolved in favor of the former as early as the final period of the World War; there then appeared the first indices of its practical resolution.

However, the World War did not yield a complete picture of the employment of the new means of attack. The exploitation of new technical means of struggle (tanks and aviation) did not achieve that effect, which one could already have demanded of them. This effect did not go beyond the bounds of their narrow tactical employment and nowhere did it grow into the operational level.

At the same time, the evolution of the technical means of struggle has moved far ahead and the modern tank and combat aircraft qualitatively represent a completely new weapon of struggle than was the case in 1918. It is sufficient to point out the evolution of the following basic indices:

Tank	**1918**	**Now**
Speed	2–4 kilometers per hour	25–40–60 kilometers per hour
Range	40–50 kilometers	300 kilometers

Combat Aircraft	**1918**	**Now**
Motor power	500–600 horsepower	3,000 horsepower
Bomb load	0.4 tons	3–4 tons
Speed	120 kilometers per hour	200–300 kilometers per hour
Range	250–300 kilometers	700–1,000 kilometers

At the same time, the modern indices are in no way the maximum achievement and have a tendency for unlimited growth.

In these conditions, the resolution of the problem of the competition between defensive and offensive weapons in favor of the latter opens up even further vistas.

In the quantitative sense, weaponry will, of course, always remain stronger in the defense than in the offense. The machine gun and the battery in the defense will always be stronger than the machine gun and battery in the offensive. This proceeds not from the quality of these fire means, but from the character of their goals in the defense and offense. The machine gun and battery in defense have as their goal the immediately attacking groups of infantry, which present easy and rewarding targets for their fire. The battery in the attack has as its goal the individual, scattered, camouflaged, and hidden machine guns and guns, which demand a large expenditure of shells and a significant amount of time and accurate fire for their suppression. These are, of course, completely different conditions.

The quantitative massing of firepower in the offensive therefore remains a binding condition.

However, qualitatively new technical means of struggle are acquiring an undoubted superiority over firepower in the defense.

In essence, the tank is not a new type of fire weapon. It carries in itself that same machine gun or cannon that called forth its appearance and serves for them merely as an armored means of movement. But, it is precisely here that the qualitative solution to the problem lies. The mobility and all-terrain capability of the machine gun covered with armor have imparted to its fire a *new quality* as to its relative degree of protection from defensive fire and its capability to directly destroy the defender's firing points with its material weight as *a new kind of strike and attack*. There's no doubt that such a machine gun on a tank is proving to be more powerful than the same machine gun, buried in the ground, in the defense. There's no doubt that a gun on a tank is proving to be more powerful to the same degree as the same gun deployed in a position in the defense.

One must recognize as true the tenet in Fuller's theory that the tank, to a certain degree, has changed the entire correlation between the means of defense and attack in favor of the latter. At the same time, one must keep in mind that the problem of mechanization is resolving yet another critical question for overcoming intolerably deep modern march columns, which represent an excellent target for aerial attack in moving along the roads.

The technical problem of mechanization is, in its tactical application, the problem of moving to *off-road tactics*, to the tactics of large-scale movement over any terrain, which altogether eliminates the former significance of the road network and the necessity for moving in deep march columns. In this way, the most rapid attack and the comparatively best passive defense against air attacks is achieved. Off-road tactics, as a completely new indicator of the actions of modern mechanized formations, has enormous significance for the evolution of the character of operations and through its influence alone conditions the shift to a new era in military art.

All this naturally increases the offensive's opportunities. Basically, an analogous qualitative evaluation of the tank as a weapon is also applicable to combat aircraft, which deliver by air the same weapons and explosive shells (bombs) that the defense employs in a static position on the ground. There is no doubt that these weapons of annihilation, when employed in the air from flying apparatuses of enormous speed, are proving to be more powerful than those same weapons employed by the defense on the ground.

At the same time, one should keep in mind that the means of air attack remain at the present time more powerful than the means for combating them from the ground. In this sense, anti-aircraft defense from the ground is undoubtedly inferior to attack from the air.

To be sure, this circumstance is a double-edged one, both for the defense and the attack: aviation, as a means of attack is an equally menacing weapon for the attacker as well. However, in this case the problem must evidently be decided by the achievement of air superiority along the axes of the decisive offensive operation. The massing of major air strength will be just as mandatory as the massing of firepower in the offensive on the ground.

Thus the degree of protection against the defense's machine gun fire, mobility, all-terrain capability, and, finally the rapid overcoming of space in the air—are the decisive factors that are conditioning the superiority of the new technical means of attack over the firepower of the defense.

The problem is basically solved through *mobility*, which imparts a new quality to firepower, which is superior to its strength in the defense.

The entire evolution of modern military equipment is occurring chiefly under the auspices of increasing and improving this mobility. Everything that raises it increases the capital of the offensive. Defensive capital may be increased only by increasing firepower. However, as far as the rapidity of fire and the introduction of the machine gun into the infantry are concerned, everything has already been achieved in the era of the World War. Only the problem of automating artillery remains an unresolved problem. Overall, defensive capital has achieved its maximum accumulation as regards firepower. Thus the search for means of counteracting the offensive is developing along other lines, such as the introduction of rapid-fire anti-tank and anti-aircraft guns.

The engineering art, chemical means, obstacles, anti-tank barriers, minefields, and the possibility of employing electricity and radio as means of resistance and destruction over large distances—in general, high technology and science—are being opposed by modern defense to the attack. It's necessary to take into account that only a certain stabilization of the front and its positional nature may facilitate the employment of these scientific-technological means. At the same time, the development of modern rapid means of struggle—aviation and motor-mechanized forces—are conditioning to a significantly greater degree the mobile development of military activities.

Nevertheless, there undoubtedly lurk definite prospects and possibilities for countering the attack in the evolution of science and technology. However, at the present stage, it is obvious that the means of attack are ahead and that the means of defense are developing as a subsequent reactive phenomenon.

That the superiority of the means of attack is beginning to seriously worry the European general staffs may be best observed from the development of modern permanent fortifications. Along the eastern border of France, this is taking the form of concrete fortifications, all the approaches to which are girded by electrified fields of death.[68] The overcoming of such a concrete belt is likely impossible for modern offensive weapons. If one keeps in mind the likely evolution of the fortification art in the sense of fast-hardening concrete, which would allow us to erect concrete fortifications at great speed and in conditions of the maneuver nature of military activities, then one must consider the possibility that military art will come face to face with a completely new problem of positional warfare at the highest scientific-technological level. One may, however, presuppose that its prerequisites are rooted in the prospect of a second imperialist war in Western Europe. Its hopelessness in these insuperable conditions of positional warfare will doubtlessly be an accelerating factor in transforming this war into a civil war on a worldwide scale.

One cannot discern the prerequisites for such positional warfare along our Eastern European theater of military activities. However, if such a front should arise along individual sectors and in another qualitative condition, it is necessary to have a sufficiently clear idea in order that such a concrete line be stormed along another plane—not on the ground, but from the air.

Airborne landings, which are sometimes approached with a certain lack of faith, have a great future in this regard, and it is hardly likely that we can fully foresee all of their great significance for the evolution of operational art.

In our modern conditions of colossal technological progress and the prospects of our second Five-Year Plan,[69] one must beware of being insufficiently progressive and far-seeing.

As a result, the competition between offensive and defensive weapons leaves an entire field for experimentation and research. However, one thing is beyond doubt: at the present stage the tendency toward the superiority of offensive weapons over defensive weapons is manifesting itself all the more clearly. And it is this circumstance, in those political conditions that determine the nature of our future war, that yield such an incontestable material justification for the possibility of overcoming the front of fire and the solid front and the organization of a deep offensive operation.

4. The Offensive's Deep Operational Formation

The means of struggle are the necessary material prerequisite for resolving the problem, although they, by themselves, still do not yield an immediate solution to it. In the history of military art one can point to a number of cases in which new means of struggle failed to yield the necessary effect, because their employment was constrained by old and no longer applicable types of combat formations and methods of activity. This was the case, for example, at first with rifled field artillery, which continued to be towed in the rear of the column.

New weaponry requires new forms for its combat employment. And if this question was resolved tactically in the transition to the group combat formation and the deep engagement, then as regards the control of large troop formations, it remains on a completely outdated historical level.

If the object of the attack is the great depth of resistance, then the operational organization of the offensive's combat formation undoubtedly requires the most fundamental changes. It is hardly likely that a single line of drawn-up armies will be capable of solving the new *problem of the deep offensive*. One may even definitely state that *a single wave of the linear strategy's operational efforts will not be able to resolve anything here* and that it must helplessly break itself against the depth of modern resistance.

This problem brings us to the central question of justifying a deep strategy in the new era.

At the same time, one must first of all discern the character of the depth of modern resistance. The strength of this resistance has a tendency to increase and reach its highest culmination point at the strategic zenith, when the attacker is already close to his goal and when the defender is forced to put everything on the table in order to save the situation. Given the mutually exclusive contradictions of the sides, when the struggle is being waged for their political and economic independence, eliminating the possibility of a compromise agreement, this resistance may manifest enormous power at the last stage.

As early as the World War, given the enormous exacerbation of imperialism's contradictions, the development of the operation in 1914 flowed along the path of the increase in combat efforts. This was completely misunderstood by the Germans, who entered into the first frontier battle with a great deal of operational intensity, but who reached the Marne very weakly prepared to meet the Anglo-French forces' increased powers of resistance.

To an equal degree, the curve in combat efforts was completely misunderstood by us in 1920, during our offensive to the Vistula. Then, following the forcing of the Neman, it was even planned to weaken the Western Front's armies and the conclusion of the campaign was considered to be guaranteed by the initial development of the offensive itself. A battle of enormous intensity along the Vistula was completely missing from our operational forecast, and this was, of course, a bitter delusion. It spoke to the profound misunderstanding of the dynamics of the development of the modern operation.

The exhaustion of the offensive has as its true cause not so much the exhaustion of the attacker's forces as the growth of the defender's resistance.

In conditions when the linear strategy essentially just pushed back with its offensive front the one who should have been engaged, in order to defeat him; when it was nowhere capable of destroying the enemy's personnel—this circumstance manifested itself with particular force, and the retreating enemy, having won a favorable operational situation for himself by his retrograde movement, proved to be significantly stronger at the culmination point in the development of the operation than in its beginning. At the same time, the attacker thoughtlessly and with no prospects arrived at this strategic Rubicon and assumed that the final moment of the operation would be the easiest. This was his cruel mistake.

It is precisely the first step, which is always supported by the preliminary grouping of forces and systematic planning that will be the easiest in this regard.

Difficulties should be expected along the path of development, which cannot be calculated beforehand in all their details. *One should expect the greatest intensity and the crisis at the end.* The art and firmness of operational leadership consists in arriving at this decisive moment with complete foresight, with a new wave of operational efforts and fully supplied with the necessary men and materiel for the final conclusion of the destructive operation.

Woe betide that commander, who in modern conditions takes it into his head to arrive at the Marne and Vistula the way it happened in 1914 and 1920. He will perish ingloriously, no matter how significant his operational achievements during the offensive. Moreover: the more significant these achievements, the more horrible will be the disaster, if his forecast has not foreseen the concluding stage of the operation.

The modern operation is the operation in depth and it must be calculated to the entire depth and must be ready to overcome the entire depth.

At the same time, it must be taken into account that this depth, in relation to the force of its own resistance, has a tendency to thicken and increase from the front line to the rear of the defense.

In these conditions, modern operational art faces, in organizing the deep offensive operation, the completely new problem of organizing the offensive combat formation. Only one thing is clear: *the linear strategy of a single wave of operational efforts cannot resolve this problem of the offensive.*

The solution must be found along the new paths in the evolution of operational art.

We will have to smash more than one tenet of the old military theory along these paths.

It is first of all necessary to finish with the tenet that *strategy achieves its decisions on the basis of the simultaneity of actions.* This tenet, which has spread its stagnant influence all the way up to our time, can be traced back to Napoleon and has long been in irreconcilable contradiction with altered conditions. It was formulated by Clausewitz in any number of places of his work, in which we read:

"In tactics, given the gradual commitment of forces into the fighting, the chief decisions are put off until the end; in strategy the opposite is true and the law of the simultaneous employment of forces almost always strives for a decision at the beginning of the overall action...."

"Tactics allows for the consecutive commitment of forces into the battle, while strategy absolutely demands their simultaneous commitment...."

"One must employ as many forces as strategically possible; their employment must be simultaneous...."

"Strategy can in no way recognize time as an ally for its own sake and with this aim commit its forces successively and gradually into the battle. All available forces, designated for achieving the strategic aim, must be employed simultaneously...."

"In strategy the dispersal of effort is antithetical to the goal; all available forces should be used in the fighting at once...."

This theory, while true for the conditions of the Napoleonic era and the dawn of the linear strategy in the age of Moltke, when the operation still led, in general, to the single-act battle, decided by a single wave of operational efforts, proved that it no longer corresponded to the new conditions of armed struggle in the age of imperialism. However, one could even earlier discern its dying out in the new conditions of the final decades of the 19th century.

During the second period of the Franco-Prussian War of 1870–1871, the French army, newly reorganized after the fall of the Second Empire, placed the Prussians in the difficult condition of not having enough forces for a new fight. Then Moltke was aided by counterrevolutionary and treasonous bourgeois France, which concluded peace with Bismarck[70] over the heads of the National Guard. It is difficult to foretell how the Franco-Prussian War of 1870–1871 would have ended in different conditions. In any case, Engels wrote about its possible outcome: "...the belligerent forces are now in greater balance than at any time since Sedan..." (Engels, *Stat'i o Voine*, 1924, p. 169).

Even then the first signs of permanent mobilization and the impossibility of achieving in strategy a decision by a single simultaneous effort had undoubtedly been revealed.

Moltke understood that he had collided with a completely new phenomenon in the history of armed struggle and thereafter more than once repeated: "This struggle; that is, the a continuation of the 1870–1871 war after Sedan, surprised us from the military point of view to such a degree that the question it put forward must be studied over the course of several years." This was undoubtedly a question worthy of study. The fact of the appearance of new armed forces after the first-line army had ceased to exist, indicated that in the future strategy will evidently not be capable of resolving its tasks with a single first-line army, deployed at the beginning of the war, and that it will require behind it the commitment of second-line forces from the depth, and possibly even a third line. This was a convincing hint at the age of deep strategy, which Moltke already dimly foresaw.

In his famous speech to the German Reichstag in 1890, he said:

"If a war, which has been hanging, like the sword of Damocles, over our heads for more than ten years, should finally break out, then no one will be able to foresee its length or its end. The great European powers, armed as never before, will enter into combat with each other. *Not a single one of them can be crushed in one or two campaigns to an extent that it would admit that it was defeated, so as to be forced to conclude peace on harsh terms, and that it could not arise and renew the struggle* (our italics, G. Isserson).

Moltke was already speaking as a strategist of a new era. However, the new views still proved helpless to overturn the old theory. Historical experience passed completely without a trace and as late as the beginning of the 20th century Foch was writing in his *Des Principes de la Guerre*: "Instead of the tactical law of the consecutive layering of efforts, in strategy there operates the law of the coincidence of efforts." This proved to be untrue as early as the war of 1870–1871 and was even more the case in 1914–1918; however, now this tenet is in flagrant contradiction with the new character of the deep offensive operation.

Moreover, what during the second period of the Franco-Prussian War of 1870–1871 revealed itself only at the strategic level was exposed during the World War and is today spreading fully into the field of operational art.

The modern multi-act deep operation is not resolved by a single simultaneous blow of coinciding efforts. It requires the deep operational layering of these efforts, which increase as they approach the highest point of achieving victory.

The deep echeloning of the resistance calls forth an equally deep formation of the offensive. This offensive must resemble a whole series of waves, rushing to the shore with increasing force, in order to wash away and crush it with its uninterrupted blows from the depth.

In this sense, the modern deep operation calls forth the dispersal of efforts in time and to the same degree conditions this at the strategic level.

The events of the World War and our civil war showed this. *Of course, it would be entirely incorrect to understand this in the sense* that the Germans did in the Battle of the Frontiers in 1914 and we in the battle on the Mnyuta River in 1920, when we immediately committed too many forces, while they should have been committed consecutively. All available forces at a given moment should be committed into the first initial operations in accordance with the sides' correlation of forces. However, the heart of the matter is that simultaneously we must plan for the deep echeloning of even more new efforts, so that at the decisive moment of the operation we arrive with such amounts of men and materiel and in such a grouping that will guarantee the final achievement of victory.

The echeloning of modern operational efforts in depth is not the same as employing them in detail or in operational packets; this is the *consecutive and uninterrupted augmentation of operational efforts, calculated to break the enemy's resistance throughout the entire depth.*

The greater this depth as to its possible resistance and the greater the possible intensity of resistance, the greater must be the echeloning of the offensive's operational depth.

In organizing the modern deep operation, it is necessary to calculate men and materiel not only in a linear dimension along the front, but also in the new dimension in depth.

The problem of the deep operational formation of the modern offensive violates yet another tenet about strategic reserves.

As long as strategy resolved its task in a single simultaneous effort, reserves were not needed. Clausewitz speaks of the absurdity of the idea of a strategic reserve and considers it superfluous, useless and even harmful. He demands that "all strategic efforts be compressed into one action and in one moment," and writes: "The idea of purposely retaining ready forces for the purpose of employing them only after the general decision cannot but be recognized as an obvious absurdity."

As long as this general decision was achieved in the Napoleonic era by a single act this was true. During the second period of the Franco-Prussian War of 1870–1871, this tenet was already in doubt, while at the dawn of the 20th century it had obviously become false.

Schlieffen foresaw this problem to a certain degree. He demanded a powerful reserve army behind the Germans' right flank during the offensive on Paris. However, his motives here were somewhat different. He demanded an operational reserve; in order to be able to lengthen his right flank, should the latter prove to be insufficient for an envelopment during the development of the offensive. Thus, in the final analysis, this reserve was supposed arrive at a single line with the attacking front.

In modern conditions, operational reserves are mainly necessary not for lengthening the front, although that still has significance for us in the beginning of a war. Overall, the flank has already achieved its maximum spreading along the front, while reserves are necessary now for augmenting operational efforts and crushing the enemy's resistance throughout the entire depth.

The very problem of operational and strategic reserves is, in modern conditions, growing into the problem of *operational echelons*, behind which in the depths for the prospects of the evolution of armed struggle, the silhouettes of just such strategic echelons can already be discerned.

This naturally leads to a further growth in the size of the armed forces, while at the same time sweeping away all sorts of theories about small professional armies, as both historically conservative and absurd.

The numerical growth of armies in the age of imperialism was conditioned, alongside other reasons, by the linear strategy's striving to occupy as much of the enveloping attack front as possible. This numerical growth at the present time is to the same degree conditioned by the deep strategy, which requires powerful operational echelons in depth and the deep construction of the offensive.

This once again points to the grandiose scope of modern armed struggle and reveals the entire evolution of the operation's nature in the new era of the deep strategy.

5. The Deep Entry into the Modern Operation

Any new phenomenon in the historical process is born not only because it is called forth by new requirements, but because it is predetermined by an entire series of new prerequisites.

When in 1866 Moltke first deployed the Prussian armies along a 400-kilometer front, this new phenomenon in operational art not only corresponded to the altered character of the armed struggle, but also had been predetermined by new objective conditions in the form of railroads. At the same time, we have Schlichting's famous words that Moltke approached such a broad deployment with great doubts.

In the same way, one may view *deep deployment* with fear and doubt; but whether we wish it or no, we must inevitably arrive at this.

The deep deployment has been predetermined in modern conditions by an entire series of objective prerequisites. This proceeds from the nature of a future war, which demands an enormous intensity of struggle. Of course, no country, having embarked upon this struggle, will limit its mobilization intensity to the single first echelon of its already mobilized cadre army. At the same time, no country is in a position to simultaneously concentrate all of the forces it is capable of fielding by the start of military activities. In order to do this, it would be necessary to significantly postpone the beginning of military activities and, in order not to expose one's deployment to the threat of defeat in detail, move it back sufficiently far into the depth of the country. And the weaker opponent, in this case, would prove to be stronger, having acquired enormous advantages at the start of the struggle. It's doubtful that anyone would decide on such a method. It's quite obvious that *consecutive and permanent mobilization* calls forth the consecutive layering of efforts in the struggle.

Second-line and third-line forces will follow behind the first-line army, which determine the army's entrance into the war in deep strategic echelons.

This inevitable outline of the deep entrance into a future war has already been foreseen by the modern organization of the armed forces in peacetime. How else is one to explain the maintenance in France of a special covering army (*l'armee couverture*) on the Rhine as a first strategic echelon, behind which will deploy and enter the fighting a second echelon of the main mass army, which will doubtlessly not be the last? The conditions of the Treaty of Versailles make this outline for the deep deployment and entrance into the fighting inevitable for Germany as well.

The deeper and more spacious the country's territory, the greater its mobilization prospects and the stronger its intensity in war—the greater will be the scope of this deep echeloning of strategic efforts. This is the situation our country is in and in this lies also its enormous power, which supports the *greatest growth of efforts at the final, decisive movement of the struggle.* As concerns our small western neighbors, with their insignificant depth of territory and fewer mobilization opportunities, the first mobilization strain at the beginning of the struggle will be quite close to the maximum and the absolutely necessary consecutive layering of efforts will be possible for them to a significantly lesser degree, if the major imperialist countries don't arrive to seriously aid them with men and materiel.

The echeloning of the armed forces upon entering the war is predetermined in modern conditions not only at the strategic level of the entire country and the entire war, but at the operational level as well. This has as its prerequisite a series of material factors in the evolution of the modern technical means of struggle.

In establishing the motif of the technical evolution of modern armaments, we see that its guiding foundation is the striving to achieve the greatest possible range and striking power. It all comes down to shifting the striking of the target to the maximum range. In the capability of rapidly overcoming space we see the entire reason for the combat employment of aviation; in this we see the value of motor-mechanized weapons.

The evolution of firearms moved along this path. It's interesting to note that in the middle of the 19th century the evolution of this weaponry advanced according to an increase in range and rate of fire. At the dawn of the 20th century, before the World War, evolution mainly moved to increase the rate of fire, leaving the range of fire pretty much at its previous level. Having achieved the maximum limit in the rate of fire with the introduction of the machine gun, technical evolution in the World War moved and is moving at the present time primarily toward increasing range. The machine gun has the clinometers and azimuth circle combined for firing at distance targets from hidden positions. The artillery has significantly increased its range of fire, achieving average ranges of 12–20 kilometers. This circumstance has close to decisive significance for the evolution of the technical forms of the engagement.

Historical evolution shows that the switch from the Napoleonic concentration of all forces before the battlefield to the meeting engagement, which unfolds directly from the march, had in the Moltke era as its main prerequisite the increase in the range of fire.

However, in modern conditions, given the significantly increased range of fire, we face the further evolution of the meeting engagement.

At the dawn of its appearance, in the second half of the 19th century, soldiers fired at the same distance they could see, which essentially brought about the start of the engagement from the march. Now we are seeing fire effects at a significantly greater distance than we see in field conditions. This means that *the modern engagement will unfold at great distances*. From this, it follows that the current system of the tactical security detachment, with its interval of 5–6 kilometers from the head of the column of the main forces, basically secures no one from the effects of long-range attack and the surprise appearance of motor-mechanized weapons, not to mention aviation, which operates at enormous distances along another plane. This security detachment is no longer capable of fulfilling its role as the vanguard and securing the deployment of the main forces for the engagement. It retains only its importance for immediate security. Alongside this, the significantly increased depth of the column requires more time and space for deploying the main forces for the engagement in the corresponding formation. As early as the dawn of the 20th century, Langlois,[71] who built his theory on the evolution of artillery equipment, wrote: "We must push our vanguards forward more than just a few kilometers, but several miles, to a distance of 1–1½ marches."

Modern combat distances have grown significantly. They require the forward movement of the march security detachment to a significant greater distance, at least 20–30 kilometers, which the main forces of a modern reinforced division can occupy in depth. The widely employed system of forward and reconnaissance detachments have essentially been created by these requirements.

This system has already pushed the vanguard, advanced to a distance of 5–6 kilometers, into the background and has adopted its reconnaissance and support functions for the march. The vanguard, moving behind the reconnaissance and forward detachments ahead of the column, naturally loses all the specific indicators of a vanguard and becomes simply the column's first echelon. However, in such a case these detachments should be called vanguards, upon making them sufficiently strong. However, this resolves the problem only on the tactical scale; that is, on the level of adopting tactical preventive measures along the given route of march.

At the same time, the modern army commander, if he wants to truly lead the modern deep operation, must first of all ensure for himself the opportunity of deploying his forces in time and committing them into the battle in a grouping corresponding to a freely adopted decision. For this, he needs his own operational support organ, represented by a powerful group of rapid and mobile formations, chiefly motor-mechanized equipment and cavalry, pushed far forward to a distance of 1–2 and more marches.

In modern conditions, we are once again returning to the Napoleonic phenomenon of the *army vanguard*, but already on the different historical basis of the new era of the deep strategy, with a completely different qualitative content. It is in this dialectical contradiction that the circle of the evolution of the operational organization of the offensive is completed. And this means that the phenomenon of the meeting engagement moves to its highest phase—from the field of tactics to the sphere of the operation. Tactically, the meeting engagement will become, as a rule, possible only for the forward operational support units. However, operationally, it is growing into the *meeting operation*, in which these operational support units carry out the functions of the army vanguard. At the same time, this means that the modern operational organization of the offensive inevitably leads to deep echeloning.

We may once again approach this new phenomenon with fear and doubt, although this is undoubtedly caused by the altered requirements of the modern operation. It will have to be approached, however, because it is also predetermined by an entire series of objective conditions.

It is necessary to take into account that as regards the capability of rapidly overcoming space, rapidity of action and range, the modern means of struggle present a very heterogeneous picture.

First place among them for range and the capability of shifting one's actions a great distance naturally belongs to aviation. The ground enemies will still have not exchanged a single shot, while this combat arm will begin its operations along the farthest trajectory in the first hours following the beginning of a war. A powerful combat aviation will undoubtedly be the first operational echelon of the struggle.

Behind it and on the ground will move forward everything that is rapid and mobile and may advance forward easily, which are naturally the motor-mechanized formations and modern mechanized cavalry. These motorized and cavalry forces, which have the highly important operational task of disrupting the enemy's concentration and occupying a favorable position for going over to a general offensive, will move forward as far as possible and will enter, together with aviation, the fighting, when the main mass of the still-deploying first-line army will still be in the grip of the complex mechanism of modern concentration. They will be the second operational echelon.

Finally, the main mass of combined-arms infantry formations will take the field in the theater of military activities. However, even it cannot immediately take the field in a single line. Modern railroads, which are growing less than the armed forces, will not be able to immediately and completely transport them. This will lead to a more extended concentration schedule for all forces in the theater of war. The greater part of the newly arrived forces will enter the fighting immediately in a drive to begin the operation as soon as possible. A certain part of the forces, having arrived later, will take the field somewhat later. Thus even this main mass of forces will enter the fray into two echelons, thus forming the third and fourth operational echelons.

When this deep system of the first strategic echelon begins to fully move, in the depths of the country the silhouettes of a second strategic echelon of mobilizing second-line forces will already be visible.

If this does is not the start of the era of the deep strategy, then it is necessary to question the very concept of depth.

The operational outline of the deep entrance into the operation should spread out over an enormous distance. Air actions will immediately reach their maximum range of 500–600 kilometers. The motor-mechanized formations and cavalry will rush forward by 2–4 marches, or about 100 kilometers. The attacking first echelon of the main mass of troops, even if each division has its own road (which is not always guaranteed), will occupy a depth of up to 75 kilometers. Finally, at a distance of a single march from it, the second echelon of the main mass of troops, which will also stretch in depth up to 50 kilometers, will advance along a broader front.

Overall, the entire first strategic echelon would occupy an enormous depth of 250–300 kilometers on the ground. Such a depth cannot be supported by modern conditions of deployment.

Schlichting, already foreseeing the conditions of the 20th century, wrote: "…the strategic deployment of the army will be separated from the first decisive battle only by a few short marches…." And Lewal[72] predicted that "in a future war contact between the sides will be established almost immediately near the railroad disembarkation stations."

In modern conditions, when the heightened mobilization readiness troops have already been moved up to the border; when the force and opportunities for covering enable us to concentrate significantly closer to it, military activities will essentially begin operationally on the spot and long marches to such a depth as 300 kilometers will undoubtedly be excluded. Moreover, the insignificant depth of a number of small states will naturally not allow for such an unfolding of military activities.

In these conditions, the potential operational depth of the offensive will remain hidden and will not achieve its full scope in depth. The operational echelons will be launched into the operation from a single line. Nevertheless, when the last of them is already peacefully marching in the deep rear, feeling only a threat from the air and the intensive work of supply and evacuation along its march route, the first operational echelons will already be involved in brutal fighting and much will have already been decided for them at the corresponding scale. It will even be difficult to define where and when this grandiose operation will begin; where and when it will draw any kind of visible boundary between the operation and the battle. We will creep into this battle, while in essence the first bomb dropped from an airplane in the deep rear and the first solitary shot along the frontier will mean that this gigantic operation has begun.

If during the age of the linear strategy the battle organically flowed from the operation, then in the *age of the deep strategy, the operation and battle are organically merging*. Any kind of boundary between them is disappearing in time and space. The modern battle will begin along the front and conclude in the depth in a single stream of increasing operational efforts. Thus wave after wave; they will fall upon the enemy's approaching front, evidently in the same operational formation. The conclusion is thus arising that *he whose operational formation is deeper will have the ultimate success*.

However, the moment is inevitable when all of these waves will merge into a single squall of fronts solidly facing each other. It's then possible that the operation's development will lead us once again to a linear front and to a linear strategy. But at this stage, which is likely to be normal in modern conditions and which may arrive quite

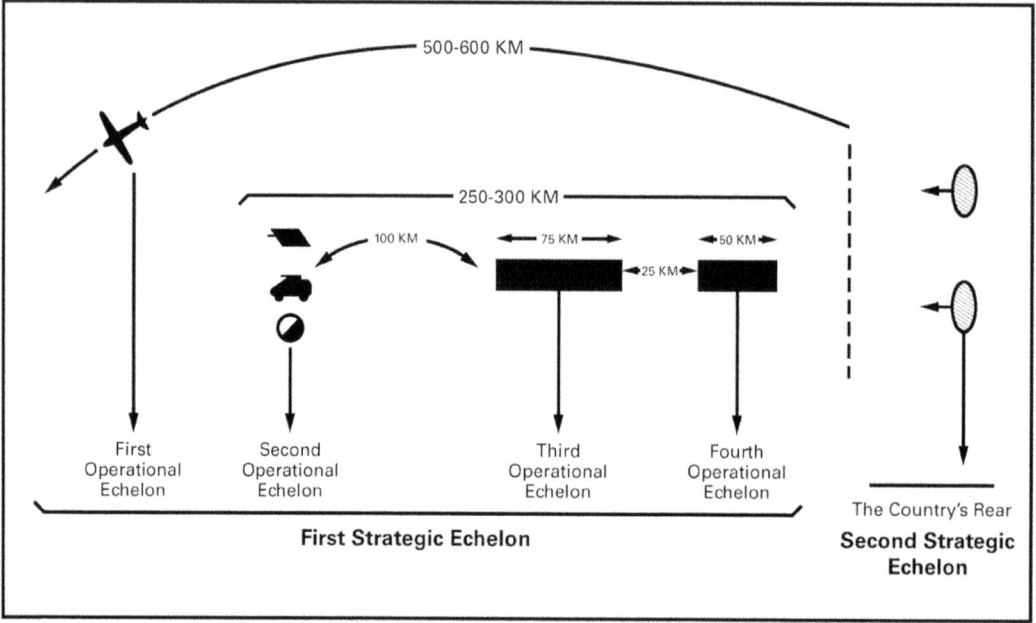

Figure 1. The Deep Entrance into the Modern Operation.

soon, the evolution of operational art will require a different decision and a deep frontal attack from the depth. It is precisely here that the demands of the deep organization of the offensive will emerge with new strength.

This will be a new operational solution to the problem of the breakthrough in conditions of the new era of the deep strategy.

6. The Deep Breakthrough and the Smashing of the Front

If operational art in the age of the linear strategy has arrived, at a certain stage in its historical development, at its antithesis and self-negation, then this has happened front has met the front and required a breakthrough.

This problem could not be resolved operationally on the basis of the linear strategy. It has brought about the appearance of new technical means of struggle; it has led to new tactical forms of the engagement; it has raised the tactical organization of offensive technique to a high level and created the prerequisites for its tactical resolution. However, on the basis of the linear strategy, it was impossible to achieve an operational decision of the problem of breaking and smashing the front.

For this, operational art had to search for new paths and had to stride into the new age. However, in the conditions of the imperialist war by attrition and starvation, this was impossible.

The new nature of our future war, with its decisive destructive operations places the decision of this central problem of modern operational art in a new light.

The front must be broken in a decisive operation. The breakthrough of the front must be achieved throughout the entire depth; it must be completely destroyed.

Here the deep strategy is undergoing the test of historical maturity, and if it proves to be predetermined by many objective conditions of our times, then at the same time it is being called forth by exactly these requirements for the decisive and complete overcoming of the frontal stalemate.

The new forms of the deep engagement, which have been conditioned by the full tactical employment of modern tactical means of struggle (tanks, long-range artillery and close combat aviation), are already resolving this problem at the tactical level. However, they only break the tactical depth of the modern defense. They do not yet result to an operational decision, although they lead that way.

A deep tactical effort must still grow into a deep operational breakthrough. This is the main problem which the operational art of the age of the linear strategy must resolve. All the achievements of the deep tactics become unnecessary if a decision is not reached on the path to the operational realization of the problem.

It is necessary that we comprehend that the first attacking echelon, which is directly breaking the front, can carry out its task only at the tactical level and break through this front. No matter how great its successes, it cannot by itself transform its tactical result into an operational one and flood into the depth through the open door. The first attacking echelon cannot carry out this task, because the broken door is still propped up from within by the powerful springs of defensive resistance, and it is necessary to continue to hold it open so that it does not once again slam shut. The fulfillment of this combat task remains the responsibility of the first attacking echelon. And if during this time, when a tactical breach has been made by it, no one arrives from the depth of the attack's operational formation, *in order to extend the attack from the depth into the depth* and transform the tactical success into an operational one—if this is not done, the breach will soon close again; all of the first attack echelon's efforts will be in vain and we will have no more than a stomach-shaped protrusion along the attack front and only wearing out and attritting the attacker's forces.

This would mean continuing the system of senseless and exhausting frontal attacks of self-attrition into which the linear strategy of 1918 degenerated.

The modern breakthrough must be undertaken not only when there are sufficient men and materiel for breaking through a sector along the front, but also when there are sufficient forces for extending this breakthrough from the depth into the depth and destroying the enemy throughout the entire depth.

It's senseless and pointless to attempt a breakthrough operation if no forces have been provided for its development; for it is senseless to knock down the door as long as there is no one to go inside.

The modern deep breakthrough requires two operational echelons as a basis for the organization of the offensive: an attack echelon for tactically breaking into the front, and a breakthrough development echelon for transferring the attack from the depth into the depth and smashing through and destroying the defense's resistance throughout the entire operational depth. At the same time, each of these echelons retains its tactical echeloning.

This deep organization of the breakthrough operation resolves the chief problem of modern operational art—the problem of the decisive and deep overcoming of the positional front all the way to its complete destruction. This preserves its significance not only for the breakthrough of defensive fortified zones, but also for any frontal attack, which is develops as the result of the frontal battle.

If one views the question from the operational point of view, only he who will dispose of the greater depth in his operational formation and will have on hand more powerful echelons in depth may count on final success.

If at the turn of the 20th century—the age of the flowering of the linear strategy—Schlieffen taught that in war he whose front is longer and whose flank is stronger will win, then now we must oppose to this teaching the tenet that in modern conditions of the deep strategy, as viewed from the operational point of view, *he whose front is deeper and whose deep echelons are stronger will win the struggle.*

And this once again returns us to the completely obvious prospect of an even greater growth of the armed forces and once again convinces us of the absurdity of any kind of small-army theories.

It remains to unfold the entire deep outline of the modern breakthrough operation.

When the crashing waves of operational efforts from the depth hit the first vanguard echelon, which has brought about the battle, and merge with it in the general melee; when front will face front, excluding the opportunity of a linear enveloping maneuver—the deep strategy's operational art will fully come into its own.

The fast and mobile *vanguard echelon* of motor-mechanized formations and cavalry will soon be squeezed from the front, because it will not be able to employ its long-range qualities. Lacking space for its actions and having carried out its role as the army vanguard, it will first contract toward the flank and then dive completely into the rear of the offensive's operational formation.

Its place will be taken by the newly-arrived echelons of the combined-arms infantry formations, which now, and with a greater basis for their useful employment, will take upon themselves the combat work along the front. Like a closed operational phalanx, supported by numerous tanks, and reinforced quantitatively and qualitatively by long-range artillery and close-range aviation, they constitute the *attack echelon.*

Behind them there will form the *breakthrough development echelon* from that group of fast and mobile formations which previously led the offensive's operational formation,. It will include major independent mechanized, motorized and cavalry formations and large masses of long-range combat aviation. He who was first when the operation began will now be last in the operational formation; and he who was last during the approach will be in the first ranks of the attack.

The operation of the modern deep breakthrough will begin in such an operational formation. It will reveal a picture of the operation's deep operational formation, which has the goal of extending and developing the attacks from the depth into the depth. It has nothing in common with the echeloning of the depth of the offensive in the breakthroughs of 1918.

Then in the March offensive the German Eighteenth Army had 12 divisions in the first echelon, eight divisions in the second echelon, and four divisions in the third echelon, while the German Seventh Army in the May offensive,[73] had 14 divisions in the first echelon, five divisions in the second echelon, and six divisions in the third echelon. But then each division was to advance only three miles in depth while the divisions standing behind were supposed to relieve and feed the units fighting ahead, continuing their attack along a single overall front line with them. The heaping up of these echelons resembled the strategy of the buffalo, who couldn't understand that in order to carry out a true breakthrough of the front, tactical efforts must grow into operational ones and that this can be accomplished only by the lengthening and operational development of the attack from the depth into the depth. However, in 1918, when independent motor-mechanized formations did not yet exist and cavalry no longer existed, the realization of such a decision had no support. The breakthrough development echelon must be, first of all, speedier than the attack echelon, in order to bypass it; therefore, it cannot consist of infantry formations. The breakthroughs of 1918 therefore could not grow from a tactical phenomenon into the operation. They could not set themselves goals that the operational art of the deep strategy has the right to set.

The modern deep breakthrough operation pursues the goals of the simultaneous defeat and breakthrough of the entire operational depth of the resistance. This simultaneity at the operational level cannot, however, be compared to tactical simultaneity. There is nonetheless a difference in time. It is determined by the smashing of the tactical depth of the first defensive zone. When this tactical task is resolved by the attack echelon and a yawning breach opens up in the enemy's front, the breakthrough development echelon will dart in from the depth of the attacker's operational formation. It will be preceded in the air by combat aviation, which from a great distance will seal the path of the enemy's reserves to the sector being broken through. Simultaneously, a landing detachment, brought in by air, will land in the enemy's rear and will serve as the first harbinger of death. At the same time, on the ground a large mass of high-speed tanks, self-propelled artillery and infantry on armored transports will pour like a broad avalanche in several waves through the breached sector of the tactical breakthrough of the front. These forces will themselves expel the last cork from the opened breach. Behind this will appear modern cavalry, the *beau sabreur*, which has been preserved by history. Finally, behind it, following the clearing of the roads will come numerous columns of motorized troops, ready to burst forth. In different conditions, each of these component parts of the breakthrough development echelon will have its own sector of the open breach. Then the development of the breakthrough will follow simultaneously along several parts of the front.

All of this will lengthen and develop the attack from the depth into the depth. The larger the breakthrough development echelon, the deeper it is aimed. In all cases, in order to accomplish the task of the operational breakthrough, it must embrace the *entire* depth of the enemy's resistance.

And when the attack echelon is still be waging a brutal struggle along the breakthrough front—in its depth, at another level, and even on several tiers of the operational depth, the breakthrough development echelon will begin its crushing actions to surround and destroy. In terms of its operational perspective, this will be a *new, grandiose, multi-tiered battle, waged on several tiers of the operational depth.*

This will lead to a rebirth of "Cannaes" on the new basis of the deep strategy. This will be an entire system of "Cannaes," with some of them beginning, others maturing, and still others concluding.

The problem of the operational breakthrough of the front will be resolved in the overall decisive crushing and destruction of resistance. The strategy of destruction has never had such brilliant prerequisites for its complete realization. And this forecast resolves still another major problem in the evolution of the character of modern operational maneuver.

The historical practice of armed struggle and the theory of military art have up to now distinguished between two basic types of operational maneuver: the maneuver along internal lines as the concentrated attack from a single point, which predominated in the age of Napoleon, and the maneuver along external lines as the enveloping attack from several directions, which predominated in the age of the linear strategy. These two types were opposed to each other and considered to a certain degree to be operational polar opposites.

Clausewitz wrote the following about them: "In strategic maneuvering, one encounters two opposites, which may seem completely special types of maneuver. The first opposite is the action along external or internal lines; the second is the concentration of troops together, or their dispersal along several points."

However, historical evolution is giving rise to something new, uniting and altering its heterogeneity.

The modern deep breakthrough operation is a peculiar combination of two types of maneuver. The attack echelon, while breaking through the front, occupies a broad continuous line and operates along external operational lines. The breakthrough development echelon carries a concentrated attack from the depth into the depth, operating along internal operational lines.

Thus the age of the deep strategy is approaching the synthesis of two types of maneuver and two historical schools of military art.

In this way the oft-repeated and certainly non-dialectical statements that in modern conditions only a frontal attack is possible, that the turning maneuver and encirclement

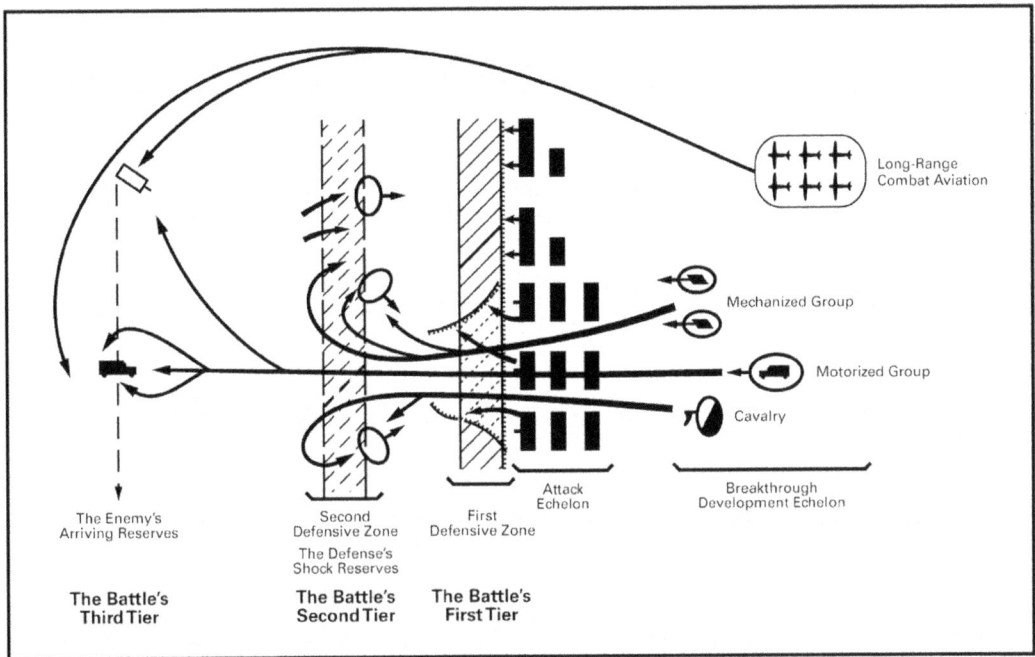

Figure 2. **The Deep Breakthrough Operation and the Contraction of the Front.**

have left the historical stage of armed conflict, is refuted. These judgments do not reflect the most basic thing in the evolution of the nature of the modern operation: they do not see its two dimensions along the front and in depth and remain conservatively on the position of the linear strategy.

The frontal attack is naturally the basic form of action for the first attack echelon. But the attack echelon alone decides nothing by a frontal attack if its tactical efforts do not grow into operational ones. And this is possible only by means of an attack along internal lines from the depth into the depth, in order to envelop, encircle and defeat.

Such a maneuver is naturally disappearing from the first line of the front of struggle, but it is shifting with new strength and enormous intensity into the depth of the front of struggle.

Here it is being fully reborn and with a new content. Here it promises a flowering of the new military art of the deep strategy as the art of magnificent maneuvers and crushing attacks in the depth.

Thus the age of the deep strategy is completing the evolution of military art.

7. *The Art of Managing the Deep Operation*

The new nature of the deep operation is naturally bringing about a new nature of its conduct.

Operational art, as the art of conducting the operation, is now facing a series of new problems. In the conditions of the new age, when the operation is organically merging with the battle, when these concepts have come to form a single whole in the modern phenomenon of deep strategy, and when they no longer have boundaries in time or space—it is first of all necessary to decisively put an end to the historically-outdated tenet that "as soon as the enemy has approached close enough for the general battle, then the time for strategy has passed and it can rest" (Clausewitz).

If modern operational art were to remove from the sphere of its competence the battle, it would eliminate itself and become nothing. This is what happened as early as 1914 and during our campaign to the Vistula in 1920. The power of conservative theory is great. The experience of the World War has by no means been taken into account. Even now, the French Col. Duffour writes:

"Strategic maneuver comes down to forming columns of various designations and aiming them toward the overall goal. At the moment when maneuver passes from the field of planning to that of execution, it is expressed only by the axes assigned to the units and the distribution of forces along these axes. Carrying out a maneuver is choosing the axes and the distribution of men and materiel between the columns."[74]

And thus according to Duffour's thinking, strategy, which the French do not always distinguish from operational art, only gathers and directs the columns and disappears when the columns begin the engagement. Further on, the operation will evidently unfold along unchanging, once they have been assigned, axes and will degenerate into the simple pressing of a solid wall, which drives from itself the one who, according to the logic of armed struggle, should be attacked. If this led to the failure of the big operations of 1914 and 1920, then in the new conditions it would simply become a comedy. One recalls the clown, who putting enormous shoes on his feet, wants to catch a ball thrown on the ground; but each time, when he gets near it he pushes it away from himself with his enormous shoe and amuses the viewers with this senseless work. In the same way, the French colonel is preparing to amuse history, for the transference of old and outdated tenets to completely altered conditions of a new age naturally becomes a historical joke. However, armed struggle, which is waged for realizing great ends, places more harsh and cruel demands on itself.

The deep multi-act and multi-tiered battle, which has come to embody the entire phenomenon of the operation during the age of the deep strategy, will be modern operational art's sphere of competence from beginning to end. Otherwise, there will be no operational art.

The gathering and aiming of columns will nearly always be its least significant part. The modern army commander, in organizing the deep operation, will already be simultaneously beginning and conducting the battle. When his combined-arms formations' main forces will still be moving by rail toward the front, his long-range combat aviation and vanguard echelon of motor-mechanized formations and cavalry will already be involved in intense fighting.

To limit the sphere of one's operational activity in these conditions to the simple gathering and directing of columns would mean declaring oneself bankrupt. Step by step, while regulating and directing actions from the depth, the modern army commander will have to continually and efficaciously manage the course of events. Each refusal by him to take an active part in this management will signify a step toward operational chaos.

At the same time, we should take into account that from the moment the battle is begun by the vanguard echelon of motor-mechanized formations and cavalry, to the moment when the succeeding combined-arms echelons enter the battle, a certain amount of time will pass. Once again a break and a pause will arise between the arrival of the succeeding echelons of the main forces and their entry into the battle. However, this phenomenon, being on the tactical scale characteristic of the Napoleonic era, is acquiring a new quantitative scale in the new conditions, and a completely different qualitative content. This pause now consists not of the concentration and placing of all one's forces in front of the battlefield, as was the case during Napoleonic times. It now unfolds not in a quiet and static situation, as was the case then, but in the intense dynamism of the rapid arrival of succeeding echelons from the depth and in brutal engagements by the already-engaged vanguard echelon. However, this pause is becoming an actual phenomenon in the beginning of the deep operation. Thus in the evolution of this phenomenon as well the circle of dialectical development is completed.

In the operational sense, this means that the modern army commander, having his aviation and vanguard echelon of motor-mechanized formations and cavalry ahead of him, will have the opportunity of grouping the succeeding echelons in his operational zone and directing them during the march to the front of the recently-begun battle in such a way that this will flow from his evaluation of the situation and the decision adopted by him. In the future it will be possible to conduct engagements, not where they will flare up along the columns' route of march, but where the commander decides.

The deep operation is again becoming manageable.

But to do this requires great skill in conducting and managing the deep operation. The various raw data of the immediate situation, which illuminate the operation in two dimensions—along the front and in depth, will demand a high level of operational art and operational culture, so that on the basis of the analysis, combination and calculation of all the data, one can arrive at a synthesis of a justified decision.

Here one will not be able to get away with flights of long-range strategic fancy, but rather specifically and consistently resolve immediate tasks, in order to defeat the enemy throughout his entire operational depth.

At the same time, the very idea of the operation will no longer comprise the main part of the art of managing and conducting the deep operation. If at the close of the last century Lewal wrote "inspiration is gradually descending from the intellectual heights to the practice of reality," and that "plans are becoming more and more dependent upon that materialization which they acquire."

The technique and calculation of the organization and realization of the operation will comprise the main subject of work for the modern army commander and his staff. The enormous scope, enormous columns, the enormous technical means, and enormous rear will all demand enormous capabilities to calculate everything, organize everything and direct everything.

The management of the modern deep operation consists, first of all, in organizing it.

Operational art itself, as the art of management, is becoming in modern conditions, first of all, organizational art—the ability to correctly calculate, precisely organize and firmly direct.

A great deal of chaos and defeat under the weight of one's own mass of men and materiel await us if this art of organization should not be top flight or fall short in even a single organizational detail.

It is becoming obvious that this art cannot rely upon organizational creativity. It has as its object a field, in which everything is closely connected, where everything demands coordination and is mutually interacting. Here everything must be resolved on some kind of definitely established organizational bases.

Modern operational art is faced with the urgent necessity of regulating the technique of organizing and conducting the deep operation within those boundaries and with that exactitude that are mandatory for any statutory tenet. We are faced with the mandatory task of creating an operational manual for the control of higher formations.

At the same time, tactical manuals must undergo significant changes, because each of the attack's operational echelons has its own tactics. In particular, we need to work out in detail the tactics of the breakthrough development echelon and organize its passage through the open breach of the tactical breakthrough of the front. A chapter like "The Development of the Breakthrough" should be included in the field manual and become one of the most important sections of that part which is dedicated to crushing the front.

However, if the organization and technique of conducting the modern deep operation have to be regulated by an operational manual, this does not yet mean that creativity, keenness and plan of a sagacious operational decision have in any way lost their significance.

In revealing the prospects of the deep development of the breakthrough, we can hardly agree with Lewal's prognosis that "imagination and creativity no longer have a place," that "the field for creative planning is shrinking," and that "the time of great preparations, brilliant combinations and magnificent maneuvers has passed."

When the time comes to unleash the breakthrough development echelon from the depth and transfer maneuver into the depth of the resistance, operational art will once again face the task of adopting a bold and smart decision corresponding to the situation.

The commander-organizer, who guides the operation in a systematic and planned way, will at that moment become an army commander, keenly and rapidly perceiving the situation in all its variety and immediately adopting a bold decision, according to his materialistically-founded inspiration. He will become at that moment a directly operating command level of authority, for sending aircraft into the air and committing the breakthrough development echelon will be accomplished through acts of direct command. Here we are talking about hours and minutes. Naturally, these decisions will not be issued from a comfortable office in the army headquarters, located in the deep rear, but from a forward-based command post near the front. At this post, the modern army commander must be in direct touch with the situation and feel the beating of the attack's pulse along the front.

Thus the modern commander will once again appear on the Pratzen Heights[75] and, surrounded by radio and telephone communications and always having an airplane at his disposal, will able to guide the modern deep breakthrough operation.

A powerful staff—the organizer and technical executor of decisions—will be at his service. Its leading operational section will be at the commander's side at all times. Another subordinate part of the staff will be located behind, managing and regulating the movements of the deep operational formation's units. Finally, its third rear part will be located even further in the depth, approximately along the railroad basing line, and will manage the entire complex mechanism of feeding and supporting the deep operation.

Such a management system for the operation is already appearing according to the deep outline, inevitably complicating the organization and coordination of the entire work. This management system must also be regulated by a series of exact statutory tenets. And even in this case, it will require a high degree of skill to force it to work in a planned and dependable way.

Finally, when the breakthrough development echelon penetrates into the depth and begins its destructive work there—management and the decision will achieve the highest level in the art of encircling and destroying in the new conditions of the deep strategy.

Following the exhausting methods of conducting attrition battles, this will be a vivid resurrection of the "great preparations, brilliant combinations and magnificent maneuvers."

Such are the prospects for the evolution of operational art in the age of the deep strategy.

8. From the Theory to the Practice of the Deep Operation

The prospects of a new era of the deep strategy, which have been outlined by us in their general theoretical forms, now demand concrete expression, so that they can pass from the field of philosophy and theory to the field of practical application.

This brings us back to the task of creating an operational manual for guiding higher formations. We have not assigned ourselves such a task here. In general outline we sought only to define the main contours of the new era in the evolution of military art.

In this, we proceeded from the chief prerequisite that our future revolutionary-class war will be a mighty act of armed struggle and, with worldwide-historical significance.

"Each great war represents an individual age in the history of military art," wrote Clausewitz.

The Evolution of the Operation's Nature

Era	Division of the Operation Along the Front	Division of the Operation in Depth	Operational Organization of the Offensive	Forms of Operational Maneuver	Type of Strategy
Napoleon, at the turn of the 18th and 19th centuries	One point	One instant	Close deep column in a single mass	Massed attack along internal lines	Strategy of a single point
Moltke, in the second half of the 19th century	A broken front of individual points dispersed in space	A broken chain of individual, unconnected battles	Broad deployment in individual groups	Concentric maneuver along external lines	Linear strategy of the broken front

Era	Division of the Operation Along the Front	Division of the Operation in Depth	Operational Organization of the Offensive	Forms of Operational Maneuver	Type of Strategy
World War, 1914–1918	A solid front of points merged into a single line	A single chain of a broken series of battles	Broad deployment along a solid front	Turning maneuver along external lines	Linear strategy of the solid front
Future war	The same	A solid chain of a merged series of battles	Deep deployment in several echelons	Frontal attack along external lines and crushing maneuver from the depth along internal lines	Deep strategy

The new age of socialist revolutions and revolutionary-class wars is predetermining such a new age in military art.

However, the new forms of military art, which have matured in the process of historical evolution, are not spontaneously born in their specific manifestation; they must be realized and studied; they must be philosophically and theoretically justified.

"Each age must have its own theory of war, or at least sooner or later take up processing it on philosophical principles" (Clausewitz).

Without a grounded theory, there cannot be intelligent practice. We began, therefore, with theory, in order to then move on to the practice of precisely calculating the deep operation.

This path has opened to us the entire evolution accomplished by military art since the beginning of the 19th century. Only in the difference of major historical ages can one establish the regularities of the development of military art and comprehend how and why this art has moved from some forms to others, and why it is concluding with the age of deep strategy.

Thus the point of the Napoleonic era broke up into a series of individual points scattered in space in the Moltke era. During the World War this series of points merged into a single solid line. And now this line is growing in depth, turning into a square and thus acquiring a new dimension in area.

We are entering a new era of deep strategy and moving from a broad linear front to a deep front.

Of course, the forms of deep strategy will not fully appear everywhere and not immediately, and not in all cases and conditions. Historical evolution does not know any such definite boundaries. In the conditions of our theater of military activities there may still be prerequisites for the linear strategy of the enveloping maneuver at the beginning of the war. The forms of the deep strategy will most likely reach their full scope in the Western European theater of war. Their complete manifestation on our front will evidently be a slower process. However, all of the principally new factors of the deep strategy will manifest themselves in our struggle with the same full force, although on a smaller scale.

It is thus why the working class is acting as a worldwide-historical progressive force and that the idea of destruction is embedded in the bases of the deep strategy and that the revolutionary proletariat will be the first bearer of the new operational art and will put forward the first craftsmen of the deep destruction operation.

"A true change in military art is the consequence of changes in politics" (Clausewitz).

Only the most immense political goals of our struggle may guarantee the historical realization of the new deep strategy.

Two

"The Fundamentals of the Deep Operation" (1933)

Part I. The Shock Army's Initial Offensive Operation

Moscow, 1933
Top Secret

TABLE OF CONTENTS

Introduction	74
I. The Basic Prerequisites of the Deep Operation	76
II. The Elements of the Deep Operation and the Composition of the Shock Army	84
III. The Place and Role of the New Means of Struggle in the Deep Operation	97
IV. Vanguard Activities in the Beginning of the War	102
V. The Operational Organization of the Approach	107
VI. The Meeting Battle	114
VII. The Fundamentals of Organizing the Deep Breakthrough	124
VIII. Carrying Out the Deep Breakthrough	137
IX. The Development of the Initial Offensive Operation	147
X. The Fundamentals of Managing the Deep Operation	151

Introduction

Two factors of the greatest importance are determining at the present time the approach to the formulation and resolution of the problems of our operational art, as the study of the conduct of the operation.

The first is the triumphant completion of the first Five-Year Plan, which has transformed the Soviet Union into one of the most advanced industrial countries in the world. The second is the accomplished technical reconstruction of the Red Army, which has become one of the first-class armed forces and has been supplied with the most modern and high-quality technical means of struggle. On the basis of these achievements, our defensive capability has reached a high degree of development; our armed forces have achieved such capabilities for their combat employment which authoritatively put forth the demand for their new employment in the engagement and the operation. In these conditions, our military-theoretical thinking can no longer limit itself to simple declarative statements on the new forms of armed struggle from the theoretical point of view. The justification for the logic of adopting the new forms of struggle [in our country], which have been given form in the concept of the deep engagement and the deep operation, to a certain degree, represents for us an accomplished fact. The basis of the deep engagement and the deep operation is already a reality for us, and it is in these conditions that we are faced with a series of new and complex tasks, which require at the present stage the elaboration of an applied theory of operational art as the basis for calculating the specific formulation of the new forms of armed struggle.

This has already been resolved to a significant degree in the field of tactics, now demanding a number of experimental verifications. However, much less has been achieved in the field of operational art, leaving here a wealth of work still to be done. Of course, it is incomparably more difficult to formulate an applied theory of operational art than to elaborate a general-theoretical draft of the new forms, which was pursued in the work *The Evolution of Operational Art*.

One must bear in mind that we are writing about an operation which nobody has ever conducted. At the same time, we are dealing with those means of struggle, the employment of which no one has tested in the engagement or operation.

Our research work in the field of operational art is substantially distinguished by these conditions from similar work in the past, when such military researchers as Schlichting, Bernhardi and Schlieffen built their operational theory {for the future} completely on [studying] the experience of recent wars, employing data that was sufficiently well-known and tested.

In the conditions of the great revolutionary age of our socialist construction, when we have created a society and army that no one has yet created; when the unprecedented growth of our productive forces each day and each hour yields material values of a completely new effectiveness—the significance of the past has for us only that significance that history has in general and always does.

However, we would have proven to be helpless to resolve the tasks of modernity, had we not gone beyond the limits of this experience, had we not reevaluated it from the point of view of the {radically} new conditions of our age, and if we had not ruthlessly thrown over everything that was intolerably out of date and obsolete.

All of our construction is revolutionary construction. It is taking place under the aegis of building the new and mastering the new. The field of our military art, as the continuation of politics, feels this extremely acutely.

In researching the forms of modern armed struggle, we are faced with completely new tasks, which were never put forward nor resolved by past experience.

This involves its own natural difficulties.

In the development of our theory of operational art, we have at the present time not yet reached that level where we can exactly and definitely regulate the fundamentals of the waging of armed struggle on the operational scale and immediately set about formulating an operational manual for commanding higher formations.

The very necessity of such a manual, which did not exist in the past, is called forth by the new conditions of the nature of the deep operation as a complex system for employing combat efforts in single, centralized and unified cooperation along the front and in depth, on the ground and in the air.

This work cannot set itself such a task. Being an expanded elaboration of lectures read by me in the operational department of the RKKA Military Academy[1] {in the spring of 1933}, it pursues more limited aims—to research the calculated norms and the specific forms of the deep operation against the background of the initial offensive operation at the beginning of a war. Naturally, at the same time the tenets put forth here cannot lay claim to any kind of a final and complete solution to the problem. This relates in particular to the calculated norms laid out here, which will naturally still be subjected to inevitable changes.

This work is also not meant to serve as a direct guide for action.

We should altogether reject the expediency of such a statement of the problems of [the modern operation] {operational art}, by which we can be guided to a ready outline and recipe.

The very concept of operational art presupposes the distinctiveness of methods and forms, which are each time employed in conformity with the given specific conditions of the situation.

We should understand each tenet put forth in the field of operational art as an orienting idea, which can achieve this or that specific form only in and depending upon a definite and real situation.

Thus this work would have a negative effect if someone took it into their head to attach to the tenets set forth therein the significance of a ready-made outline for operations. There can be no such outline in operational art.

We are speaking here about establishing the fundamental and basic features of the new forms of the deep operation at various stages of its initial development.

Only this significance should be attached to the tenets set forth by me here.

Finally, in the same way, this work is tries to solve the problem in the field of organizing the coordination of all elements of the deep operation, not in the sense of to how to do it (an outline, a standard), but only as to what to do (an idea, a directive).

It is quite obvious that all of the tenets put forth in this work are subject to intensive and all-round testing; only after this will they be able to acquire the significance of material for elaborating an operational manual for commanding higher formations.

This could serve as the entire justification for the present work, which embraces only the first part of the entire theme of the fundamentals of the deep operation. This part treats only the initial offensive operation, examining the dynamics of the development of the initial period of war in a conditionally adopted process of consecutive stages of deployment, the approach, the meeting battle, and the breakthrough.

Naturally, this does not yet embrace the entire volume of operational activities, requiring the subsequent elaboration of questions of defense, the organization of the operational withdrawal and special forms of the operation, {particularly} in relation to fortifications.

I. The Basic Prerequisites of the Deep Operation

The deep operation as a new form of employing modern weaponry in the engagement and operation, has its cutting edge directed at the front of fire and is designated for breaking and crushing this front throughout its entire operational depth. The solid front, as the opposition of front against front, is the main background against which the deep operation finds its application. The overcoming and destruction of the solid front is the chief task which is resolved by the deep operation and which called it to life at the modern stage of development. It is quite clear that if the general operational nature of the front does not condition the [element of] {continuous} fire front, then the deep operation [loses its special topicality and] does not acquire {its own} complete forms of development.

Thus the initial, practical problem of justifying the deep operation consists of studying and determining the operational nature of the front of struggle, which will be applicable to us in the likely theater of military activities.

Specifically, the problem comes down to determining whether this front will be continuous, in which case it will require a breakthrough; or will it be a broken one, allowing for the possibility of turning maneuvers and thus, at first, not requiring the organization of the kind of combat formation that lies at the heart of the deep operation as the operation of overcoming and crushing the front.

The study of this problem is all the more valuable because it finds an extremely varied interpretation in modern bourgeois military doctrines and is precisely that core around which the study of the problem of the nature of future war, at the operational level, revolves.

Overall, modern operational literature very inadequately illuminates this problem. The onetime rich German military literature has almost nothing to say on this matter. We find a certain illumination of this problem in the French literature, which has a certain interest for us, insofar as France is the country having the most concentrated system of imperialism and which maintains an army that is technically well-outfitted.

The nature of the front of struggle is primarily defined by its operational density; that is, the number and composition of those forces that can deploy along a given front.

On this score, the chief of the French General Staff, Gamelin,[2] wrote not long ago: "The chief difference between the operations we conducted in 1918 and the ones for which we must prepare, now lies in the correlation of the overall number of forces and the length of the front of strategic deployment.

"The question is how many divisions we will see along a 500-kilometer expanse, where in 1918 there were about 200 divisions on each side.

"If the front is insufficiently occupied, then empty spaces will appear, and this means maneuver."

In evaluating the modern conditions of deployment along the Franco-German frontier, defined by the weak state of the German armed forces,[3] Gamelin arrives at the conclusion that this front will be weakly occupied, presenting great opportunities for developing flanking maneuvers.

He states: "At the start of a war between groups of forces along the flanks, empty spaces will undoubtedly exist."

This evaluation of the operational nature of deployment along the Franco-German border lies at the heart of modern French operational doctrine. Proceeding from this, a French manual on the employment of large formations begins with the following introduction: "At the beginning of the war the available forces will consist of small armies, designated for securing the conduct of a general mobilization in their own territory and making the same difficult for the enemy. These armies will be called upon to maneuver along open spaces. A similar situation may arise when the exhaustion of the forces will enable us to break through the continuous front, and intervals will be created."

In this way, French doctrine in general proceeds from the conditions of the broken front, which allows for a great deal of maneuver. And this is all the more significant for the Franco-German theater, where the events of the World War first reproduced for the first time the solid front of fire and gave birth to those prerequisites, which in their subsequent development, and upon a new material base lead {us}, at the present stage, to the concept of the deep operation.

Of course, French doctrine chiefly proceeds from the Versailles system of a disarmed Germany. However, as is known, this factor has already long ago lost its stability and permanency.

In the meantime, the French are drawing further conclusions from the concept of the broken front. They are obviously opposing the solid front to maneuver and believe that modern maneuver, while remaining in its forms at the level of 1914, is completely excluded if the front should represent an unbroken wall of fire. The training of the French officer corps and the rank and file is conducted on this basis. It is of interest to note that during the large maneuvers of 1931—and these were the first large maneuvers in France since the World War—three infantry divisions turned their front 90 degrees and then conducted a concentric offensive along an overall front of 60 kilometers to encircle the enemy. To be sure, at the same time long-range weapons were represented; in particular, an air landing was employed in these maneuvers, consisting of a small group of infantrymen with light machine guns, who had the task of landing 50–60 kilometers in the enemy's rear and attacking his supply bases.

To sum up, freedom of maneuver and the possibility of broad flanking movements free us from the necessity of working on such forms of the operation that would support the breakthrough of the front of fire. Therefore French doctrine remains on the position of the linear strategy, the entire essence of which consists, in the final analysis, of acting upon the enemy in a single line of direct combat contact.

In speaking about the possible forms of the breakthrough, French doctrine examines this question only from the point of view of grouping forces for their employment along the front. At the same time, two variants are considered possible: either all forces will be employed along a single axis for achieving a decisive result, an operational form the French call the ram; or, just the opposite, where all forces will be dispersed in separate groups along the front and, through individual attacks with limited aims, they will begin to shake the front as blows from a hammer. The French call this form of operation the hammer.

Completely different prerequisites and different specific conditions lead us to a completely different solution to this important question. [Of course, first of all, we must] {First of all, it is necessary to proceed from the nature of our future war as a revolutionary-class war, which takes the intensity of the struggle to the highest level}. {Secondly, we should} proceed from a specific calculation of the operational density of our front of struggle. This calculation was formulated by comrade Triandafillov and retains, with a few qualitative corrections, its great importance.

Up to approximately 70 infantry divisions and five cavalry divisions can be mobilized against us along the most important sector of our western frontier between the Western Dvina and the upper reaches of the Dnestr rivers. Of these forces, approximately 60–65 infantry divisions could appear along our western frontier, which is 800 kilometers long on this sector. This calculation yields an average operational density of 12–13 kilometers per infantry division; that is, in general, an operational density that enables each division to construct a stable defense. However, such a general calculation has more of a theoretical significance.

Our western theater of military activities is divided into two parts, separated by the large wooded expanse of the Poles'ye.[4] Both parts have an equal length of 400 kilometers. If we consider one of these parts the main one and the other secondary, then no less than 20 divisions will be required to defend the latter. In this case, up to 45 divisions could appear along the main sector of the front, which already yields a density of 8–10 kilometers of front per division. Naturally, the deployment density is by no means equal. A certain part of these forces is echeloned in depth. Denser groups of forces are created along axes chosen for the main attack. However, in the final analysis the front may end up being completely occupied, and this excludes any kind of significant free space for the development of broad turning maneuvers in our theater of military activities.

At the same time, we must take into account that the length of the front shrinks, as it moves from our western frontier to the middle Vistula, which leads to an increase in the overall deployment density.

We will note that a normal operational density of 8–10-12 kilometers per division is not something new. Such was the deployment norm for the Russian armies along the Austrian front in 1914.

However, the same quantitative indicator in 1914 is now characterized by a completely different quality.

As is known, in 1914 a division had 24–32 heavy machine guns, but not a single light one; a modern division has 108 heavy machine guns and 162 light machine guns.

Thus, given one and the same operational and tactical density, the fire density has increased considerably.

{In the conditions of our future war as a revolutionary-class one}, this circumstance of enormous significance makes the front more stable and {more} capable of resistance, while requiring significantly greater pressure to overcome it.

In our conditions, the front of struggle is thus taking on the general nature of a solid fire front at the operational level. Of course, this does not mean that in the beginning of a war the conditions for flanking maneuvers are completely excluded.

However, we must take into account the fact that these conditions offer incomparably more limited opportunities for such a maneuver.

We should, first of all, evaluate the nature of the western theater of military activities, which is extremely specific as regards those axes that, according to the terrain conditions, it offers for the development of offensive activities. Even a cursory glance at a map shows that this specificity features a pronounced corridor system, which offers very definite axes for an offensive, which are separated from each other by significant geographical barriers.

This is particularly vividly expressed in the Baltic theater, where a number of axes are divided from each other by water basins (Lake Ladoga, the Gulf of Finland and Lake Pskov). This is true to the same degree in the Belorussian and Ukrainian theaters of military activities, where distinct axes are divided from one another by large wooded and swampy areas, offering very definite opportunities for maneuver, which is limited in space by solid barriers.

In examining the engineering outfitting of the western theater to the west of our frontier, we can clearly discern that the axes allowing for the development of maneuver are being scrupulously closed up by engineering structures in the form of modern fortified areas. This factor of the engineering outfitting of the theater of military activities doubtlessly exerts an enormous influence on the entire operational nature of the front of struggle and the organization of maneuver. It is known that one of important factors in drawing up the Schlieffen Plan for an attack through Belgium was precisely the engineering strengthening of the French border opposite the Rhine, represented by the engineer Sere de Riviere's[5] well-known engineering belt. Finally, if a flanking maneuver nonetheless proves possible along individual axes, it would encounter incomparably greater obstacles than before.

These obstacles primarily consist of the modern opportunities for operational maneuver, which may oppose comparatively quickly the flanking group with a new concentration. It is sufficient to point out, for example, that along the extremely important sector of our western frontier between the Western Dvina and the headwaters of the Dnestr, the opposing side could shift {up to four} infantry divisions in a day by the existing railroads from one sector of the front north of the Poles'ye to another to the south. Moreover, auto transport, which is one of the main means of operational maneuver, can help out in this regard, as can air transport.

Thus we should not assume that a flanking maneuver can now freely develop to a significant depth; it will doubtlessly encounter a new enemy group of forces much sooner than before.

Besides, the flanking maneuver now encounters significantly greater impeding factors in the technical means of struggle themselves. The chief of these is modern aviation as an extremely powerful means of operational significance. Proceeding from a calculation of 1% of hits out of the 1,200 fragments that 40 bombs from a single light assault aircraft can deliver, it has already been proven by experience that a single detachment of ten aircraft can delay a battalion on the approach, a squadron a regiment, and a brigade a division. If this pressure is maintained over the course of several days, a squadron, if operating intensively enough, could delay an entire division on the approach. On an operational scale, this means that the movement of a shock group of troops along a 30-

kilometer front, along which six divisions can move, calculating one road per five kilometers {of front}, can be delayed in modern conditions by six assault air squadrons, or two assault air brigades.

Thus it is by no means necessary to oppose a new troop concentration to a threat to one's flank and the appearance of a new enemy group of forces; this task is more rapidly and actively carried out by aviation.

Finally, a flanking maneuver now encounters great resistance in a new factor of enormous operational significance—the obstruction of the terrain.

It is by no means necessary, even given the absence of aviation, to block the enemy's path with a new troop concentration; this task can be carried out by the engineer-obstacle troops alone, even with a very limited number of forces.

It is sufficient to point out that in order to obstruct the terrain along a 30-kilometer front and to the same depth; that is, a territory of 1,000 square kilometers, we need the following: two sapper battalions, one chemical battalion, four infantry battalions, and one ton of explosives. It is possible with these materials, and in certain time conditions, to block an area of 30 square kilometers to such a degree that the enemy's advance will, in any case, be significantly delayed and slowed down. If one sets aside a single division for the defense of this area, then its retention may be considered guaranteed for a definite period of time.

Thus the entire sum of modern conditions—the front's operational density norms, the engineer equipping of the theater, the opportunities for operational maneuver, aviation, and obstacles speaks to the fact that the operational nature of our front of struggle contains everything necessary for the establishment of an unbroken front of fire to a significantly greater degree than was lurking in the prerequisites of struggle in 1914.

All of these conditions give rise to the front's much greater degree of resistance and its greater depth and greater stability.

We must thus consider the opposition of front to front a perfectly likely prospect for our theater of military activities and proceed from this in defining the nature of the operation.

Its basic form will undoubtedly be the breakthrough.

The breakthrough requires the greatest amount of men and materiel; this operation is the concentrated expression of the entire sum of possible combat efforts; it is the fullest expression of the deep nature of the modern battle and is conditioned by those prerequisites that lead us to the new forms of the deep engagement and deep operation.

At the same time, we should keep in mind the fact that {we} are in no way pitting the breakthrough against maneuver and the solid front against maneuverability. The deep operation has as one of its main purposes the shifting of maneuver to the new dimension of depth and thus resolving the task of breaking through the entire operational depth.

To this day, the problem of the breakthrough remains in the center of attention of modern military art. This problem was not resolved by the events of the World War; it was not resolved by the events of recent wars, particularly the Sino-Japanese War in the east,[6] which, from this point of view, has been much understudied.

One should take into account that in 1932 the Chinese front between Wusong and Shanghai was not broken in fighting by the Japanese, despite their sufficient technical provisioning with tanks and aviation. This circumstance imparts to the problem of the breakthrough [in the center of attention] an extreme acuteness and topicality.

A study of breakthrough conditions in the World War leads us to establish four main reasons for their lack of success, as seen from the tactical-technical point of view. The first reason was that the offensive's combat formation at first did not dispose of such [a] means of struggle as would be capable of withstanding a bullet and to overcome it as the chief factor of defensive fire.

The second reason was that the offensive's combat formation was not equipped with such long-range technical means of struggle, which would enable it to {simultaneously} strike the entire tactical depth of the defense. This forced the offensive to try and resolve the task of the breakthrough by consecutive, methodically broken-up attacks into the depth, during which the defender's deep tactical reserves remained unscathed and could each time reconstitute from the depth the depth of the resistance, when some seat of resistance at the front would fall. As a result, this led to a situation in which the entire depth of the defensive zone was, at best, entirely moved back, although it always retained its vitality and its overall tactical depth. In these conditions, the attacker was transformed into an ancient hero fighting with a 12-headed dragon, who would grow a new head each time one was chopped off and who could only die if all 12 heads could be chopped off simultaneously.

The third reason was that even when achieving the tactical breakthrough by a massive concentration of artillery and, in 1918, tanks, this did not yield an operational result, because the attacker's combat formation did not have in its own depth any kind of factor that could burst through the open breach of the tactical breakthrough into the operational depth and thus achieve an operational resolution of the problem. In these conditions the very tactical breakthrough proved to be, essentially, unnecessary and would rapidly die out, resulting only in a sack-shaped broadening of the attacker's front.

Finally, the fourth reason, which is being established in the possible course of events, was that even if the offensive's combat formation in 1918 had possessed a special echelon in its depth that could have rushed into the tactical depth through the tactical breach in the front, it nevertheless would not have achieved its goal, because the entire defensive sector under attack did not remain isolated from its operational depth. In these conditions, the large masses of defensive reserves had full opportunity to concentrate at the front under attack. At this stage of the breakthrough operation the offensive would turn into a defense for the attacker and the defender's resistance took on the character of an offensive.

Of course, the efficacy of these reasons may be viewed only in the light of the political nature of the World War as a reactionary and imperialist war.

However, the operational-tactical side of these reasons is beyond question. And the essence of these reasons leads us today, on a new material-technical basis, to four main conditions, which are logical and binding, so that the task of overcoming and destroying the front of fire can become truly possible at the operational level.

The <u>first condition</u> is that the offensive's combat formation be equipped with such technical means of struggle as would be able to withstand the bullet and overcome it, by carrying its own fire and attack throughout the entire tactical depth of the defense. It is blatantly obvious that the tank is such a means and that the essence of realizing this condition is the tank saturation of the attacker's combat formation.

The <u>second condition</u> is that the entire tactical depth of the resistance must be attacked simultaneously. If this is not done simultaneously, then the defender's deep tactical reserves have the opportunity each time, when some point of resistance falls along the forward edge, to reconstitute the overall depth of the resistance from the depth. Without this simultaneous attack, the depth of the tactical defense is capable of continuously maintaining itself, while only falling back in its entirety.

This condition determines the logic of the concept of the deep engagement and of the mandatory simultaneous attack against all the elements of the defense's tactical depth and conditions the entire necessity of this form of the engagement as the only one that can radically resolve the problem of the tactical breaking of the front.

The <u>third condition</u> is that the offensive's combat formation must have in its own depth such a qualitatively new factor that would be capable of pushing forward immediately, following the tactical breaking of the front, through the open breach and into the operational depth and to finish the tactical breaking of the front with the latter's complete operational defeat.

Specifically, this condition leads us to the necessity of a special operationally designed and correspondingly technically armed breakthrough development echelon, without which the task of breaking through cannot be resolved at all.

Finally, the <u>fourth condition</u> is that the entire front under attack must be completely isolated at a sufficiently great depth from the overall operational and strategic depth, denying the enemy any opportunity of concentrating new reserves to the breakthrough sector. If this is not done, the breakthrough development echelon will inevitably encounter a new front of fire {in the depth} and will be deprived of the opportunity of carrying out its mission. This circumstance conditions the place and role of long-range aviation in the modern operation and its enormous significance for resolving the problem of the breakthrough.

These four conditions lie at the heart of those new forms [of the engagement] {of struggle}, which at the present stage lead us to the concept of the deep engagement and deep operation. The first two conditions are realized at the tactical level and lie at the heart of the prerequisites for the tactics of the deep engagement. The latter two conditions—the breakthrough development echelon, which transforms the tactical breaking of the front into its operational defeat, and long-range aviation, which isolates the breakthrough front from the defense's operational and strategic depth—are at the heart of the prerequisites of the deep operation and comprise its essence.

At their core, the new forms of the deep operation are distinguished from the old and impotent forms of the linear operation by the fact that they shift the combat pressure against the enemy to his entire depth and have the following as their chief prerequisites: the first is the simultaneous defeat of the defender's entire tactical depth; the second is the immediate development of the tactical breaking of the front into the operational depth, and; the third is the complete isolation of the breakthrough front from the defense's operational and strategic depth.

In other words, if French doctrine puts the emphasis on a mixed group of forces and the distribution of forces along the front for breaking through, in a single group, or several groups, then the essence of the deep operation is that it groups its forces in depth, in order to strike the entire depth. If one will allow a certain poetic license, we could oppose to the terms ram and hammer, as [a standard] {types} of the linear operation, the awl, as a type of deep operation; for its essence is that we drill, with the new force of the breakthrough development echelon, from our depth into the enemy's operational depth, thus resolving the problem of the modern deep breakthrough { figure 1}.

It is quite clear that such an operational form is becoming possible only at our present level of industrial and technical development; for the chief factors of the deep attack against the enemy are the technical means of struggle, which dispose of sufficient long-range striking power. Without these means of struggle, without long-range striking power, all-terrain capability, and protection from bullets, the task of attacking the entire depth cannot be resolved. From this it is obvious that the chief factors in the deep operation are, first of all, aviation; second, independent mechanized formations, and; third, motorized troops and mechanized cavalry. Only on this technical base do the forms of the deep operation become possible.

However, this embraces only one side of the question, for we are now speaking not only of the possibilities of the new forms of struggle on the armed front; we are now saying that these new forms of struggle are logical and thus necessary and mandatory as the only ones capable of radically resolving the problem of overcoming the entire depth of the front of fire.

Thus the concept of the deep operation proceeds primarily from the operational background of the solid front of fire; it is directed against it and is intended to destroy it. However, by no means does this mean that in all of those cases when the operational nature of the front of struggle does not condition the uninterrupted front of fire that the forms of the deep operation become non-efficacious and inapplicable.

Undoubtedly, the modern distance between two sides entering into a war is shrinking significantly. One can easily discern this from the conditions of our theater of mil-

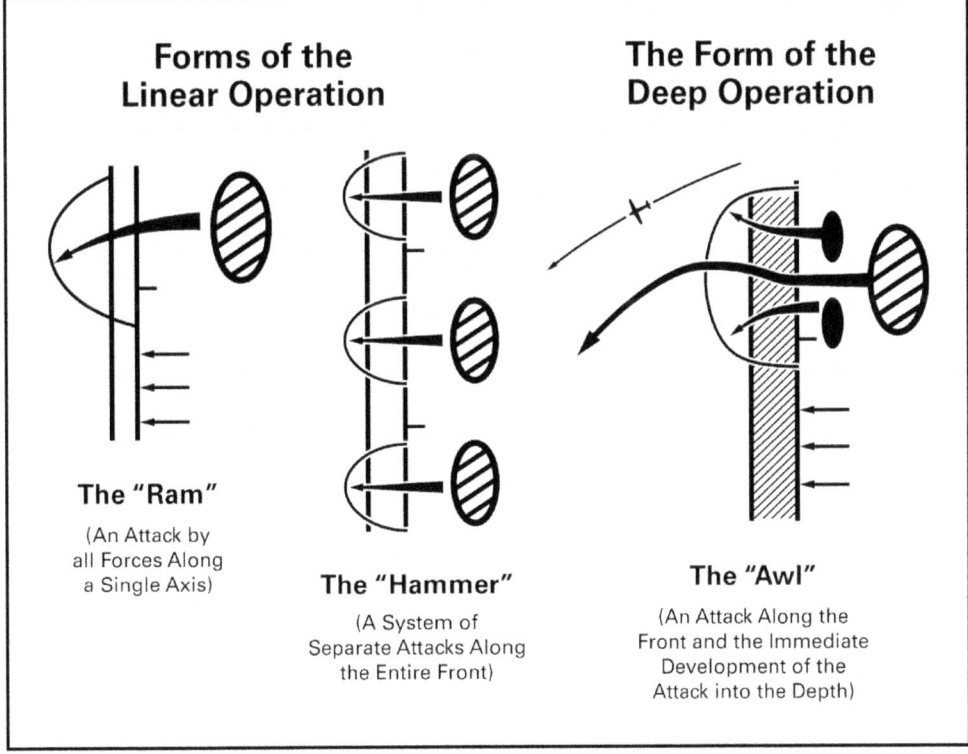

Figure 1.

itary activities; for such major locales as, {for example}, Minsk and Baranovichi, will always be a base for the sides' jumping-off position, which leaves a distance of no more than 100 kilometers between them.

Nonetheless, even this distance conditions the march, which is always linked to a certain linear maneuver. In modern conditions the latter is thus in no way excluded. However, the principles of the deep operation are extending their qualities to even this stage of the march, manifesting itself in altered forms.

The essence of the question is that during the breakthrough the deep operation's chief task is the overcoming and destruction of an already established front {; and} during the march the chief task of the deep forms of combat pressure consists in striking the advancing enemy from afar so as to destroy his front before ground contact is made with him; so that his combat formation cannot coalesce and establish itself as a result of our long-range attack and thus turn into an continuous front of fire.

If during the breakthrough we are fighting against an already-established solid front, then during the march, given the presence of an open space between the sides, we are fighting against the establishment of the solid front itself.

Thus the characteristic features of the deep operation, at the heart of which lie long-range combat pressure, retain their forms and applicability at all stages of the development of the modern operation.

An entire picture of employing the new forms of struggle unfolds, which we do not find in a single one of the foreign bourgeois armies' theories. It is completely logical that the most forward and revolutionary army is putting forth the most forward and revolutionary theory of the new forms of armed struggle.

II. The Elements of the Deep Operation and the Composition of the Shock Army

The first problem in specifically formulating the deep operation consists of defining and evaluating the elements by which this operation is carried out. This problem was not that significant in the conditions of 1914, when the entire arsenal of equipment which the sides disposed of on the operational scale consisted of uniform formations of infantry and cavalry. To be sure, there was the difference in their range of activities (in the rate of march), although this was easily reconciled.

In general, each of these combat formations was a collection of homogenously armed and homogenously organized men.

In this sense, the uniform combat formations were an element of the operation and the entire task consisted of bringing them into such a degree of cooperation along the front in which the task of defeating the enemy would be achieved.

The operational art of 1914 was, in this sense, monosyllabic and was realized on the basis of uniform means of struggle, which knew no qualitative distinction.

We are now faced with completely different conditions.

First of all, modern means of acting upon the enemy have multiplied quantitatively, according to their types; secondly, they are extremely varied qualitatively, according to their combat effectiveness, nature, tactical employment, and operational designation. An examination of the complex arsenal of the modern means of struggle, which the

II. The Elements of the Deep Operation and the Composition of the Shock Army

ATTACHMENT 1. THE ELEMENTS OF THE DEEP OPERATION

Formation	Composition of its Combat Effort	Front, Area, the Object
Reinforced Rifle Corps	3 rifle divisions, 4–6 battalions from the High Command Tank Reserve, 3 regiments from the High Command Artillery Reserve, 1 assault air brigade, 1 chemical battalion, special support units	A 10–12 kilometer attack front against an organized defense
Cavalry Corps	2–4 cavalry divisions, a mechanized brigade	A 6–12 kilometer attack front in maneuver conditions; 12–24 kilometers while holding terrain[7]
Mechanized Corps	3 mechanized brigades (T-26 and BT[8]) and a rifle regiment	A 10–12 kilometer attack front in maneuver conditions
Motorized Division	4 rifle battalions, a battalion of small tanks and an artillery regiment	A 4-kilometer attack front; an 8-kilometer defensive front
Airborne Detachment	2 motorized battalions and light mechanized equipment	
Assault Air Brigade	94 planes	Its objective is an infantry division on the march (15,000 targets)
Light-Bomber Brigade	94 planes	Destroying a vital economic area of 0.5 square kilometers, or putting out of action up to three supply stations for 12–36 hours (with 100 demolition bombs)
Heavy-Bomber Corps	144 planes	Destruction of urban and industrial structures up to 5 square kilometers (with 250 demolition bombs)

Formation	Daily Depth of Range	Operational Range in Time and space	Operational Designation
Reinforced Rifle Corps	10–15 kilometers	Unlimited	A tactical attack means for the simultaneous suppression of the entire tactical depth of the resistance
Cavalry Corps	50–60 kilometers	150 kilometers in 2–3 days. Four days of fighting and advancing, and a fifth day for rest, in normal marches	A means for attacking, seizing and holding in independent actions ahead of the front, along the flanks and in the operational depth
Mechanized Corps	100 kilometers	200 kilometers in three days	A means for attacking and seizing in independent actions ahead of the front, along the flanks and in the operational depth
Motorized Division	150 kilometers	Unlimited	A means of operational maneuver for seizing and holding
Airborne Detachment	Corresponding to the radius of the transport aircraft; on the average—400 kilometers	Corresponding to the radius of the transport aircraft; on the average—400 kilometers	A means of surprise attack against the operational depth of the resistance
Assault Air Brigade	400 kilometers	400 kilometers	A means for attacking personnel from the air
Light-Bomber Brigade	400 kilometers	400 kilometers	A means for destroying light targets
Heavy-Bomber Corps	800 kilometers	800 kilometers	A means for destroying heavy targets

deep operation disposes of for its realization, leads to the establishment of the following main elements (see attachment No. 1).

1. The chief factor of the deep operation is the troop organism designated for the immediate tactical breaking of the opposing front and the simultaneous defeat of its entire tactical depth. This is a reinforced rifle corps, which must be equipped with technical means of struggle for the simultaneous destruction of the main position of resistance, the suppression of the artillery and tactical reserves and the paralysis of the immediate troop rear; that is, for the simultaneous pressuring of the entire tactical depth of the resistance. Proceeding from these requirements, a reinforced corps must have the following: from four to six tank battalions, for an overall strength of up to 300 tanks; two howitzer artillery regiments and one artillery regiment from the High Command Artillery Reserve, for an overall strength of up to 300 guns, and; one assault air brigade, for an overall strength of 90 aircraft. This reinforcement norm is an approximate standard for the technical equipping of the reinforced corps to resolve the task of simultaneously suppressing the tactical depth of the resistance along a front of 10–12 kilometers in a day of fighting {to a} depth of 10–15 kilometers. The operational range of such a corps is essentially unlimited, requiring only the corresponding time for the troops to rest.

2. The second element of the deep operation is the cavalry corps. It is composed of from two to four cavalry divisions and, in certain cases, an attached mechanized brigade.

It is not required during the deep breakthrough that such a cavalry corps take part each time in that operational formation that is resolving this task. Cavalry may be massed in large groups and detailed as an independent operational formation for the resolution of independent operational tasks.

However, in joint actions with other combat arms, the cavalry corps is a valuable element of the deep operation.

It is capable of simultaneously attacking in maneuver conditions along a 5–12 kilometer front and [to hold] {of holding} 12–24 kilometers of ground, depending upon its composition. The daily depth of such a corps' combat range is approximately 60 kilometers and its operational reach is 150 kilometers, a distance that may be covered in 2–3 days.

Such a corps may operate up to four days in normal cavalry marches and requires a halt on the fifth day. It is mainly an attack, seizing and holding weapon in independent actions ahead of the front, along the flanks and in the operational depth.

3. The third element of the deep operation is a completely new qualitative factor in the arsenal of modern troop formations—the mechanized corps, consisting of three brigades: two tank and one rifle-machine gun, with an overall strength of up to 300 tanks.

The mechanized corps can attack in maneuver conditions along a 10–12 kilometer front; the daily depth of its combat range is 100 kilometers, while its operational reach is 200 kilometers.[9]

It is necessary to keep in mind that the duration of the mechanized corps' combat activities has a definite limit as to its technical resources and can be measured in time at no more than three days, after which it needs to be concentrated and put in order. A mechanized corps could cover 300 kilometers in three days, although this would mean pushing its technical resources to the maximum.

II. The Elements of the Deep Operation and the Composition of the Shock Army

The mechanized corps' operational designation is to attack and seize; the mechanized corps cannot operationally hold conquered territory, and for this needs to operate jointly with either the cavalry or motorized infantry. The mechanized corps operates independently ahead of the front, along the flanks, or in the operational depth.

4. Because the mechanized corps does not have the operational capability of holding ground, the following factor in the deep operation is of great significance—the motorized division. Its tactical value is not great. It consists of only four rifle battalions, a battalion of small tanks and a standard non-reinforced light artillery regiment. The motorized division is capable of attacking along a 4-kilometer front and holding an 8-kilometer front on the defensive. However, it is an extremely valuable factor in the operation, because it has great operational maneuverability along the roads, [in depth per day] {measured at a daily depth of operations} of up to 150 kilometers. Its operational reach may be considered as unlimited. As a most important means of operational maneuver, the motorized division is capable of seizing and holding and assisting in joint work the mechanized corps, which it replaces at definite stages in the fighting.

5. The airborne detachment is currently represented by the insignificant standard model of two motorized battalions and reinforced by light mechanized weapons such as the T-27 and D-8.[10]

The landing detachment is a means for launching a surprise attack against the enemy's operational depth and has less utility for employment in the tactical depth. The possibilities for the air-technical development of this weapon are colossal and depend completely upon the technical opportunities for air transport and the type of transport aircraft.

It would be least favorable to employ our ANT-14[11] transport aircraft, as it would require up to 85 of these to transport a single landing detachment. Our technical capabilities already offer us more profitable means for a landing operation.

Further, aviation:

6. The assault air brigade is the chief means of attacking personnel from the air and, based upon the calculation of being able to strike up to 15,000 personnel targets with bomb fragments; it can delay an entire division on the march.

7. A light-bomber brigade can strike an area of industrial and urban structures in an area of 0.5 square kilometers, or simultaneously put up to three supply stations out of action for 12–36 hours, depending on the effectiveness of the strike.

8. Finally, a heavy-bomber corps, with a range twice as far as light-bomber aviation, can destroy an area up to five square kilometers.

The normative indices for the effectiveness of bomber actions cited above have no direct practical value and in each separate case they must be changed, depending on those targets against which their destructive power is directed; they only indicate in general the bombing possibilities of this very powerful factor of struggle.

The above-enumerated eight elements of the deep operation are only the main factors in combating the enemy and by no means do they account for the entire arsenal of the means of struggle. We should also mention the chemical battalion, which can erect a chemical barrier along a 30-kilometer front; and the fighter squadron, which can cover the air front to a width of 12 kilometers against the enemy's reconnaissance planes.

But even this list of modern combat factors speaks to the fact that the arsenal of the means of struggle, which the modern deep operation disposes of, is extremely large in the quantitative sense and even more diverse in the qualitative sense.

Each of these enumerated combat elements has its own special quality of combat effectiveness, its own different reach, and conditions the particular nature of its employment in the engagement and operation.

It's quite obvious that modern operational art is facing new tasks of organizing the complex interaction of qualitatively different elements of struggle.

[First of all] {Most of all}, this requires comprehending their composition.

The second practical question in formulating the deep operation consists in determining the amount and composition of the elements of combat pressure on the enemy required for resolving the tasks of the deep operation.

It's necessary to keep in mind that each of the main elements of the operation listed here are individually planned to exert such pressure on the enemy in the engagement that forms an immediate fire and tactical bond.

The matter is quite different with the operation, the chief indicator of which is the unification of combat efforts not in immediate tactical and fire connection.

The essence of the operation's main indicator is in this [connection] {condition}.

We encounter the phenomenon of the operation each time we have to unite individual combat efforts, which are not immediately tactically connected, for achieving the overall goal of defeating the enemy.

From this point of view, we could define the concept of the operation as <u>a system of individual combat efforts, united in all possible dimensions for the achievement of an overall goal—the defeat of the enemy, or opposing him</u>.

From this point of view, the essence of operational art consists in unifying combat efforts that are not directly tactically connected, along the front and in the depth, on the ground and [on] {in} the air, into such a constructed system that it leads to the defeat of the enemy in a single unified effort. Consisting of a complex system of tactical efforts, this unified effort becomes operational; that is, the operation. We encounter in embryonic form very early the indicators of this unification; that is, the operation.

The division undoubtedly resolves its task by means of such a combat act, in which the entire pressure on the enemy is in immediate tactical and fire connection. The system of unifying efforts at the divisional level does not go beyond the bounds of organizing technical activities and is easily supported by a single centralized fire group of divisional artillery, the range of which may easily serve the division's entire 3–5 kilometer attack front. The division is thus the main tactical formation.

The corps, when operating along a narrow front, also does not go beyond the tactical level in the system of unifying its efforts. However, while operating along a broader front or in a turning maneuver, the corps already faces the necessity of unifying the efforts of individual divisions not directly tactically connected. In this we see the indicator of the operation. Thus we may correctly call the corps an operational-tactical formation.

However, the chief operational formation, in the full sense of unifying individual combat efforts in different dimensions, is the army. The army is the chief subject of the operation and resolves the tasks of the deep operation at its army level. The army forms and unifies individual tactical efforts into a single, general operational effort. The unification of these operational efforts, which crosses over to the strategic level, is carried out in the theater of military activities by the front[12]; the front is therefore a formation of a strategic order and resolves strategic tasks.

Thus the chief operational formation is the army and its essential indicator is that

II. The Elements of the Deep Operation and the Composition of the Shock Army

it unifies individual combat efforts along a definite operational direction. This latter factor has great significance for comprehending the concept of the army. We should in no way attach to the concept of the operational direction that content which the Leer school attached to it, understanding the operational direction as the abstract idea of the operation according to goal and direction.

By operational direction, we should understand a completely real area of terrain, which, {leads to important targets in the enemy's territory} according to its geographical conditions, allows for the unification of individual tactical acts into a single operational effort. From this point of view, the operational direction is a very specific piece of terrain, characterized by a definite unity of geographical conditions in the theater of military activities. The careful study of these conditions while still in peacetime leads to the establishment of definite operational directions, the signposts of which are important geographical areas, characterized by values of an economic, military and political order. The establishment of these operational directions, their study, the definition of their importance and operational capacity represents one of the most important tasks for the work of major staffs in peacetime.

It's quite obvious that the outline of the operational directions, which is predetermined by the overall nature of the theater of military activities, may already be established in peacetime and to a certain degree predetermines the general outline for [developing] {organizing} the offensive and thus the deployment of one's forces. This, however, in no way predetermines what kinds of forces will be deployed and grouped along the operational directions, for this depends completely on the evaluation of these directions, on the evaluation of their importance and determination of those which have major significance and become the main shock directions.

The establishment of the main operational directions and defining the men and materiel [along them] {for them} comprises the main content of strategy and the work of the higher staffs in peacetime. From this point of view, the chief essence of operational art is to employ the available forces along a given direction in accordance with the goal defined by strategy. Of course, at the same time, of great importance to strategy is the definition of the main operational direction along which the main group of forces deploys, the main component part of which is the shock army.

{A shock group of forces, as part of a front, may consist of, depending upon the specific conditions of the situation and the breadth of the front of deployment, 2–3 and even more shock armies. In any event, the task of a front operation cannot, of course, be resolved by a single shock army.}

The question of the composition of the shock groups of forces has its own, long history.

The German army's shock group along the French front in 1914 consisted of 26 corps and nine cavalry divisions, {grouped into five army formations}, thus comprising 2/3 of all the German forces deployed against France. (In all, there were 35 corps and ten cavalry divisions.)

In the subsequent course of the World War the composition of the shock groups of forces underwent enormous changes; this was the result of the new qualitative power of the front of fire, which required significantly larger forces and weapons to overcome it than was the case in 1914.

At the present stage we stand before a new resolution of this problem. The new technical means of struggle dispose of new and increased combat effectiveness, naturally changing the composition of the shock group of forces capable of resolving a task in the entire operational depth.

At the same time, we should keep in mind the fact that the increased effectiveness of combat weapons results in a broader infantry front and allows for a somewhat lesser reinforcement by artillery weapons in comparison with 1918. This tendency is perfectly logical in modern conditions. In any event, in determining the [norms] {length} of the front and its technical outfitting, one should proceed from those {norms} that have been determined for overcoming a single kilometer of the front of fire {in the entire} depth of the defense.

However, a calculation of the men and materiel must now be made not only in a single dimension along the front. Shock groups of forces must be capable not only of breaking the opposing resistance along the front, but of simultaneously striking the entire operational depth of this resistance with its long-range pressure.

From this it follows that two dimensions must lie at the heart of calculating men and materiel: along the front and in depth.

In general, the shock army, which is the chief subject of the deep operation, must be capable, first of all, of resolving its task throughout the entire depth of its assigned operational direction and, secondly, of suppressing the enemy's [operational] resistance in {his} entire operational depth with its long-range means.

Determining the men and materiel required for resolving the deep operation's tasks is a very difficult problem. The calculation listed below is only approximate and general in character; it assumes a full complement of men and materiel necessary for resolving the tasks of the deep operation, altogether. However, it is not calculated for every individual case, or each shock army. Naturally, a specific solution of this question depends in each individual case on the [tasks] {importance} of the operational direction and all other conditions in which the given shock army is operating.

In taking up the overall calculation of the shock army's composition, it is necessary to separately define the men and materiel necessary to break the front in the tactical depth and {then} the men and materiel necessary to defeat the resistance in the operational depth.

In resolving the first problem, it is necessary to proceed from the width of the front which must be attacked so that the tactical breaking supports the operational development of the breakthrough.

The modern breakthrough may be resolved only in that case if the front can be broken by a simultaneous attack along a front no less than 30 kilometers in length. This condition is determined by the range of modern artillery weapons, which can subject a smaller sector of front to flanking fire from areas of the defense not under attack. And even a breakthrough along a 30-kilometer front cannot be considered to be completely safe from flanking fire under modern conditions, while the subsequent growth of long-range artillery must inevitably increase the breadth of the front subject to attack.

Because a reinforced corps may break through a front having an overall width of ten kilometers, three reinforced corps are required for attacking a 30-kilometer front. This is the minimal force necessary for the shock army to break the front in the tactical depth.

However, it is obvious that the combat formation, both on the tactical and the operational scale, never consists wholly of a single shock group of forces; it always has a pinning group along a secondary direction, which in the shock army should be rep-

II. The Elements of the Deep Operation and the Composition of the Shock Army

resented by yet another, unreinforced, corps. Thus the normal composition of a shock army is four corps, of which three are reinforced. [Of course] In a case where the shock army is operating along an important operational direction, and if the attack must encompass more than 30 kilometers, up to but 40 {kilometers}, then it must have four reinforced corps and one unreinforced; that is, a total of five corps. Proceeding from the tank equipping of each reinforced corps with four to six tank battalions, and assuming that of four reinforced corps two will have their full complement of tanks; that is, up to six tank battalions, while two will have four tank battalions each, then the shock army would need up to 20 tank battalions. Because the artillery reinforcement of a reinforced corps will also consist of two howitzer regiments and one gun regiment from the High Command Reserve, then the artillery complement of the shock army's four reinforced corps should consist of 12 High Command Reserve artillery regiments, of which eight are howitzer and four are gun regiments. Besides this, if the shock army has to overcome a fortified area in its attack zone, reinforced with concrete, then it should have attached to it heavy artillery of no less than a regiment of 152-mm and 203-mm guns.

Thus the shock army's entire first immediate attack echelon, which is capable of breaking the front throughout the entire tactical depth, must consist, overall, of five corps (of these, four are reinforced), 20 High Command Reserve tank battalions, 12 High Command Reserve artillery regiments and one regiment of heavy artillery.

We do not dispose at present of more exact initial norms for calculating the weapons necessary for the operational development of the breakthrough and defeating resistance throughout the entire operational depth. In this regard, determining the composition of the breakthrough development echelon represents a more difficult task.

One must suppose that we will subsequently arrive at a more definite norm for calculating the density for striking the depth of the modern resistance. In any event, the breakthrough development echelon must consist of such formations that are protected against bullets and can thus pass through the front's broken tactical breach; have a high degree of cross-country performance and have long-range qualities. It's quite obvious that basically only independent mechanized formations, as represented by the mechanized corps, satisfy these conditions.

However, the development of the breakthrough into the depth is, to an equal degree, open to cavalry formations and motorized infantry.

Thus the breakthrough development echelon (ERP) should, as a rule, consist of a mechanized corps, mated in its work with a motorized division and a cavalry corps.

As has already been pointed out, it is not mandatory that the cavalry corps each time form a part of the shock army, because in some cases it will be more expedient to detail cavalry masses to separately operating operational formations of cavalry and motor-mechanized formations.

Finally, an airborne detachment must form part of the ERP.

This composition defines the strength and outfitting of the breakthrough development echelon.

The composition of army aviation, which is a factor of independent significance in the system of operational pressure on the enemy, should also be calculated on fighting in two dimensions—in the tactical zone along the front and in the operational zone in the depth.

Insofar as the complete unfolding of the outline of the deep engagement presupposes the employment of an assault air brigade, as part of a reinforced corps, for suppressing artillery and tactical reserves, each of the shock army's reinforced corps would have to have its own assault air brigade. This amounts to four assault air brigades for four reinforced corps.

Of course, this will not always be the case; some reinforced corps may not dispose of the entire arsenal of weapons necessary for the entire system of defeating the tactical depth of the resistance.

In this case, the shock army could limit itself to two assault air brigades. Besides this, the army commander must have at his disposal no less than two light-bomber brigades for attacking the enemy's operational depth.

In general, army aviation chiefly consists of light attack aircraft. Heavy aviation for destruction is chiefly strategic aviation and is centralized, as a rule, in the hands of the front. However, we must take into account the fact that in the depth of the shock army's operational reach we will often find targets (major railroad junctions, industrial and administrative centers) which will require the destructive power of heavy aviation, and in this case the shock army must be supported by a single heavy brigade.

The calculation for fighter aviation proceeds from the norms for covering the air front against the enemy's reconnaissance. The norm is 12 kilometers for one fighter squadron, against an overall frontage of 75–80 kilometers, which yields a strength of six fighter squadrons, of which three may be represented by two fighter brigades, with three designated for reinforcing the corresponding reinforced corps and defending them from the air.

Finally, the shock army should have one reconnaissance squadron and a corresponding number of corps and artillery aviation detachments.

Overall, the shock army's aviation consists of the significant amount of about 1,000 aircraft.

In defining the composition of the shock army's chemical units, it is necessary to proceed from the chemical battalion's capabilities for laying down up to 30 kilometers of chemical obstacles along the front. In order to cover a shock army's entire front {of up to 80 kilometers}, three chemical battalions are thus required.

The significantly altered quantitative and qualitative composition of the shock army naturally determines the different length of its front of operations.

The growth of this front compared to the front norms during the World War is, at the same time, a logical one and is conditioned by the increased effectiveness of the modern means of struggle.

The determination of the shock army's attack front proceeds from a reinforced corps' front during the breakthrough. A reinforced corps is capable of resolving an offensive task along a 10–12 kilometer front. This amounts to a front of 48–50 kilometers for four corps.

A non-reinforced corps, attacking along a secondary axis, may be deployed along a front up to 20 kilometers in width. This adds another 20 kilometers to the shock army's 50-kilometer front, for a total of 70 kilometers.

However, we should take into account the fact that in the army's operational zone a certain area will always be operationally dead, due to terrain conditions, and will not allow for the development of major activities.

Finally, a certain place must also be left for the maneuver capabilities of highly mobile means (the mechanized corps and cavalry corps), which, however, as a rule,

II. The Elements of the Deep Operation and the Composition of the Shock Army

must always operate at another level, and not along the overall line of the front. In calculating all of these conditions, the shock army's attack front may reach up to 85 kilometers. In general, a length of 80 kilometers will be quite normal for the modern shock army's attack front.

This army is thus an enormous organism, as regards its composition. It may consist of a total of 15 rifle divisions, approximately two cavalry divisions, three mechanized brigades, a motorized division, and an airborne detachment, having an overall weapons complement of 1,472 guns, 1,457 tanks and 1,045 aircraft. It will number 350,000 men.

It is interesting to note that the number of guns in a reinforced corps is equal to the number of tanks; only the aircraft are three times fewer.

Thus the technical outfitting of a reinforced corps may be expressed by the following formula: O (guns) =T (tanks) =3 S (aircraft).

In the shock army the technical outfitting is expressed to an even greater degree. Here the number of guns is equal, {overall}, to the number of tanks. However, there are only 1.5 fewer planes than this number.

Thus the shock army's technical outfitting is expressed by the following formula: O (guns) =T (tanks) =1.5 S (aircraft).

If the number of aircraft is less than the number of guns and tanks—which deprives this calculation of practical significance without the qualitative evaluation {of these weapons}—then it is necessary to keep in mind the fact that the weight of a single bombing by [this] {army} aviation is 237 tons, which is equivalent to a single combat load expended by a rifle division in a day of intense fighting. At the same time, aviation can deliver three such bombings in a day of intense fighting and is thus capable of delivering a fire effect equivalent to that of a single corps.

Thus the modern shock army is a complex organism, with a rich arsenal of highly effective technical means of struggle.

As regards its composition, in this sense it is fundamentally distinct from the shock groups of forces during the 1914–18 World War.

However, in comparing its indices and densities for a single kilometer of front with historical data from the {most recent} past, nevertheless something draws our attention.

We see how the composition of the shock groups of forces has been changing since 1914. There, where 37 divisions attacked along an 80-kilometer front with 1,572 guns, in 1918 75 divisions attacked with 5,700 guns; moreover with a significantly altered correlation of light and heavy calibers in favor of the latter.

In the modern shock army a significantly lesser number of infantry and artillery will attack along the same 80-kilometer front: only 15 rifle divisions and 1,500 guns. The reduction of the infantry mass is logical in modern conditions. A certain reduction in the amount of artillery, given the large number of tanks and aviation, is also logical to a certain degree. However, a density of 18.5 guns per kilometer of front in the modern shock army, compared to 80 guns per kilometer in what was admittedly the positional conditions of 1918, nevertheless forces us to assume that in a case if the shock army had encountered a heavily fortified zone and would not have had the opportunity of employing tanks in attacking the forward edge, which will often be {the case}, it would undoubtedly {have} required artillery reinforcement.

We basically are witnessing a growth in the number of tanks.

THE SHOCK ARMY'S COMBAT COMPOSITION

Composition

Rifle Divisions	Cavalry Divisions	Motor-Mechanized Formations		High Command Tank Reserve		High Command Artillery Reserve	
		Mechanized Brigades	Motorized Divisions	Independent Battalions	Organic Battalions	Howitzer Regiments	Gun Regiments
15	2	3	1	14	6	8	4

Heavy Artillery Regiments	Aviation			Fighters	
	Assault Air Brigades	Light-Bomber Brigades	Heavy Bomber Brigades	Brigades	Squadrons
1	4	2	1	1	3

Corps Air Detachment	Division Air Detachment	Army Air Detachment	Chemical Troops Battalion
5	5	1	3

Quantity

Guns			Tanks			Aircraft			
Light	Heavy	Total	Light	Medium	Total	Assault	Light Bombers	Heavy Bombers	Reconnaissance
897	575	1,472	1,157	300	1,457	372	168	50	19

Fighters	Corps and Division Air Detachments	Total	Total Personnel
306	130	1,045	352,875

O (guns) = T (tanks) = 1.5 S (aircraft)

THE SHOCK GROUPS' COMPOSITION

Shock Groups of Forces	Attack Front (in km)	Number of Divisions	Number of Guns	Number of Tanks	Number of Aircraft
German Offensive to the Marne in 1914	80	37	1,572	—	180
The German Breakthrough of March 21, 1918	80	75	5,728	—	820*
The Allies' Counteroffensive of July 18, 1918	20	18 rifle divisions 3 cavalry divisions	1,880	375	480
Modern Shock Army	80	15 rifle divisions 1 mechanized corps 1 motorized division 2 cavalry divisions	1,500	1,457	1,045

*Note. Qualitatively, the aircraft of 1918, which were primarily reconnaissance craft, had little effect on the ground.

II. The Elements of the Deep Operation and the Composition of the Shock Army

The [Evolution of the] Shock Groups' Densities per Kilometer of Front

Shock Groups of Forces	Per First-Echelon Division (in km)	Per Kilometer of Attack Front		
		Guns	Aircraft	Tanks
German Offensive to the Marne	2.1	20	2	0
The First German Offensive in Picardy, March 21, 1918	2.3	80	12	0
The French Counteroffensive (Villers-Cotterets, July 18, 1918)	2.4	65	22.5	14
Modern Shock Army	6	18.5	13.1	18.2

If the latter indices do not yield a sharp growth in density in comparison with 1918 (14 {tanks} per kilometer of front in 1918 and 18.2 in the modern shock army), the overall width of the tank attack's front is growing significantly, as is its reach into the depth and, finally, the qualitative effect of the new tank designs themselves.

The same must be said of aviation, the density of which yields somewhat smaller indices than in 1918; [because] {however}, modern aircraft cannot qualitatively be compared with the underpowered aircraft types in 1918, with their small ground effect.

In general, this is how the modern shock army looks in comparison to the shock groups of forces during 1914–18.

At the same time, [however] we {will} emphasize once again that the given composition of the shock army proceeds from a calculation of general operational requirements and in no way has significance as any kind of practical standard.

Depending on the specific conditions and the character and importance of the operational direction, the composition of the shock army will vary; it will often be more expedient to have weaker shock armies, but along several directions.

We have determined only the immediate combat composition of the shock army. At the same time, its entire composition, on the whole, should be divided into its immediate combat composition and then into combat support units and, finally, to rear support units.

Among the combat support units must be counted all of those troops and equipment which directly support the conduct of the engagement and operation. These include anti-aircraft units and communications and engineer troops.

Proceeding from the fact that a single anti-aircraft artillery battalion may cover a column to a depth of up to 12 kilometers with two layers of fire and that the depth of modern columns has significantly increased, it is necessary to strengthen a reinforced corps, aside from its authorized anti-aircraft battalion, with another two anti-aircraft battalions from the High Command Reserve. This will not always be the case and along less important axes the corps will be reinforced with only a single battalion apiece, aside from their authorized one. In general, up to six anti-aircraft artillery battalions would be required for the anti-aircraft defense of four reinforced corps.

Besides this, supply stations must be secured by anti-aircraft defense.

Overall, the shock army must be reinforced with up to ten anti-aircraft artillery battalions.

The calculation of the shock army's linear communications equipment is determined by the distance of the troops in the field from the railheads, along the line of which heavy permanent wires usually break off in combat conditions. The size of this distance should be defined as 100–120 kilometers in modern conditions. Thus line communications equipment has to be calculated to cover a distance of 100–120 kilometers by the construction of a new line and partially through stringing wire on poles. Proceeding from a calculation of six corps units (five rifle corps and one mechanized corps), the shock army must be supported by six cable-pole, five construction and four exploitation companies. At the same time, one should keep in mind that a signal communications line should be calculated for two corps formations. Besides this, in order to communicate with one's neighbors, another cable-pole company and a construction company would be required, as well as six cable-pole companies for servicing the unpaved sections of the military road. Finally, there must always be about three cable-pole companies, one exploitation company and one construction company in the operational [level]{reserve}.

Communications with airfields is not included in this calculation of equipment, as this is secured [by the latter's authorized equipment] {in a special way.}

The army headquarters itself must dispose of a communications battalion for internal use, as well as a radio battalion for organizing radio communications, and a position-finding group for conducting radio surveillance.

The army radio battalion's equipment consists of the following: a 2A type long-wave radio station, with a range of 250–500 kilometers, for communications with the front and with neighbors; a 3A type long-wave radio station, with a range of 100–200 kilometers, for communications with subordinate units, and a 12 AK type short-wave radio station, with a range of 75–150 kilometers, for communications with the command posts.

The total arsenal of the shock army's communications equipment is as follows:

a communications battalion for the army headquarters;

a radio battalion;

a position-finding group;

16 cable-pole companies;

seven construction companies, and;

five exploitation companies.

The shock army's <u>engineer equipment</u> has increased significantly in its specific weight and importance. It is necessary to keep in mind that in view of the nature of modern combat activities, supplying the troops with technical equipment and opportunities for creating obstacles requires the support of special engineer troops for each step of the offensive, particularly to such a depth as the deep operation is developing.

Proceeding from these prerequisites, each reinforced rifle corps must be reinforced with, aside from its authorized sapper battalion, yet another, which yields four independent sapper battalions for four reinforced rifle corps.

Another two independent engineer-sapper battalions are needed for the engineer outfitting of the army rear, based upon the calculation of one engineer battalion for 30 kilometers of front.

Thus, in all, the shock army must be reinforced with six independent sapper battalions, which are duplicated in their work by six construction battalions.

If in its attack zone the shock army has to force a water line wider than 120 meters, it is necessary to attach a pontoon battalion to each corps for a bridge crossing of 200 meters. Thus four pontoon battalions are required for four reinforced rifle corps.

Finally, the shock army must be supported by 4–6 camouflage companies, up to three electro-technical companies and 2–3 hydro-technical companies.

Thus the shock army's entire arsenal of engineer means is as follows:
six sapper battalions;
six construction battalions;
four pontoon battalions;
4–6 camouflage companies;
3 electro-technical companies, and;
2–3 hydro-technical companies.

The composition of the shock army's rear support units is especially significant and complex.[13] It is sufficient to point out that the shock army's complete daily requirements are calculated at 15,000 tons. It would require up to 36 trains to move this amount, of which half would consist of munitions. Of course, an army's complete daily requirement is never delivered in a single day. The delivery of up to a half of a combat load per day should be considered normal, which would guarantee the shock army 15 combat loads a month, as well as half a refill of fuel. Given the incomplete delivery of food and forage, a part of which is taken from local sources, {in these conditions} the shock army would require 24–25 trains per day.

Even greater qualitative indices arise when translating the shock army's daily requirements into auto transport units. In order to move the entire daily requirement of 15,000 tons over a 100-kilometer gap, it would be necessary to supply it with 12,000 3-ton vehicles for the round trip, or 24,000 1.5-ton vehicles, which yields 50 transportation battalions. Of course, a full load can never be delivered at this stage, nor are all the corps separated from their terminal railheads by 100–120 kilometers.

Given these amendments, the amount of the shock army's auto-transport equipment {shrinks to approximately} 20–24 auto transport battalions.

In completing here the final calculation of the modern shock army's strength, we see that despite the high level of mobility and long range of its combat means, it is an extremely complex and cumbersome formation, which presents an entire series of new problems of leadership and operational employment. At the center of these is the question of the place and role of each of the shock army's independent elements in the deep operation. This problem requires analyzing the concrete forms of conducting the deep operation at various stages of its development.

III. The Place and Role of the New Means of Struggle in the Deep Operation

In taking up the study of the specific forms of the shock army's activities at various stages in the development of the initial offensive operation, it is necessary, first of all, to take into account the diverse nature of its composition, which puts forth a series of new problems of operational leadership, the chief of which is assigning the correct place and role to each of the different combat elements in the system of the operation.

This problem is new, because in 1914 an army represented a formation of various units which, in general, were homogenous and of equal value in their combat effectiveness. The entire essence of operational art thus came down to bringing these homogenous units into a system of definite cooperation along the front. The difference in range really only revealed itself in the infantry and cavalry formations; but in the conditions of 1914, when the cavalry, for the most part, carried out a series of operationally auxiliary tasks, this difference was easily dealt with.

The problem is entirely different today. The essence of modern operational art is, first of all, to bring the individual combat efforts into a smooth system of cooperation not only along the front, but in the new dimension of depth; and, secondly, to bring into this system of smooth coordination those elements that are extremely diverse in their nature, in their combat effectiveness and their combat range. At the same time, the main indicator of the diversity of elements in the modern shock army is namely the varying depth of their combat effect. In this regard, the elements of the deep operation are basically divided into four categories: the first category is aviation, with an average reach in combat depth of 400 kilometers; the second category includes motor-mechanized formations, with a daily depth of combat reach of 100 kilometers; the third category includes cavalry, with a depth of combat reach of 50–60 kilometers and, finally; the fourth category includes combined-arms infantry formations, with a daily depth of combat reach of 12–15 kilometers when overcoming the front of fire, and a daily rate of march of 30 kilometers. The essence of the problem is to bring this differing depth of combat reach into a smooth system of a single operational effort, leading purposefully to the defeat of the opposing resistance.

The resolution of this problem requires the preliminary establishment of a definite and guiding point of view as to the place and role of each of the shock army's elements in the overall scheme of the deep operation.

<u>Aviation</u>, which we have up to now been inclined to view in the army operation as a factor for carrying out a series of auxiliary and secondary tasks for deciding the struggle on the ground, has long since grown out of this subordinate role.

We will once again remind the reader that a single bombing by army aviation is equivalent to a division's combat load and may, according to its effectiveness, have the same effect on the enemy as an attack by a corps. Each corps in the operation carries out a definite and independent task and is in no way a secondary factor for another, which is how we should view aviation now.

Army strike aviation is now becoming an independent and efficacious factor in the system of the deep operation. It would be very wrong to disperse its activities in order to carry out individual, ancillary to operations on the ground. Aviation must have a definite target, just as the corps does on the ground; it has to destroy through its activities a definite part of the enemy's combat formation. This becomes especially important before establishing direct ground contact, when the main purpose of the deep operation is to deny the enemy's combat formation the opportunity to form up frontally, to prevent his front from coalescing and to disrupt him on the march before he can come into ground contact with us. The main condition for establishing any kind of front is, first of all, its integrity. If the front is disrupted along a particular sector, it loses its stability and begins to waver.

Any front that can be outflanked loses the significance of a solid front. Taking away a particular part of the enemy's combat formation from the overall front, depriving it of the possibility of attacking and preventing its approach, is the chief task of the deep operation before the establishment of ground contact.

This task is carried out, first of all, by aviation and constitutes its independent role in the deep operation.

Following the establishment of a solid front, while the front is being broken through,

III. The Place and Role of the New Means of Struggle in the Deep Operation

army aviation acquires an even more independent significance, for it is at this stage of the full unfolding of the deep operation's forms that it must pin the enemy's deep reserves to the ground and isolate the entire breakthrough sector from his operational and strategic depth.

At the same time, one should not reduce aviation's role in carrying out these independent tasks to a single type of combat action, such as, for example, an attack against the enemy's personnel. If the aviation's target in the system of the deep operation has been defined, then it must be suppressed and deprived of its viability in all elements of its combat being. This includes: the enemy's personnel, supply routes, supply stations, and the railroad routes to them. Troop suppression, systematic attacks on his supply routes, the destruction of supply stations, attacks on railway stages leading to the supply stations—the entire system of aviation's combat reach must lead to the complete paralysis of a particular part of the enemy's combat formation and deprive it of the opportunity of supporting the front of struggle and to live and reconstitute itself.

This constitutes the independent place and role of army aviation in the system of the deep operation.

Army aviation must thus represent a massed combat force in the air and be centralized in the hands of the army command, as a very important factor of combat pressure against the enemy.

In this regard, we should unify it in the concept of the army aviation group (AGA).

<u>Independent motor-mechanized formations</u> dispose of two main combat qualities—striking power and long range.

In viewing each of these qualities separately, we inevitably arrive at {their} opposition to each other.

For in stressing the mechanized formation's striking power, we seek to preserve it for that moment when the attack matures along the army's entire front. However, this inevitably leads to a situation where we hold the mechanized formation in the rear and deprive it of its second quality—its long range.

The isolated examination of the mechanized formation's two chief combat qualities separately leads to their incorrect and mechanical opposition, in which the full combat value of this mighty factor of the deep operation is paralyzed. It is necessary to take from each means of struggle that which it can give, according to its technical possibilities.

The mechanized formation's full combat effectiveness will unfold then, when its two qualities—striking power and long range are organically united in action and employed in one and the same operational act. These qualities do not contradict each other; they may not only be reconciled, but they constitute a single organic unity in the mechanized formation's effectiveness.

Its value lies not only in its great striking power, but also chiefly in the fact that this striking power can be immediately transferred a great distance forward. And this means that the mechanized formation's place, given the presence of even the slightest space between the sides, must always be ahead of the front.

It is necessary to take advantage of any opportunity to transfer its operational effort forward and to suppress a particular part of the combat order of the enemy's front before he can come into complete ground contact with us and a continuous front takes shape. If aviation carries out this task in the air, then on the ground the mechanized formation should be the first to carry this out; for the chief condition for resolving this task is long range.

The mechanized formation's place ahead of the front sometimes encounters objections. The argument is put forth that the mechanized formation, in operating ahead of the front, may lose its combat effectiveness by the moment the attack is organized by the army's entire front. However, this does not take into account the new nature of the deep strike, in which at the moment the army's main forces are deploying for the attack a certain part of the enemy's combat formation must already be suppressed, depriving his front of the indicators of a completely solid front and stability.

This is, after all, the chief meaning of the deep operation before the establishment of the front.

Cooperation between the mechanized formation ahead of the front and the arriving main forces thus consists not of a joint attack in one line of the front, but in the coordination of the attack into the depth along different levels of the front of struggle.

This is cooperation not along the front, but into the depth.

Finally, the mechanized formation's combat capability is undoubtedly an increasing given. But even now it must be viewed, in any event, as not lower than a cavalry formation's combat capability. And by the way, no one will doubt that the place of cavalry, given a spatial distance, is ahead of the front.

This is the place to remind the reader that a great struggle was required during the age of the German Wars of Unification[14] for the cavalry to be pushed ahead of the [regiment's] {columns'} tail ahead of the front.

According to Schlieffen's apt phrase, for a long time they carried the cavalry "like a luxury item, with the wagons and transports."

Our progressive revolutionary theory must concern itself that this not be repeated with the mechanized formation. Any fears that the mechanized formation, while operating ahead of the front, may be subject to defeat in detail are not convincing. In operating along a specific axis ahead of the front, the mechanized formation will encounter no more forces than if it were to operate in a single line with the entire attacking front. And a turning movement from the flanks is least of all dangerous for the mechanized corps, due to its quick response time. Quite the opposite, possessing great freedom of maneuver ahead of the front, it will always be able to win the flank.

At the same time, while operating ahead of the front, the mechanized corps will encounter, for the most part, an enemy still on the march; that is, in march formation, when a tank attack promises the greatest effect. To defeat the enemy at this stage of the operation is incomparably easier than after complete ground contact has been established between the main forces, when freedom of maneuver is lost and the enemy's combat formation [will be] {is} deployed, presenting a continuous and closed fire formation.

Quite the opposite, as this moment in the operation's development puts a limit to the useful employment of the mechanized formation along the front of struggle, enabling us to replace it with a combined-arms infantry formation with much greater effectiveness.

Upon the establishment of a solid front and the loss of freedom of maneuver and space for long-range activities, the mechanized formation's place will be behind the front, so as to develop the slightest tactical success achieved by the combined-arms formations {along the attack front} from the depth into the enemy's depth.

And this is what defines the mechanized formation's place in the breakthrough in its capacity as the main core of the ERP.

III. The Place and Role of the New Means of Struggle in the Deep Operation 101

Thus the mechanized formation's place in the system of the deep operation is ahead or behind the front; but in both cases with the task of transferring its combat efforts ahead of the combined-arms formations' front of struggle.

The least appropriate place would be the mechanized formation along the same line of front as the latter.

If the mechanized formation should find itself in this situation, it means that the time has come to pull it out of the line and replace it with the forces of the newly-arrived combined-arms formations.

Thus the mechanized formation's place in the system of the deep operation determines its multi-tiered nature, when the battle is waged simultaneously at several levels of the operational depth.

Finally, the mechanized formation's place in the operation determines its active and independent role, which obliges us to assign it an independent and full-blooded combat task. The mechanized formation must be assigned a specific target—a part of the enemy's combat formation or an important locale (area), each time with the decisive offensive goal of defeating the former, and the capture and, if necessary, the destruction of the latter.

Only such a radical mission assignment, carried out in conjunction with aviation, corresponds to the mechanized formation's combat possibilities; it would be least correct to limit this assignment to some sort of local goals ancillary to the operation, such as seizing a line or crossings, or covering the march of the main forces. Such goals doom the mechanized formation to carrying out the passive task of holding ground, for which it is least of all suited.

However, in all cases of transferring the mechanized formation's efforts ahead of the front, a question of extreme importance is the organization of cooperation into the depth. As has already been shown, the mechanized formation is capable of uninterrupted combat activities for no more than three days. If we remain on the realistic ground of [the mechanized formation's] technical possibilities, it is always necessary to keep in mind this circumstance, which limits the operational scope of the mechanized formation's activities.

The latter's cooperation with the approaching front of the army's main forces in time and its replacement along a certain line of space by cavalry and motorized infantry comprises one of the essential tasks in organizing the deep operation. This determines the important role and place of the army cavalry and motorized infantry in the latter.

<u>The army cavalry's place</u> in the system of the deep operation is determined by [such] {those} prerequisites that have been established for the mechanized formation. Therefore its place ahead of the front, until the establishment of the complete contact of the main forces, and behind the front for transferring efforts from the depth into the depth upon the establishment of complete contact.

Proceeding from these prerequisites, it would be least correct to employ army cavalry as was the case in 1914, when it played, on the whole, an ancillary and auxiliary role in the system of the operation. This led to its dispersal along a significant front, paralyzed the strength of its attack and imparted to its actions the inevitable character of an operational screen and reconnaissance.

Given the presence of motorized and mechanized detachments in the combined-arms formations, we now dispose of other opportunities for organizing reconnaissance ahead of the army's entire front.

Army cavalry, disposing of great striking power and long range, should, together with the mechanized formation, be massed along a particular axis and carry out an independent and active task, revealing its attack.

At the same time, the cavalry's special value in the deep operation is that it is not limited by any rigid operating deadlines in time and is capable of holding ground, which determines its great significance in combined activities with the mechanized formation, and which guarantees its {relief} along a particular line.

The problem of the motorized division's place in the deep operation is resolved in exactly the same way as with the mechanized formation and the cavalry.

It is necessary, however, to take into account that the motorized division, as it transpires, is of relatively small tactical value. In these conditions, it would be quite incorrect to leave it on the front line for an extended period of time to carry out any kind of extended combat assignments. Once the motorized division has dismounted, it takes on the nature of ordinary infantry. If it is drawn into combat for an extended period of time in this way, it loses its value as a means of operational maneuver and can undoubtedly be replaced with much greater benefit by combined-arms formations.

Therefore, as a rule, the motorized division should be employed for carrying out definite, short-term tasks, which require extending our combat efforts over great distances, chiefly in conjunction with the mechanized formation, in order to hold ground seized by it or to cover its concentration.

At the first opportunity, the motorized division must be gathered once again, in order to preserve its chief designation as a valuable means of rapid operational maneuver, capable of transferring combat efforts to that area in the overall depth of the struggle where the situation will require it.

It was absolutely necessary to first establish a principled approach to defining the place and role of the deep operation's new elements, which are distinguished by the main indicator of the depth of their combat reach. Only on this basis does it become possible to examine their employment at specific stages in the development of the initial offensive operation.

IV. Vanguard Activities in the Beginning of the War

The concentration of the modern shock army in the theater of military activities represents the first stage of the beginning of the armed struggle. It is characterized by a large and complex content and is qualitatively profoundly different from those conditions in which concentration unfolded in 1914.

Then the period of concentration and deployment was linked exclusively with an entire series of organizational measures which embraced the realization of the mobilization system as a whole. These included: domestic mobilization, shipment by rail to the theater of military activities, the gathering of units in their concentration areas and, finally, deployment; that is, the adoption of a definite operational formation for carrying out the combat assignment. The beginning of operational activities itself was viewed separately as the next new stage, for in the conditions of 1914 one could still draw quite a noticeable boundary in time and space between the stages of concentration and the beginning of the operation itself. As a rule, the armies first concentrated and only then opened military activities.

A completely different picture is unfolding before us in modern conditions.

First of all, the political prerequisites for entering into a modern war predetermine the nature of entering into the modern operation.

It is hardly possible now to imagine the outbreak of such an armed conflict in which the sides will in the beginning have the opportunity of concentrating in an unhindered fashion all of their forces along the border, without getting drawn into combat collisions—and only then entering into the fighting.

Events in the east indicate the opposite.[15]

However, the entry into the operation is determined in modern conditions by a series of materiel prerequisites of the new long-range weapons. Aviation is capable of beginning its operations to a great depth immediately upon the outbreak of war. The troops located along the border, particularly the motor-mechanized formations, may also immediately open military operations and transfer their efforts far forward, thanks to their long range. The boundary between strategic concentration and the beginning of the operation thus disappears and military activities then begin, when the troops mobilizing in the depth of the country have not even begun to entrain for their departure to the theater of military activities. In these conditions, it is quite obvious that the entire stage of modern concentration is unfolding under the very powerful pressure of vanguard actions in the beginning period of the war. These activities in the air and on the ground have as their chief goal foiling the {enemy's} concentration [itself] and rendering it impossible, or at least disrupting its normal course, to push it back in time and to force the other side to pull back his deployment into the depths of his country.

In these conditions, the entire stage of modern concentration unfolds against the background of the struggle for concentration and the struggle for the right to deploy, and is filled with intensive actions, which to a significant extent predetermine the subsequent development of operational events. The length of this stage is determined by the time necessary for modern concentration. If one considers that the mechanized units, cavalry and, at a minimum, no less than one rifle corps, which comprise the shock army, are border troops quartered near the frontier—then concentration by railroad of the entire remainder of the shock army will require, in round numbers, about 1,000 trains. Assuming that the shock army will have, on the average, about 60 trains a day[16] for its concentration, we should consider that the concentration period (900–1,000 trains divided by 60) will last, on the average, {approximately} 15 days. In this period, full of enormous tension and content, the question of whether the army will deploy as was planned and whether it will have the opportunity to smoothly go over to the offensive by its assigned deadline, is decided. It is quite obvious that the very concentration of the shock army along a definite operational direction should not be viewed in isolation, outside of the {front's} entire operational deployment, within the framework in which it is carried out.

The very vanguard actions in the beginning period of the war, which are carried out before the deployment of the front's armies, thus grow out of their army confines and resolve tasks having significance for the entire front.

They are thus of a front nature and are realized at the front level.

The chief factor of these independent actions during the beginning period of war are those modern technical means of struggle, which dispose of the greatest [effect] {long range} and depth for acting upon the enemy; these, of course, are aviation, motor-mechanized formations and cavalry.

Strategic destruction aviation finds its main employment during this period. Centralized in the hands of the front, it forms the front's aviation group—in essence, an air army.

The chief and decisive achievement of the Red Army's reconstruction is the fact that we, at the present stage, already possess such strategic destruction aviation, which is an extremely mighty factor in the deep operation.

The composition of the front's aviation may vary—from one to two heavy corps, supported by a number of light air formations: assault air, light-bomber and fighter—with an overall number of aircraft from 1,000 to 1,500.

Without a doubt, during this period destruction aviation will have the greatest opportunities for effectively operating against a number of targets having decisive importance for the enemy's mobilization and concentration.

In this regard, Douhet[17] is right when he writes: "When can the air forces find their best employment if not at the beginning of military activities."

The tasks of strategic destruction aviation at the beginning of the war are; to first of all, wreck the {enemy's} concentration. This requires systematic action against [the enemy's] {his} most important railroad junctions, railroad stages, major mobilization centers, and mobilization storehouses.

The second task of strategic aviation is to suppress the enemy's important and economically vital areas, which have significance for the expansion of his military industry. This includes major industrial areas and factories; extractive industry areas, particularly oil, and other important industrial targets having significance for the country's economy.

Finally, strategic aviation's third task during this period is to terrorize the enemy's deep rear, which requires episodic activities against the country's important political and administrative centers through destruction, infection and incendiary bombing.

It is obvious that during this period the air war will take on its most developed and cruel forms. It must, to a significant degree, wreck the enemy's concentration and paralyze his country's rear.

The depth of strategic aviation's reach is measured by the flight radius of our TB-3[18] aircraft; that is, 600–800 kilometers, covering the distance to the middle Vistula.[19]

Thus strategic aviation's strike zone embraces the enemy's territory to a depth of up to 600 kilometers from his border, considering that the airfield network is located 100–200 kilometers from our frontier.

However, one cannot consider the task of wrecking the concentration to have been achieved through operations to such a depth, because in the forward theater of military activities, at a depth of approximately 200 kilometers from the border, there is already quartered a significant number of border troops, which have quite short deadlines for mobilization readiness and which concentrate along the border, for the most part, by march.

This will be completely ordinary for motor-mechanized formations. We should take into account the fact that the composition of these border troops may reach quite an impressive force, measured, for example, by up to ten divisions along a 100-kilometer front. It is obvious that these forces, which have enormous significance precisely during the beginning period of the war, should definitely be attacked. Thus suppression must be organized in the forward zone to a depth of up to 200 kilometers from the border, which is also attacked by part of heavy aviation, although, for the most part, this will be carried out by light aviation, with its range of up to 400 kilometers, as well as by ground long-range weapons, first of all, naturally, motor-mechanized formations and cavalry.

These actions, which are also carried out at the front level, take shape in modern conditions of entering a war as <u>independent vanguard actions during the beginning period of a war</u>.

The question of vanguard actions at the beginning of a war is a qualitatively new one in modern conditions. In this regard, recent past wars have offered only insignificant episodes of an invasion, which, in the final analysis, did not have any sort of serious significance for the sides' concentration and the development of operations. The only exception is the Germans' actions in 1914 against Liege, when the group of German forces sent there was, in essence, the first prototype of the operational vanguard echelon which resolved the very important task of supporting the German First Army's deployment on Belgian territory. Because, as is known, without capturing the crossings over the Meuse at Liege, Kluck's[20] First Army would not have had the opportunity to develop its maneuver to outflank the French deployment from the north.

However, in modern conditions vanguard actions at the beginning of the war are undoubtedly acquiring a broader character and have a more decisive significance.

The task of these actions is, first of all, to invade the enemy's territory, to attack and destroy his border covering units and forward troops concentrating in the forward theater and, secondly, to penetrate to the forward disembarking stations for the troops concentrating from the depth, to destroy them and then, if it is important from the point of view of securing favorable conditions for going over to the offensive, to capture a particular jumping-off point and organize its retention. At the same time, if there is an engineer-outfitted target (a fortified area) along the axis of the vanguard actions, then it will have the task of attacking it in the first days and capturing it before it can become overgrown with an entire system of engineer structures, according to its mobilization plan, and become a full-blown fortified area.

The accomplishment of such an assignment naturally requires artillery reinforcement.

Vanguard actions at the beginning of a war are carried out by a group or groups of forces, consisting of mechanized formations, cavalry, motorized infantry, and aviation. Their coordination with an airborne landing, which is made in connection with ground operations for carrying out definite diversionary acts, will be quite [active] {expedient}.

[Such a group is formed into an independent operational forward detachment, which operates at the level of the front and is its <u>independent vanguard echelon (AVE)</u>.]

Each of these groups takes shape in an independent operational forward detachment, operating at the front level along a definite axis and constituting its <u>independent vanguard echelon (AVE)</u>.

The number and composition of the independent AVE groups is determined at the front level, depending on the disposition and importance of the targets which must be attacked in the forward theater of military activities. There may be from three to four groups of varying composition along a 300–400 kilometer front. {At the same time, however, it is required that the efforts of the vanguard actions not be made along the entire front without the opportunity of achieving decisive results along a crucial direction.}

The depth of the AVE's invasion is measured by the mechanized formation's striking range; that is, an average distance of 200 kilometers, based on a calculation of retaining a 100-kilometer reserve in the mechanized unit itself.

The plan for the AVE's operations should be the subject of the most serious work in peacetime, requiring an extremely detailed study of the forward theater of military activities and the conditions of the enemy's concentration.

The outline of the AVE's {figure 2} actions allows for two variants. The first variant is when there are orders to only attack a particular target and destroy it, but the area itself does not have any value for retention from the point of view of going over to the offensive. In this case, the mechanized formation's actions resemble a raid, along the path of which the enemy's forward units and airfields are destroyed. The mechanized corps captures the assigned target (an important junction), destroys it and then immediately returns home.

The second and more complex variant is when the task is to capture a definite area and hold it to secure favorable conditions for going over to the offensive, or to even push forward one's deployment to the enemy's territory. In these conditions, following the destruction of the enemy's forward units and the capture of the assigned area (for the most part, also an important rail junction) the remainder of the cavalry or motorized division must be dispatched immediately to the mechanized corps to consolidate the

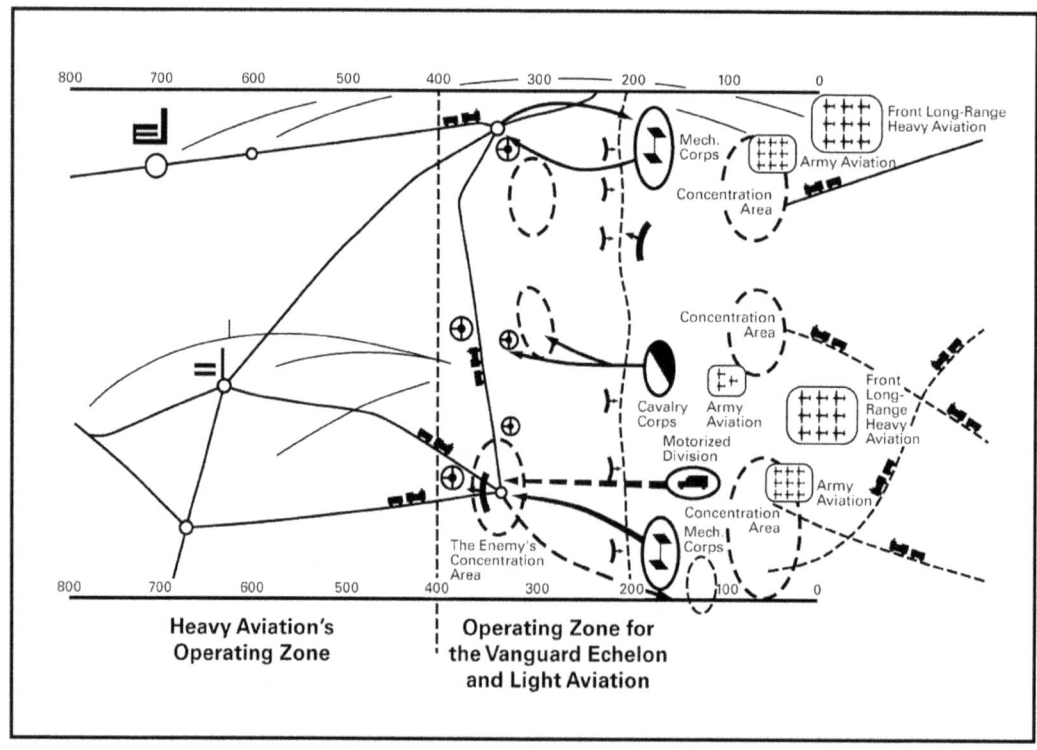

Figure 2. The Vanguard's Activities During the Beginning of a War.

captured bridgehead and organize its retention. In both cases, the mission is accomplished in conjunction with light aviation, as the mechanized corps' air group.

The cavalry corps, having its own air group, is also capable of carrying out the AVE's tasks along an independent axis; that is, to independently attack a definite group of the enemy's forward troops and to attack his airfields and reach the disembarkation stations at an overall depth of 100–120 kilometers.

It is necessary to consider that we should not count on holding the space conquered by the AVE alone for an extended period of time.

The AVE's operations may begin as early as the third or fourth day of mobilization, while the assumption of the offensive is possible, for the most part, only on the fifteenth or sixteenth day.

In these conditions, the operational plan must foresee the movement of a corresponding group of troops (a corps) from the main forces, and in certain cases the advance of the entire deployment line forward. In any case, it is necessary to take advantage of the AVE's combat warm up, so that its units retain their combat capability by the moment of the main operation's start. It would be, of course, very wrong to carry the employment of the AVE's units to such a state that they would be incapable of combat by the start of the main fighting. This question is the subject of very serious planning work for the entire operational plan.

During the time before the main forces go over to the offensive, the AVE's actions must be planned in such a way that the mechanized corps and the cavalry corps, having carried out specific tasks for destroying the enemy's forward units and having captured an important objective, be afterwards immediately covered by motorized infantry or units moved up from the main forces, which must be entrusted with the task of holding the conquered territory, if this is required by the operational plan.

Thus, the beginning period of the war is packed with activities of extremely intense activity and operational significance.

In deploying along a significant depth of up to 800 kilometers in the air and up to 200 kilometers on the ground, the beginning period of the war determines the possibility of normally concentrating and the contours of the subsequent initial operation.

V. The Operational Organization of the Approach

During the first days following the outbreak of war, as the independent activities of the AVE unfold along the forward theater of military activities and strategic aviation extends its destructive operations into the depth of the enemy's territory—the entire main mass of the concentrating troops will still be in the grip of the complex mechanism of modern concentration, and it is only with its arrival at the theater of military activities and the unloading of the railroad cars that it will emerge as an independent organism.

One should distinguish two stages in the process of adopting a specific operational formation for the beginning of combat operations: the first, or <u>concentration stage</u>, when the troops gather in specific areas following their unloading, and; a second, <u>deployment stage</u>, when the troops, having moved forward, adopt a specific operational formation for carrying out their assigned task.

In this regard, we should distinguish the <u>concentration line</u>, which usually runs through the forward detraining stations, and the <u>deployment line</u>, which is selected ahead of time in accordance with the approved operational plan and the AVE's independent activities.

In space, these two lines will be, for the most part, designated, although practically speaking, they can be neutralized because, upon detraining the troops [may] {will, as a rule}, immediately move to the deployment line.

From the moment of deployment, which is often carried out simultaneously with the movement forward, the shock army's enormous mass must move forward smoothly, carrying in its formation the deep attack against the opposing enemy.

From this moment the AVE's independent actions on the front scale will begin to directly grow into the initial operation of the shock army or a group of armies.

It is quite simple to sketch out the shock army's actions in the deep operation after the front has been established and is being broken through; but it is extremely difficult to organize the movement of this enormous mass up to the point when the front is established and to organize it in such a way that it can directly switch from a movement to waging the battle and defeating the enemy throughout his entire depth.

The shift from march formations to combat ones presents increased difficulties in modern conditions. The reason for this is to be found in the greatly increased depth of the march columns.

It's interesting to note that the very scanty material on operational art, which we find in the foreign literature, quite consciously bypasses this problem. For example, the French literature only speaks of the organization of operational activities upon the establishment of ground contact; however, it completely ignores the question of how to bring the enormous mass of the modern army to the battlefield and commit it into the battle.

As has been shown, a distance of 100 kilometers between the two sides in the initial position must be considered quite normal in the conditions of our theater of military activities.

Of course, this space does not represent an operational empty space; on the contrary, it will be suffused with the AVE's actions.

However, the march of the shock army's main mass without direct contact with the enemy is not only not eliminated, but quite the contrary, it is conditioned by this, because the AVE's mission consists in preventing the enemy's approach and throwing him back, thus maintaining the distance between the sides in the initial position.

At the same time, it is extremely important to determine the prospects of this approach from the point of view of determining whether it will encounter the enemy's oncoming movement, or an immobile front of an already-prepared defense.

The resolution of this question naturally has great significance for the organization of the approach itself.

It is quite obvious that this is determined in each separate instance, depending on the specific conditions and the operational situation. It is possible to have cases when the initial operation will begin with the arrival at the enemy's already established defensive front, leading directly to its breakthrough. However, in a whole series of cases, when both sides seek to resolve their mission along a given axis through an offensive, their oncoming movement toward each other is inevitable; in this case, the approach inevitably leads to a meeting collision. In this regard, meeting battles have by no means died out and will be quite a common phenomenon along those main axes vital for both sides. And, as the experience of past wars shows, they were the most common forms of the sides' actions in the beginning of a war.

The meeting battle in modern conditions is of special interest to us, because the organization of the movement which directly grows into the waging of the battle with an approaching enemy is a much more serious problem and, in any event, is attended by significantly greater difficulties than a march to the defensive zone, when the enemy is already opposing his immobile situation to our movement.

The problems of the operational organization of the approach and the technique of organizing the movement at the operational level must be examined from this point of view.

One should not, first of all, imagine the movement of a modern army as the movement of its entire mass of men and materiel in a single line of front on the old basis of organizing the linear operation.

[Units of the front AVE, which were operating at the front level during the concentration period, and which have been relieved by motorized infantry and by the advanced combined-arms formations before the completion of the deployment and which have thus again been brought to a condition of complete combat readiness, are now included in the stream of the army operation, joining that shock army along the axis of which they were operating.]

A specific group of front AVEs, which was operating on the front level during the concentration period and which was relieved even before the completion of their deployment by motorized infantry and advanced combined-arms formations and thus once again brought to a state of complete combat readiness—is now included in the course of the army operation, joining that shock army along the axis along which it was operating.

They must once again be employed ahead of the army' front, according to the established place of highly-mobile and long-range weapons in the system of the deep operation, with the important mission of defeating part of the enemy's combat formation on the approach, while there is still distance between the sides and it is possible to [defeat] {suppress} part of the enemy's combat formation before complete ground contact is established with him.

These weapons will now constitute the <u>army AVE</u>, which operates in organic connection with the entire army on the basis of the organized cooperation of its elements in depth.

The army AVE will thus constitute the first line of the shock army's front of struggle during the approach.

The entire main mass of the shock army's men and materiel will constitute, in relation to the AVE, its <u>main echelon</u>, forming the second line of the operational formation on the approach.

Finally, a certain part of the shock army, which is arriving to the latter's theater of military activities last, must in no way delay going over to the offensive and will thus inevitably end up in the main echelon's second echelon, forming the shock army's <u>reserve echelon</u> and the third line of its operational formation during the approach.

For the most part, the reserve echelon will include a rifle corps, arriving last, and certain heavy reinforcement weapons from the High Command Reserve. These units may not yet have a definite operational predestination during the approach and thus constitute the army's reserve.

Thus the modern shock army's operational formation during the approach basically consists of three echelons: <u>vanguard, main and reserve</u>, forming the march's overall operational depth at three levels of the approach.

It is obvious that setting into motion such a deep system of operational formation represents an extremely complex problem.

At the center of this problem is the new question of organizing cooperation into the depth of the echelons of the shock army's operational formation during the approach.

However, the organization of movement in a single line of front presents no small number of difficulties in modern conditions.

To be sure, the AVE and the reserve echelon do not yet feel this, as these elements of the approach's operational formation dispose of sufficient freedom of maneuver so as to organize their movement along a conveniently broad front.

However, the organization of the movement of the main echelon, which contains the shock army's main mass of men and materiel, is attended by other conditions.

If one takes a distance of 100 kilometers as the normal distance between the sides in the line of departure and assumes that they are moving toward each other, then this will lead to the outbreak of the meeting battle by the main echelon on the second day's march. (Both sides will cover 30 kilometers on the first day's march and thus 100-(30+30) means that they will enter into direct contact on the second day's march). On the operational scale, this means that in beginning to move from the line of departure, the main echelon must already adopt that formation in which it will enter the battle. For at a distance of one to 1 1/2 day's march from the enemy, regrouping at the operational level is excluded, because the depth of a modern reinforced rifle division's column along a single road (up to 50 kilometers, including supply trains) by itself exceeds the size of a single day's march, which means that when this column begins fighting the enemy its tail will still be along the line of departure.

This circumstance forces us to move the shock army's main echelon immediately along that front along which the units of its combat formation are capable, according to their penetrating capability and technical supply norms, of resolving their task along with the simultaneous defeat of the entire tactical depth of the resistance; we should remember that this is the basic and initial condition for overcoming the solid front to the entire operational depth and the development of the deep operation altogether.

It's quite obvious that in these conditions the shock army's main mass will move as a dense and closed phalanx and from this it follows that such a famous formula, which lay at the heart of Moltke's operational art—"march separately and fight together," is now quite out of date. Unfortunately, we now have to march together and fight together. To fight together, of course, is not difficult and necessary, but to march together is extremely difficult and this requires the careful elaboration of the technique of the movement of modern shock groups.

In order to support the breakthrough capability throughout the entire tactical depth, it is necessary to move the main echelon's reinforced corps immediately in a 10–15 kilometer zone (12 kilometers, on the average). During the first day's march this zone may be limited by converging boundary lines, in order to secure the greatest possible movement front by shortening the depth. However, because each reinforced rifle corps must occupy a front corresponding to its breakthrough capability; that is, 10–12 kilometers, then as early as the end of the first day's march these opportunities may be little expressed in practice and, in any event, insignificantly ease the technique of organizing the main echelon's movement.

Thus, in the final analysis, we have to move each of the main echelon's reinforced corps, which constitute its shock group, in a 10–12 kilometer zone. This undoubtedly difficult circumstance will be, for the most part, inevitable.

At the same time, in the conditions of our western theater of military activities, we will find no more than two through roads suitable for the divisions' movement along them in a 10–12 kilometer zone, based on the calculation of one [division] {road} per five kilometers of front.

This circumstance determines the reinforced rifle corps' movement with two divisions in the first echelon and one division in the second.

One should take into account that a reinforced rifle corps along two roads occupies a depth up to 75 kilometers and thus can fully enter the battle only on the third day.

However, these conditions may be significantly eased by the corresponding organization of the march.

At the same time, as a means of anti-aircraft and anti-tank defense, the [dispersal] {division} of the column into separate echelons into the depth basically only increases the overall depth and is [only] {quite} [a measure] relative {from the point of view} of anti-aircraft and anti-tank defense. This division is, of course, tactically necessary, although it should not cross over into the operational level, which is measured by the possibility of committing the reinforced rifle corps into the fighting in the first half of the day of its start.

Thus easing the march conditions must mainly be achieved by breaking up the column not only in depth, but along the front, by broadly employing the opportunities for moving along field roads and regular roads and, finally, over open ground.

Given a higher supply of fire weapons per kilometer of front, the march, from the point of view of anti-aircraft and anti-tank defense, also yields a significantly greater density of anti-aircraft and anti-tank fire, thereby placing its organization in easier conditions. (The simultaneous covering of the march front with a greater number of fire weapons).

In a 10–12 kilometer zone it will always be possible to find an entire series of field roads, road shoulders and open country axes, which while not enabling all three of the corps' divisions to advance in a single line, nevertheless allow the movement of two divisions, not along a single road apiece, but in a broad staggered formation along the front. At the same time, the wheeled vehicles and heavy equipment remain on the roads, while the infantry and tanks take advantage of all terrain possibilities.

In these conditions, the reinforced division, which occupies, together with its supply trains, a depth of 50 kilometers along a single road, would occupy a total of 25–30 kilometers in depth. At the same time, it's necessary to keep in mind that the divisional rear organs must halt and deploy by the end of the first day's march in order to support the fighting, and this will shorten the division's march depth to 20 kilometers at the start of the second day's march.

The division's deployment, will thus take 4–5 hours, completely supporting its entry into the fighting in the first half of the day.

The distance between the first-echelon divisions and the second-echelon division in these conditions also shrinks, securing the entry of the latter into the fighting at midday on the second day's march.

112 Two—Isserson's "The Fundamentals of the Deep Operation" (1933)

The reinforced rifle corps' movement in two echelons does not make for particular difficulties in these conditions and, in any event, is more acceptable than the movement by all three divisions in a single echelon in extremely crowded conditions along the front and, in this case, the movement of each division along one road is already inevitable.

The presence of a second-echelon division, given the possibility of its entering the fighting on the second day, is a factor favoring the development of a deep attack and enables us to augment its forces with the development of the meeting battle.

It is extremely important to calculate the schedule for the entry of the two-echelon corps into the meeting battle (figure 3). If we suppose that a first-echelon division normally moves during the day and rests at night, then upon encountering the enemy during the second day's march, its forward elements will enter the meeting engagement in the morning and fully deploy by midday of the first day's fighting. In this case, the second-echelon division, having departed in the evening {and} while maintaining its distance behind the first-echelon division's tail, will complete its first day's march at

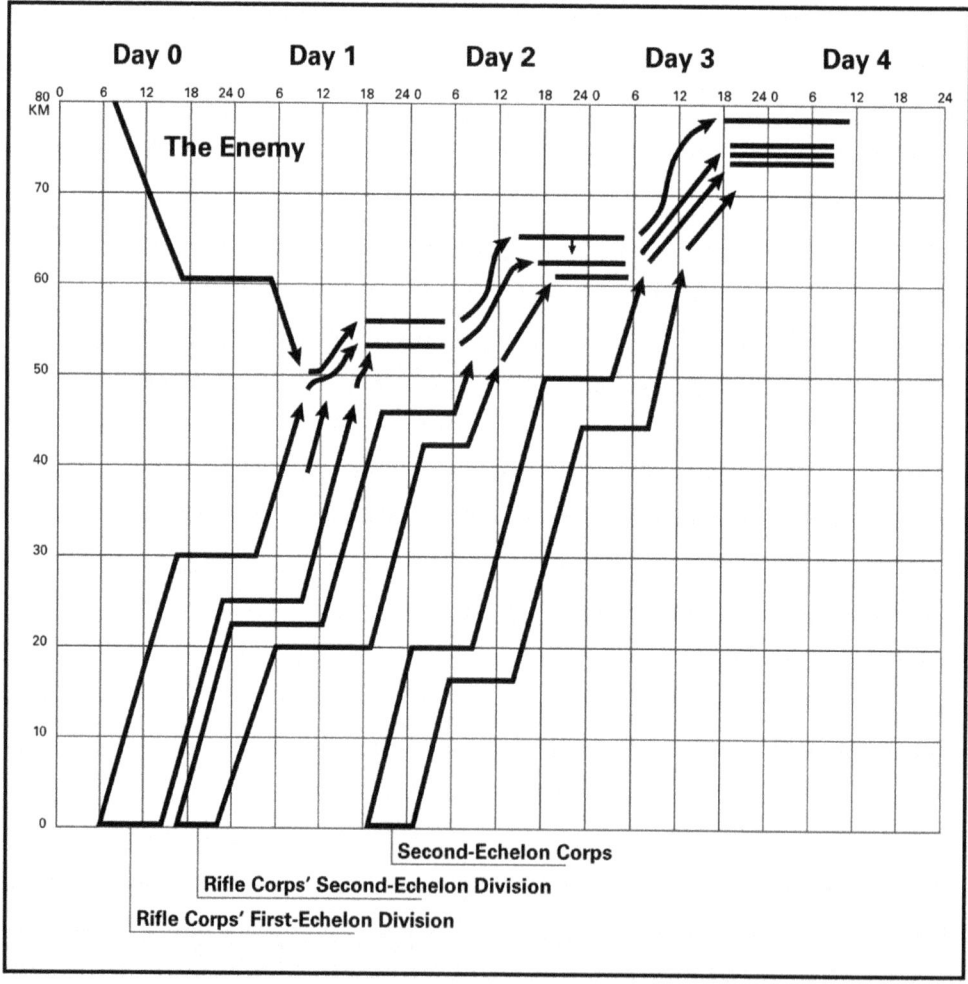

Figure 3. The Timetable for Entering the Meeting Battle.

night and will rest at dawn. In further continuing its second day's march from the middle of that day when the first-echelon divisions have already entered into the meeting engagement and, upon arriving at night in the area of the battlefield, it may enter the fighting on the morning of the following, second, day. At the same time, it should be borne in mind that the deployment of the corps' divisions naturally occurs not in place, but while moving forward. The first-echelon divisions may, as early as the first day, having thrown the enemy back, shift their efforts forward, approximately ten kilometers. In these conditions, the second-echelon division will already catch up to the first-echelon divisions during the development of the engagement and, having entered into the fighting, will reinforce the attack into the depth.

As has been shown, this method of [developing the modern meeting battle] {augmenting efforts} not only does not contradict, but actually corresponds to the requirements of conducting the modern deep engagement.

The matter is more difficult with the commitment into the battle of the army's reserve echelon, which is usually represented by a single corps.

It should be borne in mind, first of all, that the main echelon, having the AVE in front of it and well supplied with technical means, may, for the most part, itself decide the outcome of the meeting battle, without needing the reserve echelon. The latter is [more] {mainly} destined for saturating the attack front during the breakthrough.

However, if the meeting battle were to unfold in intensive and complex conditions, this would require the commitment of the reserve echelon into the fighting, which is already significantly more difficult in time.

One must keep in mind that there should remain a free space of no less than one day's march between the head of the second-echelon corps and the tail of the main echelon's corps, in order to secure the work of auto transport delivery. If this condition is not observed, then auto transport will not be able to make its way over the roads in order to deliver supplies to the main echelon's troops.

The reserve echelon's corps may therefore move out no earlier than the time the second-echelon divisions of those corps, behind which it will be moving, leave after their first night's encampment. In these conditions, the reserve echelon's corps may reach the area of the battlefield at midday on the third day.

However, bearing in mind that the main echelon may throw back the enemy another 10–15 kilometers on the third day of the battle, the reserve echelon's corps will have to catch up to the advanced front of struggle. Evidently, it will only reach the front line on the night of the [fourth to the fifth] {third to the fourth} day of the battle and thus may only exert a combat influence on the enemy from the morning of the fourth day of the battle.

It should be noted that if by this time the meeting battle has not yet been decided, then for the most part it has resulted in the establishment of the enemy's front, leading to the breakthrough.

Independent of these conditions, the reserve echelon maintains an even greater significance in the approach's operational formation, for it lends the shock army a definite stability by securing its operational depth in the case of a breakthrough by the enemy's motor-mechanized formations and cavalry into our rear. From this point of view, the reserve echelon is a complete logical and operational factor in the conditions of the deep operation.

If the shock army's operational depth remains empty during the approach, then it proves to be defenseless upon the enemy's appearance in its rear; one must always consider this possibility in modern conditions.

Thus the operational staggering of the shock army's march in depth into the vanguard, main and reserve echelons is quite logical in the conditions of the deep operation.

This conditions the significantly increased depth of the approach's operational formation, which can potentially be measured at 120–150 kilometers.[21] The start of the modern meeting battle thus presents a significantly different picture than in 1914, when, in general, the Battle of the Frontiers on the Franco-German front unfolded as a one-time, single-act event, which flared up in a single day and afterwards burned along a single definite line.

In modern conditions the meeting battle unfolds in a lengthy process of layering operational efforts, introducing significant qualitative changes to the nature of the meeting battle itself and to the dynamics of its development.

VI. The Meeting Battle

The modern meeting battle, unfolding from the march's deep operational formation, is acquiring a new character and requires the definite organization of the operational echelons' cooperation into the depth—particularly of the AVE and the main echelon.

This cooperation determines the entire development of the meeting battle, revealing the AVE's extremely important role in it.

The essence of the meeting battle is that, having defeated a definite part of the enemy's combat formation before it comes into full ground contact with us, to prevent him forming his front and create a continuous front of fire, requiring a breakthrough.

A certain unity of contradiction in the development of the deep operation here is that at this stage its forms essentially negate themselves. They struggle against the enemy's front forming and requiring a breakthrough as a full-blown form of the deep operation. The resolution of namely this task should be striven for in the meeting battle. If the enemy has been obliging and has exposed his combat formation on the approach, then all our operational aspirations should be aimed at destroying the enemy's approaching front to such a degree that it cannot be established and transformed into that continuous front of fire which requires a breakthrough. The achievement of this is the main goal of the deep operation's forms in the meeting battle.

The entire attack in this battle must already be organized on the approach, precisely in order to resolve this task.

If, during the breakthrough and in organizing the deep attack against the established front, long-range operational activity must be basically directed against the deep factors of the enemy's defensive system, then on the approach, upon the outbreak of the meeting battle, its long-range activity must be mainly directed at his lead approaching echelons, for they carry within themselves all the prerequisites and conditions for the establishment of the future solid front. It follows, of course, that deep pressure must also be exerted on the approaching enemy's depth; but, if we limit ourselves to this, having thus paralyzed the enemy's rear and suppressed his deep reserves, we will still not deprive him of the opportunity to form and establish his front. The fight against this is the main task in entering into the meeting battle. On the approach it is mainly necessary to operate against the enemy's leading echelons, so as to render them incapable of erecting and establishing a fighting front. The logic of establishing any front is that it has to be complete and unbroken, losing the opportunity to form, should any part of it be destroyed.

Thus on the approach the resolution of the problem comes down to pulling out by the roots that part of the enemy's approaching formation, at the loss of which his front will begin to waver and lose any sort of stability.

The AVE, along with the AGA, carries out the task of destroying a particular part of the enemy's approaching front before full ground contact can be established. This constitutes the enormous importance of these [fronts] {factors} of the deep operation at the approach stage and the entry into the meeting battle.

The AVE, as the name itself indicates, is to a certain degree the army vanguard, but already with a completely different qualitative content than the vanguard that was {once} employed by Napoleon. First of all, the AVE does by no means cover and should not cover the entire front of the advancing army. It would be very wrong to disperse the AVE's units ahead of the entire front. The AVE must be directed against a specific part of the arriving enemy's operational formation that conditions the integrity and stability of his front. This requires the unification and massing of the AVE's efforts along a specific axis and against a specific target.

In this way the army vanguard's tasks; that is, operational preemption, securing the deployment and the creation of favorable conditions for entering into the battle are resolved, because this is achieved at the operational level along a single selected axis of the main attack.

It is a task of the utmost importance to correctly determine the target of the AVE's operations. This is its essential part, which flows from the overall operational plan. The problem is resolved in each individual case depending on the specific conditions of the operational situation.

In the overall plan for organizing the army operation, the AVE must, as a rule, always operate along the main echelon's main attack axis.

If the main attack is launched by the left wing, then the AVE must operate ahead of it, so as to secure the unhindered development of the enveloping attack from the left (figure 4).

At the same time, the AVE may be directed either at the enemy's main group of forces, or at his secondary group.

If the enemy has preempted us in deploying and attacking, then the AVE should be directed against his main forces. In this case, the AVE's mission is to delay the enemy's offensive against our shock group through its active operations.

However, this would be the worst case of not fully taking advantage of the AVE's designation. Should events unfold normally and we are not late in going over to the offensive, it will be more favorable, given the sides' deployment shown in figure 4, to launch the AVE against the enemy's secondary group.

In this case, the AVE will obviously not be facing tanks and will enjoy the greatest advantage in its tank attack.

The enemy's secondary group may thus be destroyed easiest of all and ripped from his advancing front.

In turn, this will most likely lead to a situation in which the enemy's front will be deprived of its point of support and will begin to wobble.

As a result, the main echelon's shock group will have the opportunity of developing its enveloping attack against the flank, while the enemy will lose the opportunity to form and establish his front. The achievement of this is decisive in the meeting battle.

The AVE's actions must be organized not only in space, but along the axis of its attack, in close organic connection with the plan of the entire army operation and issue from it; they must also be planned at the operational level in time. It would be completely unrealistic to believe that the AVE's units, having been fed into the operation, can uninterruptedly take part in it to the end. The 3-day technical norm for the mechanized corps' operations to a depth of 200 kilometers lies heavy upon it and brings up two mandatory conditions: first of all, the mechanized corps can operate ahead of the front only in definite operational leaps, measured in time at no more than three days and in space to a depth of no more than 150–200 kilometers. Secondly, upon the completion of its three days of work, the mechanized corps must be pulled back into an assembly area and be put in order, while requiring during this time cover and relief by those units capable of holding terrain. The mechanized corps' joint work with the cavalry corps or the motorized division, or the combined work of all of these three elements, is therefore a mandatory condition for the AVE's actions, as a whole, while conditioning the possibility of its operational employment in the deep operation in general. At the same time, it is extremely important to plan the AVE's activities in time so that they form a definite system of cooperation with the army's main echelon.

This condition is of enormous significance in the deep operation and puts forward

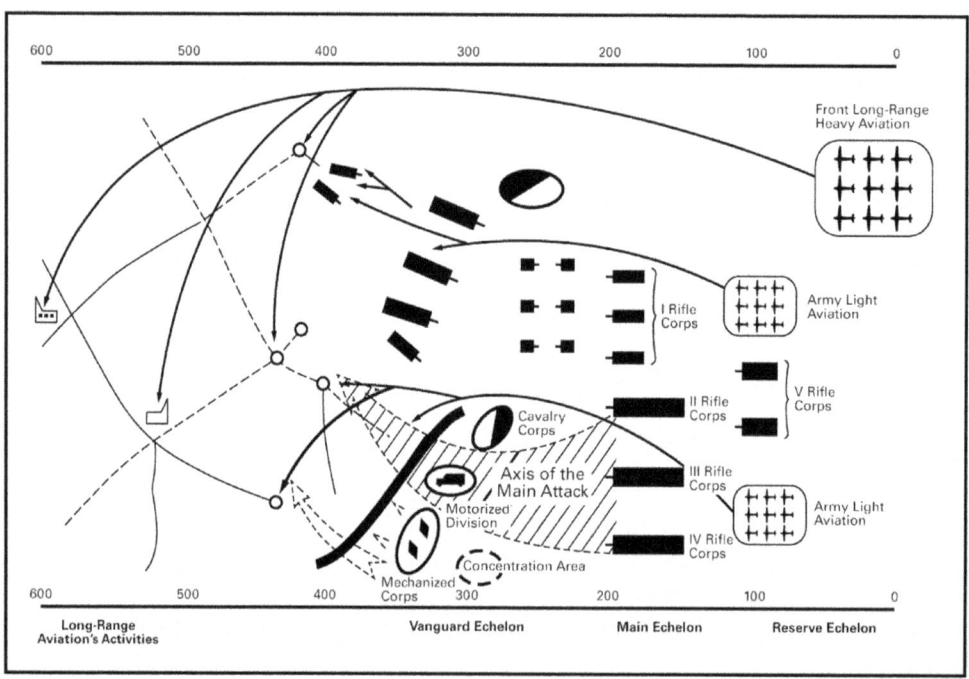

Figure 4. The Operational Formation of the Approach and Entry into the Meeting Battle.

the question of organizing cooperation between the AVE and the main echelon in time and in depth.

If while organizing the main echelon's approach we chiefly encounter indices that have significantly grown in the quantitative sense (the expanded depth of the columns)—then in organizing the AVE's cooperation with the main echelon, we encounter a qualitatively new index of measurement in depth, which determines the nature of the deep operation to a significant degree.

The essence of the question is that if the mechanized corps can operate ahead of the front for no more than three days, then its separation should not exceed that distance which the main echelon's arriving units can cover in the course of these three days; that is, the AVE should normally operate ahead of the front at a distance of no more than three days' march by the infantry, or an average of 100 kilometers.

This condition, which allows for the AVE's separation up to 200 kilometers in a more favorable situation, should lie at the heart of organizing cooperation between the AVE and the main echelon in depth. However, at the same time, we should take into account the fact that if the deep operational pressure being exerted on the enemy exceeds a certain distance, then it leads to the dispersal of efforts in depth, does not yield a concentrated attack and essentially reduces cooperation itself to nothing.

If in the conditions of the operation's linear character we have up until now spoken of the dispersal of efforts along the front, then in the conditions of the deep operation we must establish a new concept of dispersal in depth, which expresses itself in scattering combat efforts to a great depth, in which they dissolve and drown, without having the opportunity to concentrate for a massed attack along the depth in cooperation with the same type of attack along the front.

Normally, this cooperation in depth is achieved at a distance of up to 100–150 kilometers, insofar as a mechanized corps operating ahead of the front may be supported by the main echelon's units at this distance upon the conclusion of three days of combat activity, even if it is not operating together with a cavalry corps or motorized division.

However, still another condition must be observed in organizing the cooperation of the AVE and the main echelon in depth.

This condition is that the mechanized corps must be employed ahead of the front in such a way that by the moment of the meeting battle's full development by the main echelon it retains its combat capability.

It would be most incorrect to place the unfolding of the meeting battle in such conditions that the main echelon enters upon the attack while the mechanized corps, being at the end of its strength, must leave the game.

Thus the activities of the AVE and main echelon should be planned in such a way that their attack can be unified in time during the development of the meeting battle and is one of the most important facets of organizing cooperation in depth. Of course, in each separate instance this must be resolved corresponding to the situation's specific conditions.

However, the mechanized corps' firm technical resource norms enable us to establish an overall plan for resolving this problem. (figure no. 5).

As early as the first day of the start of the army's offensive the mechanized corps must be pushed forward to the entire depth of its daily combat range; that is, 100 kilometers. It will often already be this distance ahead of the army's front. However, in this case it must, having been relieved beforehand by the cavalry or motorized division, so

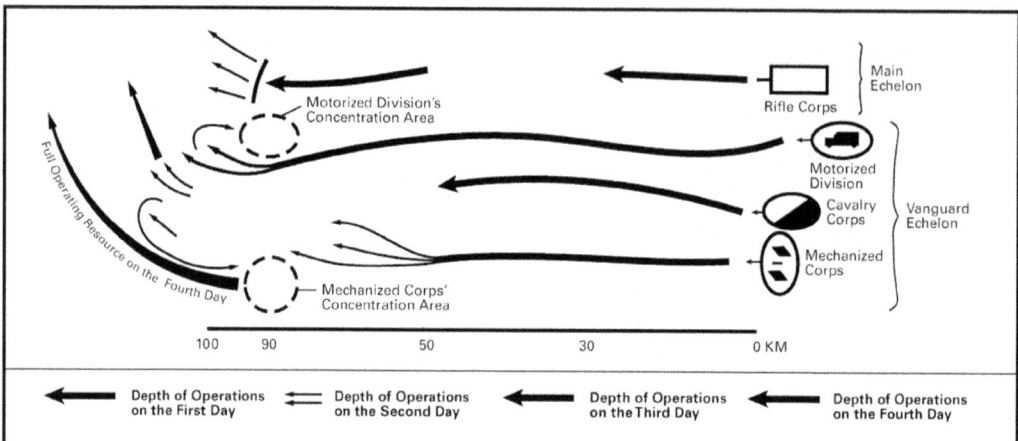

Figure 5. The Vanguard and Main Echelons' Cooperation in Depth.

as to begin its work on the first day of the overall beginning of the offensive with its full 3-day combat resource. If we assume that the enemy is also beginning his offensive on this day, then it is obvious that the mechanized corps, without having covered the entire 100 kilometers, will encounter his already moving units and will very vigorously attack him on the march and from the approach.

In operations ahead of the front, such a meeting engagement will be most typical for the mechanized corps, placing its weapons in the most favorable conditions for employment. As early as the first day the mechanized corps may attack the enemy's approaching group, consisting of no more than two rifle divisions and, in any event, completely paralyze the possibility of its further offensive. In this way a success of enormous significance is already achieved, because the enemy's approaching front is deprived of a specific group, loses its integrity along the front and uncovers the flank of that group that will continue the offensive.

On the first day, when the mechanized corps will be trying to resolve its task in depth up to 100 kilometers, the cavalry corps, if it was not located forward and has departed from the army deployment line, will cover 60 kilometers, while the main echelon's forward units will advance 30 kilometers. On the operation's second day the mechanized corps, while continuing its activities for the purposes of wrapping up the defeat of the enemy already under attack, and attacking his newly arriving units, will be supported by the newly arrived cavalry corps, reinforcing the attack in combined work with it. This will render it quite possible to completely suppress a specific enemy group along a 15–20 kilometer front as early as the second day of the operation or, in any case, paralyze its offensive along this axis.

During these days the AVE carries out its mission together with the AGA, which selects as the object of its suppression activities that same enemy group which it has been decided to rip out from his advancing front by the roots.

The most serious moment in planning cooperation between the AVE and the main echelon in depth arrives on the third day, because the latter, having carried out a march on the second day, will come into contact with the enemy with its forward units and on the third day will fully begin the meeting battle.

At the same time, it is obvious that the enemy, having been attacked along the axis of the AVE's operations, will inevitably have to delay his offensive along a neighboring sector, which will immediately impart to the line of his advanced front a broken outline, with its wing bent back toward the threatened axis.

In these conditions, the main echelon's shock group, which is attacking behind the AVE, will have the opportunity to push forward and win an enveloping position in relation to the enemy's broken front.

On the third day the decisive flank attack will be able to unfold fully, making the question of the AVE's and main echelon's cooperation particularly important.

The fact of the matter is that the operation's third day is also the day that the mechanized corps' technical resources are expended along one operational stage of its activities.

If we force the mechanized corps to operate on the third day as well, then on the fourth day of the meeting battle's complete development, when the main echelon's attack will reach its highest degree of intensity, the shock army will be deprived of its most long-range weapon of struggle on the ground, which, naturally, will not be able to lead to a decisive result for the battle. This question must be resolved depending upon what kind of combat effect {will be} achieved along the AVE's front during the first two days. If the success is significant and the enemy's front along the axis of the AVE's activities is so suppressed that on the third day there is an opportunity to penetrate into its depth, then, pushing the matter to the most decisive result, it is necessary to demand the mechanized corps' full exertion on the third day as well, [for the purpose of deeply penetrating] {in order to penetrate into the depth} and into the flank and rear of the enemy's group attacked by the main echelon and thus cause his entire front to waver. In this case, by the close of the third day the mechanized corps must be relieved by the cavalry corps and motorized division, which have been dispatched behind it.

This will be the most complete resolution of the problem of defeating the enemy in the meeting battle.

However, another variant is possible, when the AVE's activities in the first two days have not achieved a complete result and the enemy under attack manages to create an anti-tank front along a favorable line. We must unconditionally consider this possibility.

In this case, if the mechanized corps has no real prospect of operating in the depth on the third day, it should be relieved by the cavalry corps and motorized division and pulled back to an assembly area. At the same time, for the most part it will be possible on the third day to push forward units of the main echelon's shock group to its front.

When, on the fourth day, the main echelon will fully develop its attack, the mechanized corps should be once again ready for action; this time from a deep operational formation.

Thus by the moment of the meeting battle's development, it must preserve its attack's entire power and long-range reach.

It's quite obvious that in each separate case this problem must be resolved in accordance with the specific conditions of the situation; however, in all cases, the basis for this decision must be the maximum planning calculation of the AVE's activities in depth for no less than three days, which amounts to the normal technical resource for the mechanized corps' combat activities.

Otherwise, something not foreseen by the plan may unexpectedly demand the greatest efforts when there will be none at hand.

The actions of army aviation, which has been united, as has was noted, in the AGA, must be in the closest possible contact with the AVE when entering into the meeting battle, having one and the same object of attack.

The essence of the problem lies in completely pinning to the ground and suppressing that enemy group which is being attacked by the AVE. As a massed force, the AGA must be fully employed against this target, while playing an extremely important role in resolving the tasks of the vanguard's activities upon entering into the meeting battle, and from this the resolution of the latter's missions as a whole.

The AGA's actions against the chosen target must comprise an entire system of pressure from the air against the [designated] {selected} enemy group of forces—its personnel, supply routes, supply stations and railroad intermediate stages; finally, this enemy group, which is also under attack by the AVE, must be pinned to the ground in the fullest sense, paralyzed in its viability and deprived of any opportunity to enter the engagement and offer resistance.

In these conditions of the AGA's and AVE's joint operations, the main echelon's approaching shock group must be opposed along a particular axis by an enemy who has been rendered incapable of waging the engagement before the full unfolding of activities on the ground, thus offering broad opportunities for an enveloping maneuver and bringing the meeting battle to a decisive conclusion.

Characteristic of the development of this battle is the fact that the enemy's approaching front, which has been suppressed by the AGA and the AVE along a particular axis, loses its integrity and takes on the form of a broken line and thus offers the main echelon's shock group, which is advancing behind the AVE, the opportunity to win an enveloping position and immediately deploy for the flank attack with a decisive aim.

In any event, this may be achieved and must be the [army command's] main goal in the modern meeting battle. [It] {The latter} thus loses its former character of operationally frontal and exhausting actions and becomes extremely decisive according to its forms and possibilities. However, it requires great decisiveness and purposefulness of action from the AGA, AVE and the main echelon's flank shock corps.

Insufficiently energetic and non-purposeful actions upon entering into the meeting battle must inevitably lead to the appearance of the front of fire, which will have to be broken through by the enormous efforts of the deep operation's developed forms. One should not consciously try to achieve this while trying to achieve decisive operational results as early as the meeting battle, when the enemy, while attacking toward us, has exposed his forces to an attack on the march and has not yet established a front, anchored in the earth.

Thus the entrance into the meeting battle already displays all the indicators of long-range pressure, immediately brought forth against the enemy's approaching front. This reveals the nature of the deep operation, which is already capable of leading to decisive results as early as the meeting battle.

However, because the possibilities of modern defense are undoubtedly great and the organization of an anti-tank front, does not present any particular difficulties, under favorable terrain conditions, then we should take into account the fact that the meeting battle may turn into a frontal melee between the sides, localized, in the final analysis, along a single line. This will happen if the AVE has not achieved its full effect along its axis. In this case, it will at first hold the approaching enemy front with cavalry and motorized infantry, after which it will be forced to fall back and, finally, it will end up on the flank of the main echelon's newly-arrived shock corps along the same line of front with them. This is the moment of crisis in the development of the meeting battle. If the AVE has ended up along the same line of front as the main echelon's newly-arrived shock corps, then this means that one of the most important elements of deep pressure on the enemy has fallen by the wayside. In these conditions the question arises before the army command: to leave the AVE's units in the line of the new front of struggle, or, having relieved them with the main echelon's newly-arrived corps, pull them back to the rear.

Naturally, we must take advantage of every opportunity to immediately break through the enemy's front throughout the entire depth. However, if the battle along the sector of the AVE, which has now already ceased to be the vanguard echelon, breaks into flame and the prospects for immediate long-range action are lost, the time will come to pull back the AVE's units from the front. Now the newly-arrived shock corps will carry out their work of tactically overcoming the front of fire, with incomparably greater success.

However, one should take full account of the fact that this moment in the development of the operation is the first step toward the establishment of the solid front.

Does this mean that the mobile meeting battle immediately grows into a breakthrough? No, and it would be wrong to maintain this. The solid front, which forms in the process of the sides' meeting actions, still possesses all the signs of maneuver; it is quite distinct from the solid front of a defense that has taken root in the ground. The maneuver solid front, which arises as a result of the flaring up of the meeting battle, has not yet hardened and is soft and subject to great fluctuations. The latter may be very common and unexpected. The newly-arrived shock corps, having taken advantage of the strength of their attack for the simultaneous suppression of the entire tactical depth of resistance, may drive a breach into this solid front incomparably more quickly and often than in breaking through an established front, thus creating many favorable opportunities for the operational development of the success into the depth.

Enormous flexibility and skill of control will be required at this highest stage in the development of the modern meeting battle, in order to take advantage of the slightest fluctuation along the front and each open breach for the immediate development of the attack into the depth and the defeat of the entire front.

Thus as early as this stage of the meeting battle, the AVE automatically turns into the success development echelon (ERU), which develops the attack from the depth into the depth. In the meeting battle the ERU is already the ERP in embryonic form; however, its activities are naturally distinct from the character of the ERP's activities in developing the breakthrough of the organized defensive front.

They are characterized by significantly greater freedom of maneuver in penetrating the front and in actions in depth, which in this case transpire in easier conditions.

If the depth of the AVE's activities during the approach is generally determined by the distance between the sides, then the depth of the ERU's activities in developing the meeting battle must each time be determined depending on the specific conditions of the situation, the depth of the enemy's operational formation and his basing. The essence of the problem is to shake up the enemy's entire operational depth through the ERU's activities and completely defeat him. Proceeding from this decisive goal of the operation, the enemy's deep (army) reserves and his supply stations feeding his forces must be subjected to the ERU's pressure.

Depending upon specific conditions, the depth of this reach can be defined as a distance of 60–100 kilometers, and up to 150 kilometers in some cases. However, it is here that cooperation in depth for the purpose of securing a massed and purposeful attack has enormous significance for the achievement of the true defeat of the enemy.

The ERU's activities should not be pushed beyond a definite depth supporting this cooperation; otherwise, they will lead to the complete dispersal of efforts in the depth.

It's quite obvious that these activities must be accompanied in the air by a simultaneous mass attack by the AGA.

At this time, part of the latter's assault air brigades will have already been transferred to the most important shock corps as their air groups.

The entire remaining part of army aviation will push forward its activities against the deep factors of the enemy's resistance as a centralized AGA.

Thus the AGA's trajectory during the approach, as the sides move closer, contracted at first, being defined by the line of the enemy's advance, and, upon the establishment of full ground contact it once again begins to lengthen in depth.

The development of the meeting battle's success once again gives rise to a picture of it being conducted at least two tiers of depth (the main echelon along the front and the ERU in the depth), thus displaying the characteristic forms of the deep operation.

Its nature consists in the fact that it should not be waged along a single line of front. If such a situation has arisen and operations have latched onto a single line of front, this means that a linear melee has occurred, without the prospect of achieving any sort of operational result and with the inevitable and rapid consequences of the birth of the continuous front, and then of positional warfare.

The lack of understanding of the nature and requirements of the deep operation, the listless control of the operation and the lack of skill in employing the new long-range means of struggle will always be guilty of this. The well-known unity of the contradiction of their employment consists in the fact that they either operate as part of the AVE ahead of the army's approaching front, or as part of the ERU—from the depth of the main echelon which has entered the battle into the enemy's depth.

In both instances, the object of these means does not lie in a single line of the front of struggle with the army's main echelon, but in the second operational tier of struggle, which has been moved forward.

In these conditions, the employment of the entire long-range capability of the modern means of struggle and pressure on the enemy's entire operational depth may lead to a decisive result as early as the meeting battle.

The latter's development may, however, at the same time reach a high level of intensity and crisis. An energetic enemy will himself strive to penetrate into our depth with his mechanized formations and cavalry. This often coincides in time with the moment the ERU penetrates into the enemy's depth and may create an acute crisis during the battle.

At this time a great deal will depend on the energy and decisiveness of the army command, which must possess a firm operational will and carry the fight to a decisive conclusion, often under the weight of a difficult situation.

In the opposite case, the entire battle will be scrapped.

It is precisely at this moment that the entire enormous significance of the reserve echelon, which takes upon itself the struggle against the enemy who has appeared in the rear, and thus securing for the army's operational depth the necessary stability and capability of defending itself, will become manifest.

VI. The Meeting Battle

Given the absence of a reserve echelon in the operational organization of the march, the appearance of the enemy in our rear may have harsh consequences and lead to the scrapping of the entire battle, as well as to a catastrophe.

One should keep in mind these, most likely inevitable, frictions in the process of the meeting battle's development, the planned and uninhibited course of which is in obvious contradiction with the entire nature of the deep operation. This once again speaks to the enormous demands placed on the modern army command.

A question of particular importance is defining the duration of the meeting battle in time and the pace of its movement in space.

One should consider that the complete deployment of the main echelon will take about two days. On the third day the attack will take on the complete and organized nature of a blow into the entire tactical depth. Often on this day the opened breach will enable us to push the ERU forward, and on the fourth day operations in depth may decide the outcome of the battle.

Thus from the beginning of the main echelon's deployment, the meeting battle may play itself out in approximately four days in time.

To what depth will these activities develop?

It should be assumed that during the first two days the deployment of the main echelon and the movement of efforts forward will yield a somewhat lower rate of advance of up to ten kilometers per day; that is, 20 kilometers in all. During the second two days this pace, given the complete development of the deep attack, will grow to 15–20 kilometers, yielding an overall advance of 30–40 kilometers. Thus in four days of fighting the main echelon, will advance its efforts forward by 50–60 kilometers, while destroying the oncoming enemy, which yields an average rate of advance in the meeting battle of about 12–15 kilometers per day.

However, at the same time the ERU's activities will be thrust forward in to an average depth of 100 kilometers, defining by this distance the zone in which the modern meeting battle plays itself out.

Ammunition expenditure in such a four-day battle is measured at approximately three combat loads: during the first two days, when two-thirds of the main echelon is taking part, 1 1/2 combat loads would be required, while on the third day of the complete development of the main attack one combat load will be needed, and another half of a combat load on the fourth {day}.

The revealed character of the modern meeting battle fully shows its inherent indicators of the deep operation, of which it is one of the forms.

The modern meeting battle represents a large and dynamic process: it is suffused with a rich content and characterized by a high degree of tension.

It should be emphasized once again that it has as its goal the overcoming of the solid front before it can be established. It is conducted so as to prevent the solid front from being established. One of the decisive factors for achieving this is long-range pressure, which is carried out by the AGA and the AVE. The true defeat of the enemy can be achieved with the correct and energetic employment of these factors in the meeting battle; this should be pursued in the meeting battle and this comprises its basic designation.

However, at the same time, other possibilities should be kept in mind. It is quite clear that the meeting battle, as the first act of the initial offensive operation, will play out in immediate proximity to the state border. This puts a special stamp on its development. The fact of the matter is that the forward zone of the theater of military activities now contains in peacetime an entire series of designated defense lines. In these conditions, the enemy will enter into the meeting battle, having prepared defensive lines in his immediate rear. He will thus be supported by the latter, while maintaining the opportunity of [relying] {falling back} on them in case of the battle's unfavorable development.

Naturally, in this situation the task of the AVE, and then of the ERU, will consist of stepping over such a defensive line before the enemy can use it for support. However, it is impossible to count on achieving this in every case, because the enemy's prepared defensive line will be, first of all, inaccessible to tanks, for the most part; secondly, given a deep deployment, it can be occupied beforehand by his second-line forces and prepared for defense. In these conditions the enemy will be able, for the most part, to rely on the prepared line behind him and occupy it, and then the most decisive conduct of the meeting battle may be faced with the necessity of breaking through an organized defensive front. It is at this stage that the deep operation will find its most complete development.

At the same time, one must keep in mind the fact that, for the most part, the defensive front will arise gradually and unnoticed, and will by no means immediately reveal the new quality of the solid front with its roots in the ground.

However, one should by no means imagine that a breakthrough automatically assumes the enemy's initial deliberate retirement to a previously prepared line and that our march to it will take place in conditions in which we will know of its existence and outline beforehand.

The reality of the course of events is incomparably richer in its content and freer in its real-life variants.

Often, while believing the meeting battle to be at its height, we will actually be already fighting along the forward field of the already-organizing defensive front and will not sense the exact border in time and space which divides the meeting battle from the breakthrough. In modern conditions this will quite often be an inevitable and logical stage in the overall development of the initial offensive operation. Then the shock army must be ready to restructure the forms of its operational organization and the method of its operations in such a way so as to move from the energetic conduct of the meeting battle to the deep breakthrough.

This will be the natural growth of one form of the deep operation into another and into its most complete and developed form of the breakthrough.

VII. The Fundamentals of Organizing the Deep Breakthrough

The situation that leads the shock army to the breakthrough does not arise immediately, but is rather created in a complex process of combat events. The establishment of the immobile front of struggle from the very beginning of military activities can be expected only along those axes where the enemy has decided to assume the defensive. And in this case, he will often adhere to active forms of defense, while attempting to resolve his mission through decisive attacks with a limited goal. The course of action by which the Germans defended East Prussia in the beginning of the war in 1914 will and often be employed, particularly by small armies in defending a limited territory.

Naturally, the planned disposition to defend, or the planned withdrawal to a previously prepared line places the breakthrough in the most normal conditions for the timely adoption of the corresponding operational formation.

However, in the process of developing the initial offensive operation, which assumes the sides' decisive meeting activities along a given important axis—the situation obliging us to undertake a breakthrough increases in more difficult conditions.

The struggle along the front of the meeting battle will still appear to be in full swing, when individual bits of data will begin to accumulate, pointing to the establishment of a front along certain sectors. The first indicator of this will be the prolonged nature of the main echelon's actions along a particular line. The second indicator will be: the loss of spatial opportunities for the AVE and the ERU's unsuccessful attempts to win them by piercing the front.

Along some sectors of the front of struggle the enemy, having been overthrown by a decisive attack, will pull his forces back during the night, covered by an obstruction zone. Direct signs of an immobile defensive system will be established along other sectors. And then the operational prospect of the contours of the enemy's forming defensive front will begin to appear in outline.

Undoubtedly, this will become known earlier in general form through aerial intelligence.

However, one should not think that the command will always know the exact outline of the defensive zone. Nor will this information arrive immediately, because contact with the defensive zone will by no means be established simultaneously along all [areas] {sectors} of the front.

Finally, a complete picture of the enemy's defensive zone and its depth will appear in a definite process of the situation's maturation.

However, the essence of the problem is that measures cannot be adopted on an operational scale only when they will be directly forced by the situation that has arisen. Foresight has decisive significance for the correct organization of the process of command and control at the operational level. This condition requires that at the first signs of the front's possible stabilization, the shock army's entire combat formation be adapted so that the approach operation may directly grow into the deep breakthrough operation. This acquires particularly great significance for the organization of supply, which must plan ahead of time the accumulation of the amount of munitions necessary for the breakthrough.

The entire initial operation must be organized in such a way that the necessity of a breakthrough does not force us to make special halts in time. This is all the more important, because even when fully foreseeing and adopting all the necessary measures ahead of time, the preparation for the army breakthrough will require an average of up to two days.

However, this interval must in no way remain dead space for the attacker's activities and enable the enemy to qualitatively strengthen his defensive system.

Should the preparation of ground activities for the breakthrough force us to make an inevitable halt, then our aviation, even while rebasing its forces forward, can continue its work, while engaging in the corresponding planning of its flight resources.

Just as in the meeting battle, in which aviation was the first factor in combat pressure on the enemy during the march, in the breakthrough pressure on the enemy should open in the air.

This will acquire enormous significance during the preparatory period for the breakthrough and must wear out the enemy even before the ground attack begins.

The resolution of this task is achieved through an entire system of pressure from the air.

Chemical pressure from the air against the defense's garrisons in the defensive zone itself will be the most effective.

The defense's deep operational reserves must be subjected to assault air strikes and the supply stations feeding the defense to bombardment.

The enemy's defensive garrisons must be worn out, his operational reserves suppressed, and his supply paralyzed by the entire system of these actions that purposefully attack a specific chosen sector of the defensive system to the entire depth.

The enemy must thus, on the whole, be significantly deprived of his combat capability by the day of the breakthrough.

Aside from this, a very important mission of army aviation during this period is the struggle for air superiority, which is realized through the complete suppression of all the defender's main airfields. The essence of this task is that by the time of the breakthrough the enemy's aviation is to be basically destroyed on the ground. It would be highly incorrect to attempt to resolve this task during the course of the breakthrough, once it has already begun, when all of the AGA's forces have to be free for carrying out combat tasks in the system of the deep operation.

Thus if before now the suppression of the defensive system called only for the necessity of the artillery preparation of the breakthrough, then now this suppression must include in its system the new concept and new period of the <u>aviation preparation of the breakthrough</u>.

Naturally, the expenditure of flying resources must, at the same time, be strictly calculated, because the chief flying time is during the breakthrough itself and its development. Thus no more than one-fourth, or a maximum of one-third, of the available flying resources should be expended during the aviation preparation.

Nonetheless, the aviation preparation of the breakthrough is becoming quite an independent period of activity in the air, opening the modern deep breakthrough operation.

The breakthrough operation basically consists of two acts: the first act consists of overcoming the immediately opposing front of fire; this is achieved by the breaking of the defensive zone throughout its entire tactical depth and is carried out by the reinforced corps' combat formation, which forms the attack echelon (EA). The second act consists of defeating the defense's entire operational depth; this is achieved by the development of the tactical breaking of the front from the depth into the depth, and is carried out by the operation's long-range elements, which form the breakthrough development echelon (ERP).

The first tactical act must grow directly into the second—the operational act, for without this the breakthrough operation can in no way be realized. Both of these acts comprise a single organic whole; unified and coordinated in the dynamic of the breakthrough, in which the means is the tactical breaking of the front and the goal is the development of the breakthrough into the operational depth.

Both of these acts lead during their realization to a single, unified and multi-act deep breakthrough operation, which is conducted along several levels (tiers) of the operational depth.

VII. The Fundamentals of Organizing the Deep Breakthrough

Here we see the chief indication of the modern deep operation, which finds its most complete expression in the breakthrough.

[Thus] {The} main thing that characterizes the deep breakthrough operation is the defeat of the entire depth of resistance.

This requires, first of all, comprehending {what} the operational depth of the modern defense [represents] {is}, according to its spatial dimensions and its combat content.

It's quite apparent that this concept cannot have a single standard scale for all cases, being dependent upon the many conditions of the enemy's operational doctrine and {the nature of} his theater of military activities.

By the operational depth of the defense we should understand that zone, which is bounded from the front of struggle by the forward edge of the line of fire resistance, and from the rear by the main distribution station junctions, to which defensive supplies stream from the depth of the country and from which they are distributed among the defending troops, feeding their viability and supporting their combat capability.

Naturally, the depth of this zone cannot be defined according to any permanent scale.

However, the operational elements of its parts lend a basis for measuring the latter.

The operational depth of modern defense may be basically divided into three zones, each of which has a definite significance and character in the entire system of organizing resistance.

The first, or tactical, zone has an overall depth of 15–20 kilometers and embraces the location of that combat formation of the defense which is designated to put up immediate tactical resistance to the attacker. This zone will, for the most part, consist of two defensive zones: the first to a depth of 5–6 kilometers, and the second at a distance of 12–15 kilometers from the forward edge of the first. The resolution of the defensive task itself is determined by the retention of the tactical zone, particularly of the first zone. This zone, particularly its first defensive zone, is thus heavily infused with elements of direct tactical resistance, which represent a front of organized fire defense, usually occupied by one division per 10–12 kilometers of front.

The second zone is determined by the depth of the location of those final rail delivery locales; where the supply stations are located that directly feed the tactical zone's combat formation.

One should keep in mind that the enemy, who has been forced to go over to the defense on his own territory, will always be in the most favorable conditions as regards his rear. For the most part, he will have the opportunity to base himself on the railroads leading immediately to his front. However, even in these conditions, given the long range of modern weapons, supply stations may approach the front no closer than 50–60 kilometers. This distance may, on the average, define the depth of the second zone. It thus basically embraces the zone of the dirt roads from the supply stations to the border of the troop supply sector.

If the tactical zone is suffused with the defense's combat formations, then the second zone contains elements of operational significance.

Usually in this zone, along the line of the supply stations; that is, at a distance of up to 50 kilometers from the first defensive zone, are the artillery reserves, which form at this depth the skeleton of the army rear defensive zone.[22]

The enemy carries out his operational maneuver from the depth in this zone, which is the main element in the operational stability of the defensive depth. The army aviation's airfields are echeloned here. Finally, the operational center of the army headquarters is usually situated at a distance of 50–60 kilometers.

The second zone thus contains the main operational elements of the depth of the modern defense, which is why this zone may be called the <u>operational zone</u>, as opposed to the tactical one.

Besides this, all of the support elements are located within the operational zone, which are represented, first of all, by the rich complement of transportation equipment, mainly motorized; then engineer troops, communications troops, and different types of support troops and equipment that service the operational rear.

Finally, within the bounds of this zone, in the area of the supply stations, materiel supplies will inevitably be accumulating, measured, on the average, by 1–2 daily requirements of the troops being supplied.

It's quite obvious that this zone has enormous significance for the entire defensive front's operational stability and its materiel supply.

<u>The third zone</u> embraces the railroad supply sector to the supply stations from the main distribution stations, where the more capacious materiel depots are located.

This zone is significantly less suffused with elements of operational significance.

Here will be located the reserves of front (strategic) designation, heavy aviation airfields and the main centers of army and, possibly, front, command and control.

On the whole, this zone is characterized by its rear significance for the defensive front, which is why it may be called the <u>rear</u> zone. It connects the front with the country's rear along railroad arteries and within its bounds localizes the flow and distribution of supplies for the fighting troops. While containing within its borders a significant norm for their needs, it acquires significance as a base upon which the defensive front relies at the operational level. Besides this, however, it is also a bridgehead for operational maneuver along the railroads and, while containing in its borders the major railroad junctions, may always be an area for concentration a new group of forces being transferred from the rear from another axis.

In this sense, it acquires great operational significance at the front level and of course cannot be left unmolested within the confines of the ongoing army operation.

The depth of the rear zone is the least constant one and depends on the outline of the railroad network. Major railroad junctions are located at an average of 100–120 kilometers from our state border, along the meridian Vilnius—Lida—Volkovysk—Kovel'—Krasne—Stanislavov.[23]

Thus proceeding from the possible variant of the establishment of a front along the forward theater of military activities close to the frontier, we may define the depth of the operationally organized defense, calculated as the distance from the first defensive zone to the central distribution station, as, on the average from 100 to 120 kilometers. A the same time, we will emphasize once again that such a definition of the depth is by no means standard for all cases, and in each individual and specific situation it depends on the overall structure of the theater of military activities along a given direction.

Thus the operational depth of the modern defense consists of three zones: tactical, operational and rear, and is calculated, on the average, at approximately 100–120 kilometers.

Naturally, the entire depth of the resistance at the strategic level, which is defined as the depth of the entire theater of military activities, is not limited by this. In the given case, we are speaking only about the operational depth of the defense, organized along a particular line.

The problem of the deep breakthrough of such a defense is that the breaking of its first tactical zone should grow directly into the development of the breakthrough throughout the entire operational depth of the operational and rear zones.

The breaking of the tactical depth of the resistance, which is carried out by the EA, does not go beyond the boundaries of organizing the deep engagement at the level of a reinforced rifle corps and is resolved entirely within the confines of tactics.

At the operational level, the problem chiefly consists in developing the breakthrough by the ERP, which, on the whole, is the main and decisive factor in the complete and radical resolution of the problem of the breakthrough.

The development of the breakthrough requires the defeat of the entire operational depth of the defense.

This problem of organizing the deep defeat stands at the heart of resolving the problem of the breakthrough at the operational level.

The organization of the deep strike, which is carried out {by the ERP as the most important} factor in the breakthrough, must first of all be built upon the careful study of the composition and disposition of the defense's army reserves.

The definition of the correlation of forces in depth has the same significance in the deep operation as the calculation of men and materiel required for the tactical breaking of the front. It's quite obvious that if this correlation in the depth will not develop in favor of the ERP, then it should not be pushed through the broken defensive front. However, this means that the entire breakthrough operation cannot be undertaken.

The composition of the ERP, which is being directly pushed forward through the open breach in the front, will be defined in the shock army, as a rule, by the mechanized corps and the cavalry corps.

A motorized division may be thrown in behind them.

The composition of the defense's army reserves will naturally be different in each case. Usually {along an 80-kilometer front, defended by 5–7 divisions in the first line}, there will be a reserve of 2–3 infantry divisions and cavalry units, stationed in individual groups.

In this composition, the ERP's units, taking advantage of their great mobility, will be fully capable of coping with this force.

The mechanized corps is capable of attacking two divisions in a single day. The cavalry corps is capable of attacking more than one division.

Thus such a correlation of forces supports the successful resolution of the ERP's tasks in the depth. This may be provided for against an even larger group of defensive reserves, but only in the event they are scattered and unable to simultaneously enter the fighting in the depth. In this case, great rapidity and maneuver will be required of the ERP's units, in order to rout one group of reserves before another can arrive to join it.

In any event, the careful determination of the composition and disposition of the enemy's deep reserves is one of the most important factors for our intelligence during the preparation period for the breakthrough, predetermining its capability and the success of the deep defeat of the entire defensive system at the operational level.

The deep defeat must be organized in time so that the elements of the defense's operational depth come under pressure from the attacker by no means later than that moment when they begin to manifest themselves as combat factors capable of opposing the breakthrough.

The army reserves usually begin their movement from the depth of the operational zone no earlier than when the breaking of the tactical depth becomes clear, which will take place in the second half of the first day of the breakthrough. Deep reserves from the country's depth will begin to concentrate in the rear zone of the defensive no earlier than the second or third day of the breakthrough, when it will already be evident at the operational level. Of course, this may vary in each separate case; however, on the whole, these considerations should be kept in mind in defining the depth of the attack pressure in the breakthrough.

The essence of the question consists in defining to what depth and in what sequence the elements of the entire depth of the defense should be subject to the attacker's pressure.

It is quite obvious that one must enter the depth with a definite task and aspiration, aiming for definite targets and putting them under pressure in that sequence in which they take on the operational significance of a factor capable of opposing the breakthrough.

Only given a large and very powerful ERP may we pursue the goal of immediately flooding the entire depth of the defense. However, in this case the concentration of forces for a massive attack against a definite target in the depth also retains its binding force. Naturally, it would be most dangerous to try and develop the breakthrough in depth in general, scattering one's efforts in depth without a concrete objective. It has already been shown that cooperation in depth has a definite limit in its range. If during the development of the breakthrough, the ERP is pushed forward further than the distance at which this cooperation can be achieved with the EA, then its efforts will inevitably be dispersed in depth and lead to the overdevelopment of the breakthrough, which can guarantee the operational resolution of the task to the same small degree as its underdevelopment. Thus it is critically important to resolve the problem of to what depth and in what sequence the deep strike efforts should be pushed forward.

According to the distance of its ground effect, the main factor in developing the breakthrough—the mechanized corps—may immediately push its efforts forward into the defense's rear zone as far as the distribution station, if it is located at a distance of about 100 kilometers from the front line. However, what kind of significance can this have for the actual resolution of the breakthrough mission? The distribution station may be captured; it may be destroyed and, in the most favorable case, even be forced to stop its work. But to what degree will the operational stability of the enemy's entire defense suffer from this if it has reliable garrisons in the tactical zone and combat-capable reserves in the operational zone, which one should count on?

The paralysis of the distribution station for receiving and distributing supplies during the first days of the breakthrough will not materially be reflected in the matter of the defending troops' materiel supply, for, first of all, a certain part of the supplies are with the troops and will be, for the most part, unloaded with the defense within the bounds of the tactical zone; secondly, a certain part of the supplies is concentrated at the supply stations. Thus pushing the ERP's efforts immediately forward into the rear zone and the capture of the distribution station cannot materially and immediately influence the defense's supply. At the same time, the elements of important operational significance for the defense, which are located within the operational zone—the army reserves and an enormous number of support troops—should, in this case, retain complete freedom for opposing the development of the tactical breach of the front, which, according to the experience of 1914–18, always led to the complete elimination of the breakthrough and which guaranteed, in the best case, the sack-shaped widening of the attacker's front. Besides, overall, the delivery of supplies along dirt roads from the supply stations to the troops remains, in this case, undisrupted. Thus the ERP, when passing immediately into the rear zone of the defense, passes through it like an awl and drowns in the operational depth, without exerting a material effect on the stability of the defense.

Obviously, this problem must be resolved in another way.

The defense's main operational elements, which support its operational stability and oppose the development of the tactical breaking of the front, are mainly located in the operational zone of the defense. These include the army reserves, supporting troops, army air fields, the command and control center, supply stations, and dirt road delivery routes.

These elements must first of all be put under pressure in the depth.

The ERP's main and immediate task is first, having penetrated into the depth of the operational zone, to begin here its decisive work on the ground and in the air.

The army reserves and support troops must be attacked and defeated; the supply stations and airfields destroyed, the command and control center suppressed, and work along the dirt road supply routes completely paralyzed.

Thus the defense must be deprived of any kind of operational stability; those factors that directly feed and support the enemy's tactical capabilities of resistance must be destroyed in the depth.

The deep effect efforts thus at first spread to the depth of the defense's operational zone; that is, to the line of supply stations at a depth of 50–60 kilometers. It is only after all of the main defensive elements in this zone have been suppressed that the deep strike efforts must be pushed further into the rear zone as far as the line of the distribution station, which acquires significance only in the following days; for the deep reserves from the country's rear will begin to arrive no earlier than when the breakthrough becomes evident on the operational scale.

Thus the sequence of the deep defeat of the defense's operational depth consists, first of all, in the attacker's pressure embracing the tactical and operational zones to a depth of 50–60 kilometers, and only secondly is the defense's rear zone, at an overall depth of up to 100–120 kilometers, engaged.

This is how the problem of sequencing deep pressure on the ground is resolved.

In the air, however, the isolation of the sector undergoing the breakthrough must be organized immediately through the entire operational depth, barring any access by the enemy's reserves to the rear zone and preventing their concentration by railroad and by auto transport.

Putting the distribution station out of action will be very important in this regard. Representing, for the most part, a major railroad junction, this station will require attacks by heavy destruction aviation.

During the breakthrough period, the shock army must thus be serviced by no less than a single heavy aviation brigade. However, the resolution of the task of isolating the breakthrough front is not limited to this, requiring simultaneous activities by front strategic aviation to a significantly greater depth, in order to completely paralyze in the enemy's deep rear his capability of forming up reserves and their dispatch to strengthen the defense.

The isolation of the breakthrough front must therefore be understood significantly broader than merely preventing the concentration of reserves in the defense's immediate rear.

Thus in the air deep pressure is pushed forward immediately throughout the entire operational depth of the defense.

However, on the ground the attacker's deep blow must, along with the main mass of the AGA, engage with its efforts, first of all, the defense's tactical and operational zones. Insofar as possible, pressure to this depth must be organized immediately.

Simultaneity retains its significance here as well, while the opposite case leads to a series of consecutive and disjointed attacks in time, which do not prevent the creation of a new front in the operational depth of the defense.

However, in the operation this simultaneity is naturally measured according to another scale than in tactics.

If in the organization of the deep engagement the penetration by the long-range tank groups into the defensive depth is measured in minutes, then at the operational level these [scales] {deadlines} naturally increase.

In general, they are measured by the period necessary to break the defense's tactical depth.

Of course, the ERP may [be thrown] {break} into the defense's operational depth only upon the creation of a definite breach in the tactical depth.[24] The ERP's passage through the defensive front and into its operational depth must be tactically supported in a corresponding manner.

This condition raises a very important question in the system of the deep operation of determining when the moment has come to commit the ERP into the defense's operational depth. This moment, when tactics grows directly into the operation, is particularly crucial; it determines the unfolding of the true forms of the deep breakthrough operation, and thus its success at the operational level.

To unleash the ERP prematurely, when the tactical depth of the defense has not yet been disrupted, means pulling it into the fighting along the front, for which it was not intended. To commit the ERP too late, when the tactical breach has been reinforced by the defender's reserves is to doom to failure the very penetration into the operational depth. In both cases, this signifies the elimination of the forms of the deep strike and, it follows, the collapse of the deep breakthrough operation itself.

As a rule, the development of the breakthrough must follow immediately upon the breaking of the defense's tactical depth along a specific sector; that is, following the overcoming of the first defensive zone to a depth of 5–6 kilometers. It is completely unnecessary and wrong to wait for that moment when the entire depth of the tactical zone is tactically overcome to a depth of 15–20 kilometers, because, for the most part, the second defensive zone does not have its own independent garrison and is constructed in order to accommodate the troops of the first defensive zone during their retreat and give them the opportunity to halt on it. Thus the second defensive zone must therefore be seized before the first defensive zone's forces can fall back on it, which is the ERP's task.

At the same time, we must in no way await the moment when the tactical breach yawns open like a totally empty space, completely free of the presence of any kind of elements of enemy resistance. To wait in expectation of this moment would inevitably mean stretching out all deadlines and missing the time for developing the breakthrough. A completely empty space in the tactical breaking of the front may form only when the enemy has already completely abandoned the first defensive zone and fallen back on the second. However, it is the ERP's task to prevent this.

It units must therefore be committed into the breakthrough only when the defensive system has been disrupted by the pressure of the deep engagement throughout the entire tactical depth of the first defensive zone. The ERP's mechanized formations will themselves knock out the last cork in the defense's tactical depth through their great strike force and will completely clear a passage into the operational depth. The securing of this, however, remains the task of the EA's corps. Nor is there any necessity of waiting for {that moment} when the broken tactical breach will open upon along such a length of front that will fully secure the passage of the ERP's units against flanking fire from still-resisting defensive pockets. This would also lead to missing all timetables.

The passage of the mechanized corps or the cavalry corps through the front, which have been organized by tank brigades and cavalry divisions into a single echelon would require the breaking of the tactical depth of the defense to a width of no less than ten kilometers. Once again, however, the breaking of the defense along such a front is possible only after the enemy's complete withdrawal to the second defensive zone.

Thus the smallest formation capable of leading the passage through the front is the tank brigade or cavalry division, having a mechanized regiment in the lead.

The breaking of the defense's tactical depth along a length of {from} 3-{5} kilometers satisfies the conditions for the passage of each of these formations through the front.

At the same time, one must keep in mind the fact that the passage of the ERP's lead units must inevitably lead to the immediate broadening of the breach along the front. The width of this breach must therefore never be viewed statically, as something required for the given moment during the fighting. It is a given of the dynamics of the engagement's and changes along with its development. Its length must therefore always be evaluated through the lens of its inevitable further broadening as the ERP's lead units pass into the breakthrough.

Sometimes the penetration by a group of tanks through the breached defensive depth along an insignificant front may immediately broaden it to a scale supporting the passage of all the ERP's units. Thus the slightest breach, which has opened in the defense's tactical depth, must be employed for beginning the echeloned penetration of the ERP's units through the second defensive zone into the operational depth of the defense.

At this stage the engagement must grow directly into the operation, erasing any boundary between them in time and space and organically joining them in a single deep breakthrough operation.

However, one should bear in mind that the enemy will, for the most part, try to locate his second defensive zone behind a powerful local obstacle.

In any event, this is the [required] {sole radical means} for opposing the forms of the deep operation, because it can seriously paralyze the ERP's activities. Under favorable terrain conditions, the second defensive zone will always be located behind a line difficult of access—a water line, a wooded and swampy zone, or an obstructed area impassable for tanks.

If this zone also has its own garrison, which has already constructed an anti-tank front in the depth, which is {also} [also mandatory] the {single radical means} of opposing the forms of the deep operation, then the ERP's activities may encounter serious and sometimes insuperable obstacles following the penetration of the first defensive zone's depth. To unleash the ERP in these conditions, which do not secure its penetration into the operational zone, would be incorrect. In such a situation, the possibility of which must always be considered, the outline of the deep breakthrough is complicated by the expanded scope of the EA's actions.

In this case, the EA must overcome with its deep attack the defense's entire tactical zone, including the second defensive zone—normally to an overall depth of 15–20 kilometers.

The EA's corps must thus also be ready to attack the second defensive zone immediately upon capturing the first zone, thus finally securing the penetration by the ERP into the operational depth of the defense. In this case, the competence of the deep engagement expands throughout the entire tactical zone to a depth of up to 15–20 kilometers and its conquest will inevitably be prolonged, requiring a great deal of speed and uninterrupted activity.

Thus the task of the deep engagement, which is normally limited in depth by the breaking of the first defensive zone, cannot yet be considered to have been achieved if the second defensive zone represents a fully-formed factor of resistance in the defense's tactical zone. In this case, the outline of the deep engagement is complicated by a second stage, requiring the new organization of simultaneous deep pressure against the second defensive zone.

However, under maneuver conditions, when the breakthrough grows directly from the approach and meeting battle and when the defense has not had time to construct its entire depth, the attacker may be spared this stage, having the opportunity to develop the breakthrough immediately upon the breaking of the first defensive zone.

In these conditions, the ERP's units may be committed into the breach 4–5 hours following the start of the attack against the forward edge, because during this period the deep engagement can count on overcoming the tactical depth of the first defensive zone.

Thus the ERP has the opportunity of engaging the defense's operational zone to a depth of 50–60 kilometers as early as the first day of the breakthrough.

Having accomplished here its mission of suppressing the combat factors in the enemy's operational depth, the ERP, under favorable conditions, can as early as the second day push forward its efforts into the realm of the rear zone to an overall depth of 100–120 kilometers.

Thus under favorable conditions the elements of the deep operation can in two days engage the entire operational depth of the modern defense to an average depth of 100–120 kilometers; within two days this depth may be ignited and inflamed by the attacker's deep pressure.

The problem of the simultaneity of deep striking at the operational level is resolved in this approximate time scale, measured thus not by minutes and hours, as in the tactics of the deep engagement, but by hours and days.

The operational organization of the breakthrough's combat formation basically consists of two factors—the attack echelon (EA) and the breakthrough development echelon (ERP).

VII. The Fundamentals of Organizing the Deep Breakthrough

If the operational organization of the approach thus consisted of three echelons (AVE, main and reserve)—among which the reserve echelon had a specific significance as a combat factor, supporting the operational stability of the army's rear in the meeting battle, then the operational organization of the breakthrough basically consists of two echelons.

Of course, given a surfeit of forces, the reserve echelon can be part of the operational formation of the breakthrough, although this is by no means mandatory for it from the point of view of securing the operational stability of the army's rear, which, in the case of the enemy's assuming the defensive, may already be considered guaranteed against any kind of threat.

The EA's combat formation consists of a single echelon of corps of a shock and holding groups.

Only given the presence of a second defensive zone, difficult of access and already occupied, may the EA have individual divisions in its second echelon, or a corps, which are designated for immediately continuing the attack against the second defensive zone, following the capture of the first.

The main breakthrough sector, as has already been shown, must be no less than 30–40 kilometers in width, which is why the EA's shock group is determined at 3–4 reinforced corps. The EA's holding group will contain no more than one regular corps. Under favorable conditions, the main front under attack by four reinforced corps may be increased to 45–50 kilometers. At the same time, it should be borne in mind that the overall operational width of the breakthrough front is always greater than the tactical breakthrough sectors, which, due to the terrain conditions, are limited to specific areas. Thus an increase in the operational width of the main breakthrough sector should by no means be reflected in the tactical density of the attack. The regular corps of the army's holding group may be assigned a breakthrough sector of up to 15–20 kilometers in width. Thus the operational width of the breakthrough front of a 5-corps shock army may reach 70 kilometers. {However, given the necessity of breaking through the defense along a continuous attack front, its width may be no more than 50–60 kilometers}. [However] {Normally}, it is by no means required that the attack be carried out along the army's entire attack front to resolve the breakthrough mission. Certain sectors of the defensive zone may be tied down through passive activities. Finally, specific terrain conditions (wooded and swampy areas, which are especially common in our western theater of military activities), for the most part exclude the possibility of launching an attack. In these conditions, the shock army's entire operational zone along the front may be measured at approximately 80 kilometers, having, in this case, specific axes which, due to terrain conditions, remain completely untouched by active operations. At the same time, however, it should once again be emphasized that an attack along the main breakthrough sector must be, in all cases, secured by the following tactical density—normally ten kilometers of front per reinforced corps. The width of the main breakthrough sector will enable us, for the most part, to launch the main attack only along one general axis, which secures, as a rule, the creation of a single overall breach or a series of individual smaller breaches, although along a single axis. Only in special cases, when the terrain conditions will force us to launch an attack along individual sectors, separated by local obstacles—wooded areas, inaccessible areas, or water lines— may the attack be organized along two (or several) axes; in such a case, the passage of the ERP's units through the breakthrough front is supported along different axes. This has material significance for organizing the entire plan for developing the breakthrough. If the ERP's breach is secured only along a single overall sector, then its activities inevitably develop in depth along internal lines from one axis, for the task of launching a massed attack with all of its forces against a certain number objectives in the defense's operational zone. If it proves possible to develop the breakthrough from two sectors, then the ERP's units may launch an attack in two separate groups along converging axes, thus seizing the operational zone in their pincers. In this case, more effective results may be achieved, all the way up to the independent encirclement of the defense's operational reserves in depth; however, a larger ERP is required for this.

Normally, one must assume that the task of encircling the defense's deep reserves on a broader operational scale is achieved not on the scale of a single army, but at least by two neighboring armies, whose ERPs [cooperate] {are aimed} into the depth along converging axes. In this case of the most developed form of the deep operation, the ERP's activities grow out of the scope of the army operation and are resolved at the front level.

The development of the breakthrough by two adjacent shock armies thus yields a fuller outline of the front deep breakthrough operation. However, this by no means excludes the possibility of accomplishing the deep breakthrough's missions within the confines of a single army.

The ERP is the chief factor in the resolution of this task.

As was already mentioned, its power on the ground is measured by the composition of all the mechanized, mounted and cavalry formations attached to the shock army. For the most part, these will be: the mechanized corps, a cavalry corps and a motorized division. The AGA and ADO reinforce this force from the air.

Depending on the organization of the EA's attack along one overall sector, or two, the ERP will consist of one overall group, or will be divided into two groups. The first case will be the more common and normal case for a single shock army.

The entire operational organization of the breakthrough's combat formation takes on a deep form. (figure 6).

The EA's combat formation will be {from} eight [and sometimes] to ten kilometers in depth.

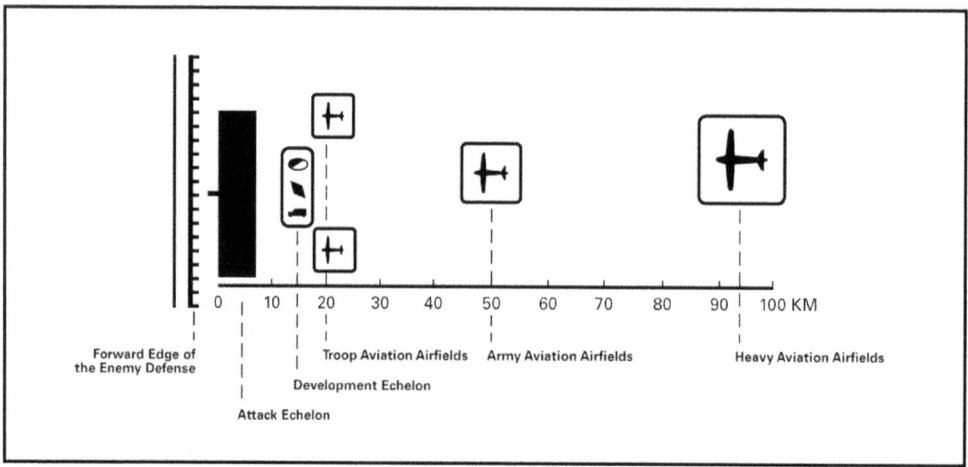

Figure 6. The Operational Organization of the Breakthrough's Combat Formation.

By the start of the breakthrough, the ERP is located in the intermediate position area at a distance of 12–15 kilometers from the EA's front, behind its shock group along the axis of the expected breaking of the first defensive zone. With the development of the EA's attack, the ERP's units move forward to occupy the jumping-off area at a distance of 5–6 kilometers from the front line of struggle, from where they are unleashed into the defense's operational depth.

The aviation's airfields are echeloned in the depth of the breakthrough's combat formation: troop aviation at a distance of 20 kilometers, army aviation and the ADO at a distance of about 50 kilometers, and heavy front aviation at a distance of up to 100 kilometers.

Thus the operational organization of the breakthrough's combat formation is measured by an overall depth of up to 100 kilometers.

The modern breakthrough operation begins with such a deep formation.

VIII. Carrying Out the Deep Breakthrough

The deep breakthrough operation basically consists of three periods, each having its own content and character.

The first period is the aviation preparation of the breakthrough, which has as its mission the exhaustion and suppression of the most important elements of the resistance even before the beginning of the ground attack. The radius of action during this period must embrace the defense's entire operational depth to 100–120 kilometers.

Its length is determined, depending upon the conditions of the ground preparation for the breakthrough, at an average of 1–2 days before the beginning of the breakthrough.

The main attack factor during this period is the AGA, the trajectory of which is extended into the depth of the enemy country by the front's strategic aviation.

The latter's sphere of activity, and that of the AGA, is demarcated in depth by the boundary line, indicated by the front and running, as a rule, through the distribution station that feeds the defense along the breakthrough front.

The second period embraces the tactical period of the struggle for the tactical depth of the defense, which is normally determined by the depth of the first defensive zone. Only given the presence of a highly-inaccessible and previously occupied second defensive zone, which does not allow for the ERP's breakthrough into the operational depth before its fall, must the tactical period embrace the overcoming of the entire depth of the defense's tactical zone, including its second zone.

In this case, it will naturally be more extended in time. However, given the ERP's penetration into the defense's operational depth immediately following the breaking of the first defensive zone, the length of the tactical period is determined at 4–5 hours, which is sufficient, employing the methods of the deep engagement, to break the first zone of the defense to a depth of 5–6 kilometers.

This period concludes with the passage of the ERP into the operational depth of the defense, which determines the direct growth of the engagement into the operation.

It would be wrong to allot the passage of the ERP itself through the opened breach in the front to a special period. This act organically grows into the tactical period, unfolds fully within the tactical bounds of the struggle in the depth of the first defensive zone and comprises, with the period of this struggle, not only a single organic whole, but its main goal. This is accomplished in the closest direct tactical connection with the actions of the EA and ERP, of which the first secures the passage of the second through the front. Only after the ERP has broken out of the struggle's tactical depth will its actions take on independent operational significance. The ERP's passage through the breach in the front into the tactical depth must thus be included in the content of the tactical breakthrough period.

At the same time, if only a single breach is forced in the defense's tactical depth, then the mechanized corps will try to break through first, followed by the cavalry corps.

If two breaches are formed along a single common axis, then the mechanized and cavalry corps may simultaneously break through different sectors of the front.

The passage of [these] the ERP's units through the front is one of the most complex and important acts in the breakthrough operation, requiring the precise organization of its tactical support. The entire responsibility for this organization lies on the EA's shock corps, which are obliged to create the conditions for the most unhindered breakthrough possible by the ERP through the breach in the defense.

For this, the attack by the EA's shock corps must, as a rule, develop [along] {not by} the entire front in a single general direction, which was characteristic of all the linear breakthrough operations during the World War, but from the captured bridgehead in the breach along diverging axes, in order to develop this breach toward the flanks, thus securing a place for the passage of the ERP's units. (figure 7).

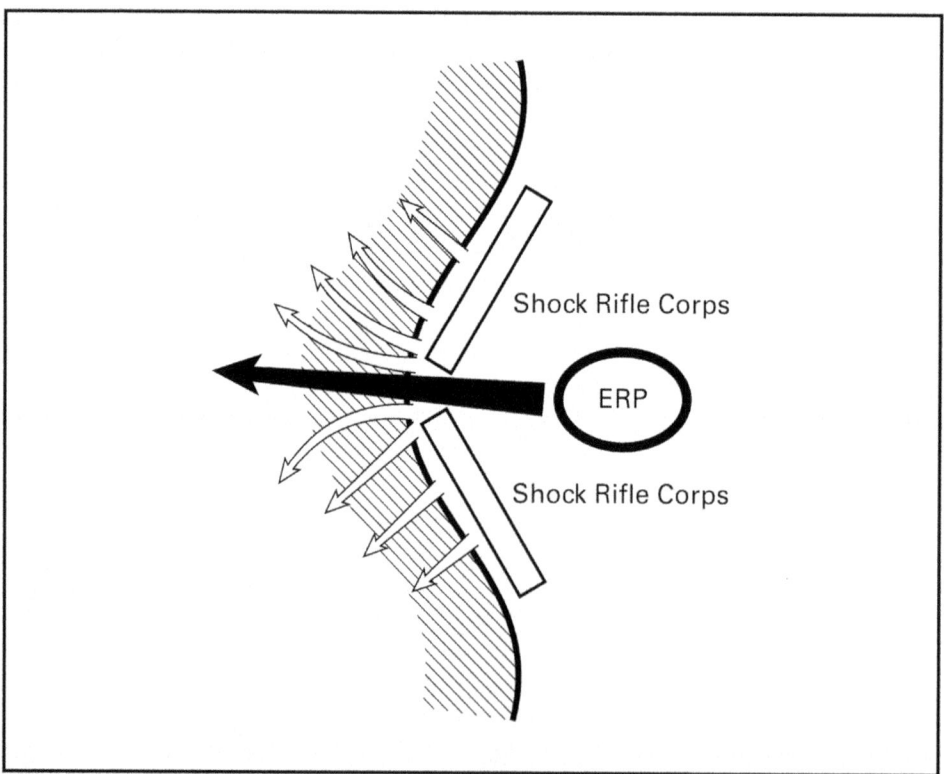

Figure 7. The Development the Breach for the Passage of the ERP.

The third period begins at the moment of the ERP's penetration beyond the bounds of the defense's tactical depth into its operational depth.

This is already the operational period, which carries out the highly important task of developing the breakthrough throughout the entire operational depth for the complete rout of all the elements of resistance on an operational scale.

As has already been shown, the pushing forward of all the ERP's efforts is not immediately required throughout the defense's entire operational depth.

The first task is the outflanking of the defense's operational zone to a depth of 50–60 kilometers, with the outflanking of the defense's rear zone to an overall depth of 100–120 kilometers only as a second task.

Thus the operational period is lengthier, up to the time of the complete spreading of long-range pressure throughout the defense's entire operational depth.

The defense's operational zone must be engaged by the ERP's actions as early as the first day of the breakthrough.

The mechanized corps must immediately push forward its activities to the depth of the location of the defense's army reserves, while not being distracted by any kind of secondary objectives.

The main task consists of attacking and defeating them as early as the first day of the breakthrough, and penetrating to the enemy's supply stations. In this way, the depth of the defense must be deprived of any operational stability and, the garrisons in the tactical defense zone, of any support.

According to territorial conditions, this makes it possible to simultaneously suppress the army's headquarters as well.

However, any kind of plentitude of missions for the mechanized corps in the depth of the operational zone, particularly on the first day of the breakthrough, must be unconditionally excluded.

The mechanized corps must break through into the defense's operational zone with a very definite and purposeful task, the main goal of which is the enemy's closest and most important group of army reserves.

The cavalry corps is thrown forward on the first day, either to catch up with the mechanized corps for joint actions with it, or it is directed to outflank an important group of the closer reserves [from the depth], which are located behind the second defensive zone. In this case, its activities will have a shallower depth of approximately 30–40 kilometers, which, however, will require an overall march intensity, counting the movement from the intermediate position, of 50–60 kilometers.

At the same time, the overall plan for developing the breakthrough should call for the mechanized corps' and cavalry corps' actions along a single axis of the breakthrough's operational turning into the flank of the still-resisting sectors of the operational zone, in order to immediately widen the length of the broken front and to achieve the decisive results of an encirclement.

Thus the ERP's actions in the defense's operational zone must take on the character of the front turning in depth in a particular direction, into the enemy's flank and rear, while at the same time sweeping away his army reserves.

The axis of this turning movement will be one of the shock corps that has made a breach in the defense's tactical depth; the cavalry corps will be the center of this front, with the mechanized corps its extreme turning flank. (figure 8).

At the same time, the ERP's units must dispose of complete freedom of maneuver in the depth and, while having quite definite objectives, they do not need to be assigned boundary lines.

Of course, such an outline is only one of the possible variants for developing the breakthrough and leading, in a particular situation, to the most decisive result.

However, in all cases it is necessary to keep in mind that the very nature of the ERP's actions is defined by its extremely decisive character.

The ERP, which has broken into the defense's operational depth, is itself encircled. In this situation it will either destroy the deep factors of the enemy's defense, thus bringing the matter to a decisive result or it will itself be destroyed if its actions prove to be insufficiently energetic and the enemy in depth sufficiently active and bold. The ERP's return behind the EA's front through the breach which, given the unsuccessful development of operations in depth, will close again before long, will, for the most part, be unlikely to succeed.

In these conditions, the ERP is faced with the problem of "all or nothing."

From this arise the enormous demands that are put on the ERP's entire rank and file and the enormous responsibility that lies upon it.

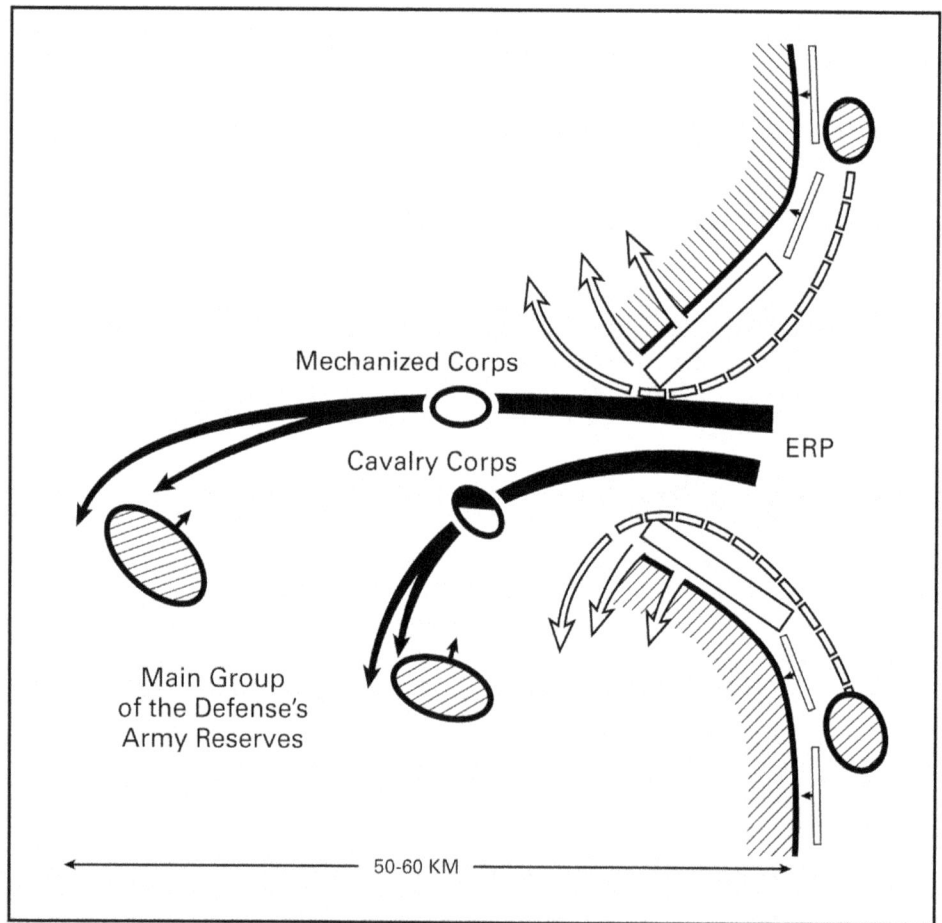

Figure 8. The Development of the Breakthrough into the Depth of the Defense's Operational Zone.

A totally different composition of forces and completely different training for commanders are required, so that the ERP's missions may be truly accomplished.

All of the ERP's extremely intense actions will begin to tell from the first moments of its penetration into the defense's operational zone; that is, as early as the first day of the breakthrough.

At the same time, it is necessary to keep in mind that on the first day the mechanized corps cannot as yet be supported by the motorized division, the passage of which through the breach in the defense requires the restoration of the roads on the battlefield that have been made impassable. Moreover, its passage requires the greater expansion of the breach for the purpose of securing it against flanking fire from those areas of the defense still holding out. This will become possible, for the most part, only on the night preceding the breakthrough's second day.

One can only approximately define the length of time required for the ERP to carry out its task in the defense's operational zone.

In any case, immediately following the ERP's breakthrough, the EA's shock corps must continue their activities without pause in order to fully turn the breach toward the flanks, particularly toward that flank, against the rear of which the ERP's deep turning movement is directed.

This should result in the arrival at the second defensive zone as early as the first day of the breakthrough, after which the turning flanks of the EA's corps must be immediately extended (broadened) into the depth of the defense, employing troops dispatched from other sectors of the army breakthrough. At this stage it will often be more expedient to halt the frontal attacks against still-resisting sectors of the defense and to concentrate all possible men and materiel (especially tank forces) at the breach in the defense, in order to push them into the depth of the breakthrough and to lengthen the flanks of the shock corps developing it toward both sides. (figure 8).

Thus on the second day of the breakthrough the EA's units, which have been pushed forward, may already appear in the defense's operational zone. On this day the motorized division will also arrive in the defense's operational depth.

In favorable conditions this will give the ERP the opportunity, while blocking up with one part of its forces the deeply outflanked and encircled units of the garrison's defense from the rear, to push forward with another part into the area of the enemy's rear zone as far as the distributing station, in order to capture it.

Thus in favorable conditions, the operational period will require up to two days for the complete spreading of its long-range pressure throughout the entire operational depth of the defense and is itself divided into two stages: the stage of outflanking the defense's operational zone and the stage of outflanking the defense's rear zone.

A great deal of intensity will be required during these days from aviation, whose activities comprise an organic part of the entire system of the deep [breakthrough] {attack}.

The AGA, having carried out the aviation preparation of the breakthrough, must retain the main part of its flight resources on the ground by its start.

The most important shock corps, which make the breach for the ERP, must be supported by assault aviation during the tactical period for suppressing the defender's artillery and reserves.

Having put pressure on the elements of the tactical depth of the defense, for which in the system of the deep engagement one sortie per plane will be required before the beginning of the infantry attack, the assault air brigades are once again united with all of the army aviation in the AGA.

Henceforth, the AGA's main task becomes the suppression of the defense's army reserves that may delay the ERP's actions in the depth.

With the beginning of the passage of the ERP's units through the breach in the defense, the entire AGA rises into the air. The task here is, through overwhelming pressure, to reduce the defense's army reserves to a state of complete inability to fight by the time they are attacked by the ERP's units.

It is necessary to comprehend the extreme importance of this mission.

Should the defense's army reserves remain untouched by the time the ERP's units reach them, they will always be able to organize an encounter, having previously occupied an anti-tank front. In this case, the ERP's actions may end up being tied down in the depth and unsuccessful.

The essence of the AGA's activities at this stage is not only that they assist the ERP from the air, but that they are also a necessary and independent factor in resolving the task of developing the breakthrough.

During the operational period the AGA resolves its tasks in the defense's operational depth along with the ERP's units in the role of their AG. At the same time, air activities must each time anticipate actions on the ground, so that the enemy's arriving deep reserves are always deprived of the opportunity of encountering the ERP's units with an organized anti-tank front. Thus during the second stage of the operational period; that is, on the breakthrough's second day, the AGA's pressure must be chiefly directed against the distribution station, which is the main junction for the arrival of new reserves by rail.

At this stage the AGA's main task lies in the complete isolation of the breakthrough front from the depth of the enemy's territory.

Of course, the struggle against the enemy's aviation, which is chiefly waged during the aviation preparation period, must be continued throughout the entire time of the breakthrough, should a threat or obstacle to our aviation's work be detected.

The ERP's actions, in particular, should be always be covered by fighter aviation, the army's entire available mass of which must create a dense and truly impenetrable curtain in the air during the ERP's passage through the breach.

It has already been pointed out that the struggle in the air is not limited to army aviation's operating range.

The front's <u>strategic aviation</u> simultaneously puts under pressure the depth of the enemy's territory, while resolving the task of isolating the broken defense by preventing the gathering and concentration of new reserves from the country's rear.

The resolution of this task requires the bomber destruction of major mobilization centers {and} depots; industrial and economic targets and railroad junctions.

Thus the system of deep pressure in the breakthrough operation spreads its striking force throughout the entire depth of the enemy's territory.

The airborne landing is an important factor in this system of pressure.

The <u>ADO</u> is basically a factor for developing the breakthrough into the operational depth. It operates together with the ERP, anticipating in the air its appearance in the defense's operational depth. Thus the depth of the ADO's landing must be determined by the depth of the ERP's activities on the first and most decisive day of the breakthrough; that is, at the line of the enemy's supply stations, which are assumed to be at a distance of 50–60 kilometers from the front.

VIII. Carrying Out the Deep Breakthrough

The ADO's main target must be the enemy's command center, as its guiding brain, the destruction of which, of course, has enormous importance from the very start of the breakthrough.

Thus having landed in a remote place at a certain distance from the enemy's unearthed control center, the ADO immediately launches a surprise attack against it, after which it occupies a particular locale important for supporting the arrival of the ERP's units in the depth. At the same time, a number of separate diversionary attacks are organized against important targets in the operational depth of the defense.

Naturally, in each separate case the plan for the ADO's activities must be drawn up in accordance with the specific conditions of the given situation.

Nor is such a situation excluded in which the ADO may be immediately dropped at a depth of 100–120 kilometers, at the line of the distributing station, for the purpose of organizing a diversionary attack on the latter.

This is possible if the defense's operational zone is poorly supplied with elements of resistance and does not cause particular difficulties for the ERP's activities.

The time of the ADO's landing may be decided, depending on the situation, in different ways. Under more favorable conditions in which the defense's operational depth is poorly manned, the ADO may be dropped on the eve of the breakthrough, without, in this case, being subjected to the danger of being destroyed separately.

However, a more normal case will be the ADO's landing at the beginning or right before the start of the breakthrough on the ground. In this case, the ADO, having carried out its immediate task of attacking the defense's headquarters, may establish contact during the second half of the day with the ERP's units that have broken through into the operational zone.

After this, the ADO is subordinated to the latter and, being motorized, operates together with it.

Thus all the elements of the breakthrough's operational formation and the deep striking of the defense have their definite place, in time and space, in the deep operation. {figure 9}

Their interaction requires precise and specific organization, which is expressed in general terms in [the attached table] supplement 2.

The organization of this cooperation is calculated on the complete spread of long-range pressure throughout the defense's entire operational depth; that is, to a depth of 100–120 kilometers, over two days.

In two days the entire operational depth of the defense must be ignited by deep strike factors and burst into a bright flame in an entire series of pockets to an overall depth of 100–120 kilometers.

By the close of the second day of the breakthrough, the operational depth of the defense will thus resemble a series of pockets burning in the flame of combat.

The entire question is now how much time will be required for the enemy's resistance to burn up in this flame. As is known, this depends on the refractoriness of the material; that is, the strength and stability of the defense's combat formation.

Thus the breakthrough operation is by no means completed on the second day of the complete engagement of the defense's entire depth. More likely, the battle for the destruction and crushing of the entire depth of the opposing front is here only beginning in its developed forms.

Thus the breakthrough operation concludes with a <u>fourth period of the defense's</u>

144 Two—Isserson's "The Fundamentals of the Deep Operation" (1933)

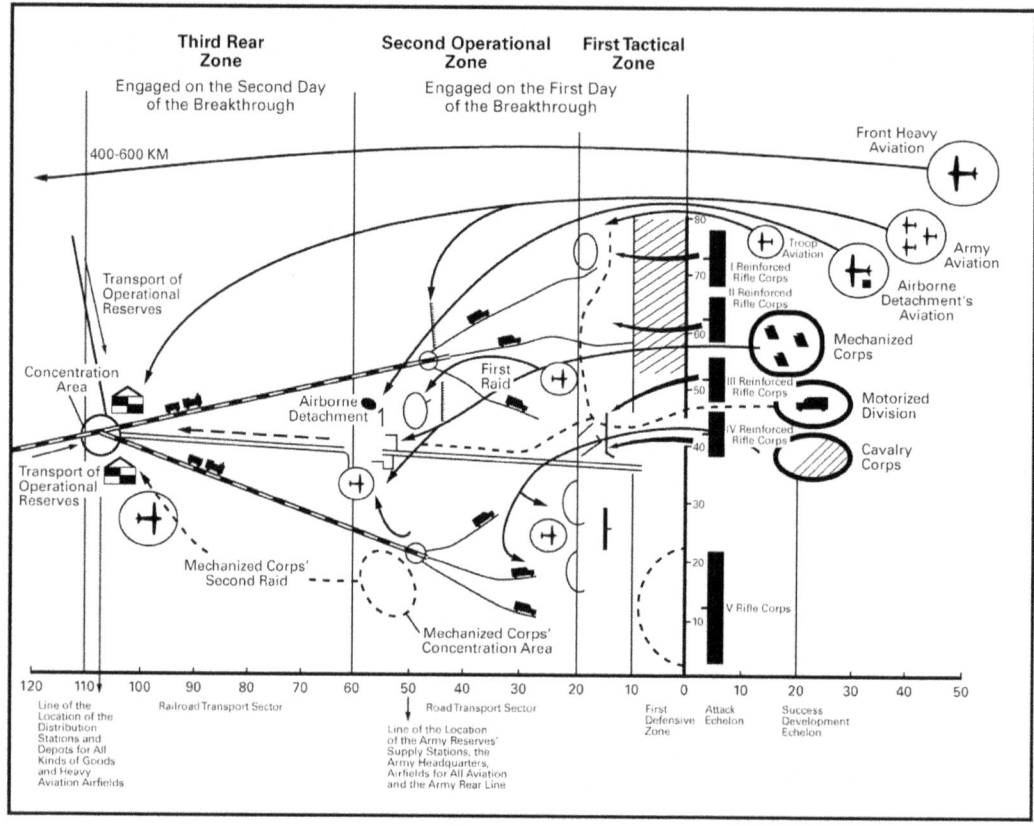

Figure 9. The Outline of the Deep Breakthrough.

defeat. The content of this period, in the sense of regulating all of the deep pressure elements cannot be precisely established beforehand, depending completely on the situation that has arisen in the depth.

The entire struggle will now take on extremely distinctive forms. Like whey, it will curdle toward pockets, around which the fight to encircle and destroy will be waged.

Attachment 2. The Deep Breakthrough Operation
(Up to the Time of the Complete Spread of Long-Range
Pressure Throughout the Defense's Entire Operational Depth)

Periods	The Aviation Preparation Period	The Tactical Period
Combat Content	The exhaustion and suppression of the elements of resistance in the defense's entire operational depth	The breaking of the defense's tactical depth and the passage of the ERP's units through the front
Depth of Pressure	The defense's entire operational depth, to 100–120 kilometers	The first defensive zone, to a depth of 5–6 kilometers*
Length	1–2 days before the start of the breakthrough	4–5 hours during the first day of the breakthrough
EA	Prepares for the breakthrough and occupies its jumping-off position	Captures the first defensive zone through the simultaneous striking of its tactical depth and secures the passage of the ERP's units through the breach in the front

VIII. Carrying Out the Deep Breakthrough

Periods	The Aviation Preparation Period	The Tactical Period
ERP	Prepares to develop the breakthrough and occupies the intermediate area	Moves to the jumping-off area and breaks through the breach forced in the defense's tactical depth
AGA	Tries to achieve air superiority and systematically exhausts and suppresses the elements of the enemy's defense, being particularly active on the eve of the breakthrough	With the beginning of the ERP's passage through the breach in the front. Its entire mass suppresses the enemy's army reserves
ADO	Prepares for the landing**	Lands in the depth of the defense's operational zone, for the purpose of attacking the army command center and securing the arrival of the ERP's units into the depth
The front's strategic aviation	Operates against the enemy's deep rear, for the purpose of preventing the gathering and concentration of new reserves at the breakthrough front and its isolation from the country's rear.	

Notes: * Given the presence of the defense's inaccessible and previously occupied second defensive zone, the depth of the deep pressure during the tactical period reaches throughout the entire tactical zone, as a result of which the length of this period increases.
** The landing may be made on the eve of the beginning of the breakthrough.

Periods	The Operational Period	
Combat Content	The development of the breakthrough throughout the defense's entire operational depth and the defeat of all the elements of resistance at the operational level	
Depth of Pressure	The defense's operational zone to a depth of 50–60 kilometers	The defense's rear zone to a depth of 100–120 kilometers
Length	The second half of the first day of the breakthrough	The second day of the breakthrough
EA	Following the ERP's passage into the depth of the defense, it seizes the second defensive zone and deploys its front toward the flanks, in order to outflank and encircle the enemy holding out in the tactical zone	Eliminates the enemy encircled in the defense's tactical zone and moves forward to cooperate with the ERP's units in the depth
ERP	Pushes its efforts into the defense's operational zone and attacks and destroys the defender's army reserves, headquarters, airfields, supply stations, and support units	Concludes the elimination of the enemy in the defense's operational zone and, upon the arrival of the EA's units, pushes its efforts into the defense's rear zone, in order to capture the distribution station and attack the newly-arriving reserves
AGA	Operates jointly with the ERP's units in the role of their AGA	Suppresses the distribution station and prevents the concentration of new reserves, while operating in conjunction with the ERP's units
ADO	Operates on the ground in conjunction with the ERP's units	
The front's strategic aviation	Operates against the enemy's deep rear, for the purpose of preventing the gathering and concentration of new reserves at the breakthrough front and its isolation from the country's rear	

The task is now that individual groups of the enemy's combat formation be completely encircled and completely destroyed or captured, having been deprived of any opportunities of breaking out into their own rear.

At this stage the intensity of the struggle will increase even more, because the enemy is faced with the choice of life or death.

It should be assumed that the fourth stage of defeating the defense will take another three days. During these days the line of the fighting will advance all the more into the depth, and one may assume that on the third day of this period; that is, the fifth day of the breakthrough operation, the offensive's entire combat formation will overcome the defense's operational depth and reach the line of the distribution station.

At the same time, the EA's forces ceaselessly advance on auto transport to the ERP's front line, thus strengthening the attack into the depth.

Thus the entire operational depth of the defense may be overcome to 100–120 kilometers in 4–5 days, which yields an average operational offensive pace during the deep breakthrough of 20–25 kilometers per day.

However, this should not be understood as the rate of advance of the EA's forces, which, of course, will overcome a depth of no more than 10–12 kilometers on the first day of the attack.

The calculation made here speaks to the speed of the overall coverage of space in the operation and is conditioned by the fact that following the overcoming of the defense's tactical depth, the ERP immediately pushes forward its efforts into the operational depth and, as early as the second day of the breakthrough, engages it to a depth of 100–120 kilometers.

This offers an opportunity to the EA, by taking advantage of the freer maneuver conditions, to immediately push its efforts forward, broadly employing auto transport. The expenditures of the deep breakthrough operation are determined by the significant demand for munitions.

The first day of the breakthrough, including the artillery preparation and the overcoming of the defense's tactical depth, will require a particularly great expenditure of munitions, defined as an average of 2.5 combat loads.

The no less decisive second day of the breakthrough will require, on the average, 1.5 combat loads.

The third day of large-scale and intensive fighting in the depth will require, on the average, a single combat load. Finally, the fourth and fifth days of finishing the breakthrough will also require half of a combat load each.

Thus the entire deep breakthrough operation, calculated at five days, will require a total of up to six combat loads.

Given a regularly organized delivery of half a combat load per day, the concentration of these supplies would require 12 days.

However, the troops will have on hand two combat loads, which are continuously replenished, at the moment they approach the defensive zone. Besides, during the two days of preparing for the breakthrough, we will be able to bring up one combat load. Thus by the beginning of the breakthrough, three combat loads will already have been concentrated, which will completely cover the first day's fighting.

However, on the second day of the breakthrough, given the delivery of only half a combat load, there will be a shortage, which demands the intensive delivery of no less than one combat load during one of the preceding days.

In this case, the remaining required munitions may be concentrated during the subsequent days of the breakthrough.

However, given the slightest difficulties in delivery, the concentration of munitions for the breakthrough must be planned ahead of time, in anticipation of the impending establishment of a front.

In the opposite case, the preparatory period for the breakthrough must be considerably broadened in time, which will give the defense the opportunity to strengthen itself qualitatively and place the accomplishment of the deep breakthrough in incomparably more difficult conditions.

The organization of supply must thus guarantee the timely accumulation of the necessary munitions norms; for the greatest possible uninterrupted switch to the breakthrough of the front, upon its establishment, is one of the most important conditions of the deep operation.

This operation has as its goal to rout the opposing front throughout the entire operational depth, in order to deprive it of the opportunity to solidify in dead forms of positional warfare, with its roots deeply embedding in the earth.

IX. The Development of the Initial Offensive Operation

Given the normal development of events during the beginning period of the war, when along a given important axis the enemy has attempted at first, to attack, and then as a result of a meeting battle, was forced to go over to the defensive, the initial offensive operation concludes with the breakthrough of the latter, as a definite beginning stage of military activities at the front level.[25]

The subsequent course of events may be foreseen only in a general way. However, Its content determines the increase in the intensity of struggle characteristic of the new age of the deep strategy.[26]

Of course, the fate of a campaign, much less of the war, cannot be decided by a single initial offensive operation. The breakthrough of the front in the forward theater of military activities, while linked to the defeat of a particular enemy group, by no means yet signifies its ability to continue the struggle. The attacker {still} has a long and intensive road to the final goal of achieving complete victory.

Thus the initial offensive operation is only the first stage in a large and deep front operation, which must be calculated to the entire depth of the operational line defined by strategy.

At the same time, the prospect of the development of modern armed struggle is such that the initial offensive operation must directly grow into a new operation and that the completion of the first must directly lead to the beginning of the second.

The breakthrough of the front at the operational level is such a serious event for the enemy that he will naturally begin to immediately concentrate all his available reserves from his country's depth and other sectors of the front along the threatened axis.

The isolation of the defense being pierced, which is carried out by aviation, must interfere with this, being a necessary element of the entire deep strike system in the breakthrough operation.

However, it is impossible to count on the complete fulfillment of this task. The great opportunities for operational maneuver will always enable the enemy to concentrate a definite group of reserves in the depth of the front being pierced.

For the most part, this will coincide in time with that moment when the ERP will engage the defense's rear zone. Approximately on the fourth or fifth day of the breakthrough operation, when it will already be coming to a close in an intensive struggle at an overall depth of 100–120 kilometers, this new group of enemy reserves will make its concentration known beyond the line of the distribution station, which has already been occupied by units of the ERP.

A new hotbed of struggle will break out at this depth before long. At first little noticed and only gradually bursting into flame, it will actually connote the beginning of a new operation. It will be impossible to establish any kind of more or less noticeable boundary in time and space when this occurs; for the ERP's first report about encountering a new enemy will already define the moment of the immediate growth of the breakthrough into a new operation.

This will be evident significantly sooner if the enemy decides to immediately organize a counterblow. He may, however, having renounced this intention due to a shortage of forces, immediately construct a new defensive front in the depth.

In the first case, the situation will lead to a meeting collision; in the second, it will require a new breakthrough.

However, in both cases the operational formation of the breakthrough will once again immediately grow into the operational formation for the approach.

The sooner this approach is organized, the more decisive will be the new attack launched, and the more possible will be the rout of the new enemy group before it can establish its front. Any delay in this situation and indecisiveness of actions would enable the enemy to organize himself for a counterblow or to build a new front in time. Besides, if the breakthrough is not immediately organized, then all the prerequisites for the growing prospect of positional warfare in the depth are realized.

Thus here the struggle against the establishment of a solid front acquires particular significance. The immediate organization of a new approach is therefore necessary for resolving the task in the entire depth of the operational line.

The completion of the breakthrough in these conditions immediately grows into a new operation, the entry into which requires, first of all, the organization of a new approach.

In the outline of this approach, the ERP, which has already established contact with the new enemy, will automatically be transformed into the <u>AVE</u>.

The <u>EA</u>, which is moving by individual corps to the depth of the former distribution station, will once again become the <u>main echelon</u>.

And inevitably, one of the corps, which has, for the most part, been delayed along the enemy's former defensive line to completely eliminate its garrison, will automatically become the <u>reserve echelon</u>.

Thus the operational formation of the breakthrough automatically grows into the operational formation for the approach, and the breakthrough itself—into a new operation.

In this lies the dialectic of the development of the deep operation, as its forms alternate in one continuous process of one growing into the other.

The approach with a meeting battle grows into a breakthrough, the breakthrough once again into the approach, and the latter, for the most part, once again into a breakthrough, etc.

This dynamic determines the place of the long-range elements in the development of the deep operation: the AVE becomes the ERP, the ERP once again into the AVE, and the latter again into the ERP, etc.

All of this takes place in the process of the uninterrupted alternation of the deep operation's forms and their direct growth of one into the other.

It is only in this sense that one can speak of a series of consecutive operations.

Naturally, one should not imagine modern military activities as a single operation in the entire depth of the operational line.

Naturally, these activities break up into a series of operations, according to their forms and organization. However, the essence of the question is that, as opposed to the nature of the development of military activities in 1914, these operations grow directly from one to the other, that they alternate in an unbroken process of the growth of some into others and that their movement from some forms to others does not know any kind of noticeable boundary in time and space.

In this lies the major qualitative difference in the nature of the development of the modern deep operation and the nature of the development of the linear operations of the past, which consisted of separate stages, divided in time and space by definite distances and presenting a picture of a broken chain of consecutive operations.

This brokenness, due to the pushing forward of striking power immediately to the entire operational depth, is now disappearing, leading to the direct growth of one operation into another as a continuous chain of operational efforts into the depth.

However, at the same time this nature of the modern deep operation is raising the question of its scope into the entire depth of the operational line, with incomparably greater urgency than in 1914.

As is known, this problem, which has its own rich history, depends on the opportunities for delivering the materiel support means for combat and is dependent upon the restoration of the railroads.

The depth of the likely operational line in our western theater of military activities, from the state frontier to the line of the middle Vistula is a straight-line distance of 400 kilometers, and 600 kilometers in a roundabout fashion.

According to the meaning of the deep operation's nature, this length of the operational line must be overcome in a single unbroken process of the growth of one operation into another.

Naturally, this requires the organization of supply, in which the delivery of materiel supplies would keep up with the pace of the operation's development into the depth. And this is possible, insofar as the delivery of the required materiel supplies will keep pace with the advance of the front of struggle.

At the same time, we are most interested in the depth of the deep operation's scope of 400 kilometers, because this length is the shortest one to the line of the middle Vistula and also overcomes each of the sectors of the western border's main area to the north and south of the Poles'ye—to a depth equal to their length.[27] To a significant degree, this determines the overcoming of the depth in the entire western theater of military activities.

If one proceeds from an average daily norm of advance of only 15 kilometers, then a depth of 400 kilometers could be overcome in approximately 25–26 days (400 divided by 15). However, one should never count on the forward movement of the front of struggle every day.

High paces of advance must inevitably alternate with short halts of 1–2 days, which are necessary for the preparation of a breakthrough.

Thus one should count on the possibility of overcoming a depth of 400 kilometers approximately over the course of a month.

The problem is thus to what length may the railroad, upon which a given shock army is based, be restored in this period.

If one proceeds even from the most cautious and quite realistic pace of eight kilometers per day for restoring the railroads, which has actually been exceeded in our time, then in the course of a single month 240–250 kilometers of railroad (8 km × 30) can be restored.

Thus upon the expiry of a month the gap between the advanced front of struggle from the terminal railroad supply station will be defined at a distance of 150 kilometers (400–250). Of this, the automobile delivery sector, which works in shuttle fashion, covers 100 kilometers, while the divisional sector accounts for 25 kilometers.

Thus there remains a gap of 25 kilometers, which, however, is easily covered by the growing pace of restoring the railroads, or the expansion of the auto transport sector to 125 kilometers, which is quite possible for modern auto transport.

Thus a scope of 400 kilometers for the deep operation, from the point of view of the restoration of the railroads, should, in general, be recognized as possible [figure 10].

However, the essence of the problem lies not only in making sure that the restoration of the railroad keeps pace with the rate of the shock army's advance; but also that the restored railroad should have a capacity that guarantees the delivery of all the necessary materiel supplies to the shock army.

However, in this sense the prospects appear to be less favorable. As has been shown, the delivery of the shock army's daily requirements is calculated at 24 trains.

But because the railroad must retain as a minimum one-third of this number for its own purposes and restoration work, then the railroad's capacity upon restoration should be measured as no less than 32–35 pairs of trains.

But such a capacity norm is technically impossible for a restored railroad and is achieved only after a lengthier period of time.

A restored railroad yields, on the average, no more than 15–16 pairs of trains per day, and this norm cannot fully satisfy the shock army's supply needs.

Does this mean that such an obstacle arises in the development of the deep operation to a depth of a 400-kilometer operational line in the delivery of supplies that inevitably forces us to break off the offensive, thus giving the enemy the opportunity of securely establishing his front {and}, [bringing] {to bring} it to a positional condition?

It is possible that in certain very unfavorable cases, such a situation will prove to be inevitable, subsequently placing the deep operation before the necessity of manifesting the greatest intensity in organizing the deep attack.

However, the rise of such a situation is by no means necessary.

The available number of supply trains will always be able to support {with supply} such a part of the shock army that may be designated for directly extending the attack into the depth against an enemy who has already been significantly broken during the proceeding period.

This designated group from the shock army should consist of all the long-range elements of the operation (the mechanized corps, cavalry corps, the motorized division) and units from the motorized combined-arms formations. It will now form a larger and more independent AVE, the main task of which consists of launching a decisive attack and preventing the broken enemy from reconstituting his forces and building a new front.

The deep operation is thus continued at this stage by a reinforced AVE, as an independent factor, with the support of the entire mass of the AGA.

Up to a certain time, this may yield a great result.

However, one should bear in mind that the growth in the intensity of the struggle, which increases as one draws near the final goal of the operation, to its highest culminating point, will require of the shock army the employment of the entire sum of its combat efforts.

The lesson of 1920, when our army ended up during the march to the Vistula without supplies, should serve as a particular warning in this sense.

At the final and concluding stage of the deep operation, the shock army, as the chief factor at the front level, must be fully supplied, and this places extremely important tasks before the technology of restoring railroads.

In any event, the activities of the independent AVE, which has been detailed from the shock army because the latter cannot be supplied in full, must continue just as long as is required to bring the carrying capacity of the restored railroad to the necessary norms.

The further improvement in the technology of restoring the railroads must fully reduce this undoubtedly difficult and important period in the development of the deep operation, having secured at the final and decisive moment of its completion the highest and maximum intensity of all of the shock army's combat efforts.

Thus the modern deep operation must count on the uninterrupted growth of one operation into another throughout the entire depth of the operational line and the equally unbroken growth of combat efforts toward the culminating moment of achieving its final goal.

This operation is thus deep not only because it embraces the entire operational depth of resistance with its striking pressure, but because it is capable of uninterruptedly pushing this deep attack into the entire depth of the operational line assigned to it.

X. The Fundamentals of Managing the Deep Operation

The prospects of the deep operation's development point to the entire extreme complexity of its organization and conduct.

This places before modern operational art with particular urgency the problem of management, which is growing into an entire problem of enormous importance.

The management of the deep operation is determined, according to its nature and content, by essentially different conditions than the management of linear operations during the World War of 1914–18.

When the struggle was waged through the employment of all combat efforts along a single line of front, which characterized the age of the linear strategy, the essence of management consisted of distributing forces among particular groups and directing them along specific axes. Subsequently, operational events were left to their own devices and management consisted, in the best case, as a result of an objectively arising situation, of indicating a new axis and redistributing one's forces along it. However, the latter, given the absence of deep reserves (such as was the case during the offensive toward the Marne in 1914 and our offensive toward the Vistula in 1920), was always achieved with great difficulty, or could not be achieved at all.

{However}, management methods are sometimes more tenacious than operational forms.

And now, given the completely altered conditions of armed struggle, it is proclaimed from the rostrum of the French military academy that "the essence of operational leadership consists in forming columns of various designations and directing them toward a common goal."[28]

The nature of the deep operation has long grown out of such a definition of the functions of management.

Modern operational art, as the art of conducting the deep operation, must in no way limit the volume of its activity to the formation of columns; that is, to the determination of groups of forces and their direction to the goal; that is, the indication of axes.

This sphere of work more likely relates to the competence of strategy, which at the front level must group men and materiel along definite operational axes according to their significance.

At the army level operational management's center of gravity shifts under conditions of the deep operation to the sphere of organizing the coordination of combat efforts along the front and into the depth and, chiefly, in the latter dimension for the purpose of deeply striking all elements of the resistance.

To limit in modern conditions the competence of management to the definition of groups of forces and indicating their axes would be to completely eliminate the very forms of the deep operation and renounce the organization of any kind of coordination of the strike elements into the depth.

The entire essence of the deep operation consists in that it is not conducted through the employment of combat efforts along a single line of the front, but in the deep zone of the entire operational depth of the enemy's resistance.

In these conditions, the task of management consists least of all in forming columns and indicating their axes; it consists primarily of <u>unifying in organized coordination all of the strike elements throughout the entire operational depth</u>.

Thus operational management in modern conditions consists of the fact that <u>activities, which are the result of the formation and direction of groups of forces, are waged and directed in the process of the struggle</u>.

In essence, only in modern conditions of the deep operation does operational art achieve its true content as the <u>art of managing the operation during the course of the fighting</u>.

In this it essentially distinct from the operational art's content during the age of the linear strategy.

There are three chief characteristic features in managing the deep operation.

<u>The first is organization</u>. The management of the deep operation receives its new organizational content.

X. The Fundamentals of Managing the Deep Operation 153

To manage the deep operation means to organize the cooperation of the elements of the deep attack at each separate stage of the struggle.

One should note, in general, that in modern conditions the organizational [battle] {content} suffuses management at all command levels.

We find signs of organization in embryonic form as early as the management of a group of two sections.

This manifests itself to an even greater degree in the management of the joint activities of unified combat arms as, for example, at the level of a reinforced battalion.

This is conditioned by the composition of the various technical means of struggle, which are part of the armament of small infantry formations and other combat arms attached to them. A new technical base imparts a new organizational character to management in battle.

On the whole, the very organization of the engagement, at the level of unifying fire and shock in one directly connected tactical action, has a tendency to rise to an increasingly higher level.

In 1914 it was thought that the division organizes the engagement; the corps was only accorded the competence of generally organizing the engagement, which was expressed in indicating axes and lines.

Now the organization of the deep engagement is realized at the corps level, for all of the chief factors for striking the tactical depth—the long-range tank group, long-range artillery and battlefield aviation—are in the hands of the corps, which directly organizes their cooperation.

At the level of the operation, this organizational indication shifts to the level of the army command.

And in certain cases, such as during the development of the breakthrough by the ERPs of two neighboring shock armies, when the organization of cooperation in depth will be required at the front level—the indicators of organizational content will be manifested in the management of the front.

The deep strike at the operational level consists primarily in the organization of cooperation between all suppression elements through the entire depth of the resistance.

The main elements in the deep operation are long-range means of struggle, such as aviation, the mechanized corps, cavalry, and the motorized division.

These means are in the hands of the army commander, who is their direct manager.

Any employment of these means requires the assigning to each of them specific tasks and the organization of their cooperation throughout the operational depth.

It would be completely empty, from the point of view of the requirements of the deep operation, to reduce management to just the indication of the axis and line to which the combat efforts must be pushed.

The question is now to indicate, how the cooperation of all the attack elements should be organized at a given operational depth.

Three

"The Fundamentals of the Defensive Operation" (1938)

Brigade Commander Isserson
(An Outline)
Educational Section of the RKKA General Staff Academy, 1938

TABLE OF CONTENTS

Author's Note	156
Introduction	156

Part One. The Fundamentals of Organizing the Defense at the Operational Level

I. The Role and Place of Defense in Modern War	157
II. The Problem of Defense	159
III. The Evolution of Defensive Forms in the World War	163
IV. Prerequisites of the Modern Organization of the Defense	166
V. The Fundamentals of the Operational Organization of the Modern Defense	169
VI. The Engineer Outfitting of the Army Area of Defense	173
VII. The Operational Organization of Forces in the Defense	174
VIII. The Composition of Men and Materiel in the Defense	175

Part Two. The Fundamentals of Conducting the Defensive Operation

I. The Operational Conditions for the Assumption of the Defense	179
II. Covering the Occupation of the Defense	180
III. The Organization of the Withdrawal	181
IV. The Fundamentals of the Decision to Defend	182
V. The Evaluation of the Situation	184
VI. Drawing up a Plan for the Construction of the Army Area of Defense	187
VII. The Distribution of Men and Materiel in the Defense	189
VIII. The Counterblow in the Defensive Operation	194
IX. The Employment of Aviation in the Defense	198

X. The Development of the Defensive Operation	200
XI. The Rear in the Defensive Operation	204
XII. Anti-Aircraft Defense	205
XIII. The Organization of Command and Control	206
XIV. Political Support for the Defensive Operation	208

Conclusions 208

Author's Note

This work on the defensive operation is being published only as a summary of lectures and is in no way a finished and systematized work. It should least of all serve as some kind of template.

Nor does the work lay claim to an exhaustive and final resolution of all of the defensive operation's problems, and is only of a research nature.

All the specific and applied tenets regarding the organization and conduct of the defensive operation are put forth as a means of framing the question as a result of specific scientific research, and are **unofficial in nature**.

The summary is prefaced with an applied elaboration of the theme of "The Defensive Operation," based on a specific example and has as its purpose offering a few initial data for this.

The author hopes that the applied study of this theme will introduce an entire series of addenda and correctives to the resolution of the problems of organizing and conducting defense at the operational level, after which it will be possible to impart to the summary a more developed, systematized and finished character.

—Brigade Commander Isserson

Introduction

1. The conduct of modern defense at a major operational level is linked to a large volume of measures for preparing the terrain and consists of a complex system of combat activities that have a multifaceted and diverse content. These activities often occupy a lengthy period of time and are conducted over large areas and require a very purposeful combination of the struggle in time and space. They are directed, on the whole, toward repelling an enemy attack, halting his offensive, exhausting his forces, and then inflicting a final defeat on him through an active offensive blow.

This system of combat activities, which are conducted along specially prepared terrain and unified in time and space by the general idea of repelling the attacker and defeating him, should be called the defensive operation.

The World War of 1914–1918, our civil war of 1918–1921 and the ongoing wars in Spain and China offer bounteous examples of such defensive operations on an enormous scale and of enormous intensity. These operations are gaining a growing share and place over the entire course of historical events.

Their study opens before us a large and complex field of military art.

2. At the same time, it is necessary to admit that this field has been least of all studied by theory and least of all highlighted by military literature. There are definite reasons for this.

The defense is strong due to qualities exclusively inherent in its nature. The offensive, on the other hand, is a difficult task, which requires an enormous superiority in men and materiel.

It is namely the correlation of the defense and offense that has always forced us to give full preference in studying the latter and treat the study of the former; that is, defense, with significantly less attention.

And it seems that the historical course of events has justified such a tenet.

The World War of 1914–1918 concluded, in general, under the banner of the victory of the defense. Nowhere did the offensive enjoy a complete and decisive result.

The enormous significance of our civil war of 1918–1921 for the development of military art is that it triumphed under the banner of the offensive, which, overall, was conducted sequentially from beginning to end, until the final rout of the enemy.

However, all subsequent events showed once again all the difficulties of the offensive and the strength of the defense, *per se*, even independent of its forms and methods.

This, in general, is the experience of the Italian-Abyssinian war,[1] and the wars in Spain and China.[2]

3. As a result, the problems of the offensive now occupy the center of attention for military theory. All of modern military literature is basically dedicated to the offensive.

Much less is written about the defense and it remains, in general, in the shadows, lacking profound theoretical research.

Nonetheless, one should note that this situation is primarily felt in the operational field. Tactically, the organization and construction of the defense are regulated by all modern manuals, alongside the offensive. However, operationally the organization of modern defense at the army level is not being definitely resolved. And the defense during the World War has been viewed only at the tactical level.

The study of defense at the operational level presents a series of new problems, which require scrupulous and all-round study.

Part One—The Fundamentals of Organizing the Defense at the Operational Level

I. The Role and Place of Defense in Modern War

1. In embarking on the study of the problem of defense as a whole, it is first of all necessary to comprehend the place and role of this kind of struggle in modern war.

Our operational doctrine is the doctrine of the offensive.

The entire historical meaning of our just war imparts to it the character of the most decisive offensive, which will lead the Soviet people against the attacking enemy, while pursuing the goal of his complete destruction through crushing attacks on his own territory.

"Our army exists not to attack, but only up to the moment of the enemy's attack on our Motherland. The army will be the most offensive of all attacking armies that ever existed, should the enemy force it to do this."

These profound words of comrade Voroshilov define the fundamentals of our operational doctrine as the art of conducting destructive offensive operations with the most decisive goal of completely overthrowing the enemy.

2. However, it would be a serious mistake and an obvious failure to understand all of the conditions of waging a future war if we were to understand our offensive doctrine in the sense that a future war will turn, along all sectors of the front {and during all periods of waging the war}, into the single and unbroken offensive conduct of military operations, knowing no breaks or deviations, or a temporary or lengthier adoption of the defensive.

"Such wars as began and ended in an unbroken and triumphant offensive have not taken place in world history, or if they have taken place, then as exceptions," says Lenin (vol. XXVII, p. 69).

It is quite clear that a future war, given its grandiose scale, will abound in an exceptional wealth and variety of types, forms and methods of conducting military operations, among which defense, in a specific situation, cannot be an exception.

3. The more decisively an offensive doctrine is expressed, which always concentrates the force of the attack along selected axes and forces us to consciously leave a minimum of men and materiel along the secondary axes, the more prerequisites there are in conditions of the front's great length for the adoption of a defensive form of activities along certain axes and at particular periods of time.

No one ever disposes of such enormous amounts of men and materiel so as to equally conduct offensive operations along the entire length of one's strategic deployment front.

In this regard, the defense may be a component part of the strategic deployment's overall offensive plan. However, it may also be inevitable along certain secondary axes, given a shortage of forces along them for the offensive resolution of the mission.

{Finally, a situation may arise at certain times in the overall strategic situation, when the continuation of an offensive at the present time and under present conditions is impossible or inexpedient and requires a temporary or more prolonged halt, which is always linked to the assumption of the defense. Thus} defensive actions, depending upon the conditions which brought them about, may be temporary or lengthier and, finally, quite stable, resulting in a positional struggle along certain sectors of the strategic front.

4. Strategically; that is, on the scale of the armed struggle as a whole, one should examine three general cases of defense at the operational level:

a) **the first case** is when the defense of a certain axis or important area, where an offensive is in no way justified by overall strategic conditions, is a component part of the strategic deployment plan.

This defense is strategically intentional, consciously decided upon beforehand and is the most consistent strategic expression of its destiny. In this case, it is strategically a goal.

b) **the second case** is when during an offensive operation along a secondary axis a further offensive is impossible, as a result of shortage of men and materiel and, as a consequence of the overall operational situation, it is more expedient to go over to the defense here, so as to, for example, put an attack by the main forces in a more favorable outflanking position.

This prospect should always be kept in mind in any development of an offensive operation along a secondary axis. In this particularly common case, the defense, on the scale of a front operation, is a means for securing the resolution of the main task of defeating the enemy along the main axis.

c) **the third case** is when the development of the offensive operation along the main axis may require a temporary break (a so-called pause) for the accumulation of new men and materiel. Naturally, any such operation, by the very course of events, inevitably leads to the adoption of a defensive formation and a defensive method of operations.

In many cases, given the forced breaking off of the operation for a lengthier period, the defense may acquire the significance of an operational goal at a given stage and inevitably lead to the birth of positional warfare.

These three general cases may still be substantially differentiated and are far from exhausting the profusion and variety of those conditions of the situation which may render the defense natural.

However, these also show what a large and broad place the defense can occupy in modern war and what a significant role in conducting military operations it is acquiring.

In all of the enumerated cases, the defense acquires its expression at the operational level. In this regard, it requires scrupulous theoretical study and operational justification.

II. The Problem of Defense

1. The problem of defense has a rich theory on questions of its purpose and goal.

Clausewitz, in one of his tenets, calls the defense a form of struggle "by means of which we strive to achieve victory, in order to, upon achieving an advantage, to assume the offensive; that is, to achieve the war's positive goal."

The most substantial thing in Clausewitz's tenet is that "**they strive to achieve victory**" through the defense; that is, to inflict a physical defeat on the enemy. Clausewitz does not speak about holding terrain as the chief essence of the defense. In the defense the retention of terrain is a derived element, expressing the practical methods of conducting it. It is impossible to put up resistance without holding terrain and impossible to exhaust and halt the attacker. This is what distinguishes the defensive from the offensive that the former's expression is the retention of terrain and the latter's expression is conquering terrain.

Strategically, the retention of an important area (political, economic or strategic) is often the main goal in the defense. Then the retention of terrain acquires the strategic significance of a goal. Such is the case, for example, in the defense of Madrid.[3]

However, one must distinguish the aim being pursued by the defense as a whole, and the goal of combat operations, by means of which the defense resolves its mission.

In all cases, the infliction of a defeat upon the enemy remains the main goal of operations in the defense, for without having inflicted a defeat upon the enemy, it is impossible to hold terrain. In this regard, it is necessary to clarify for oneself the internal goal, inherent to its meaning, of the defense. This goal consists of defeating the attacker, who, given the modern power of defensive fire, must be put in the condition of a "mass shooting" or "mass murder."

It is especially important to emphasize this now, because during the World War the concept of defense was substantially distorted.

Falkenhayn,[4] for example, expressed the entire strategic essence of the defense in 1915–1916 with the words: "**Hold what you have and don't give up an inch of what you have gained.**"

For the twilight of military art in the World War and the bankruptcy of the German General Staff, this was the most characteristic emasculation from the defense of the very essence of its destiny, which in all cases must first of all consist of the physical destruction of the attacking enemy.

2. If there is a shortage of men and materiel for defeating the enemy through offensive means, or if this is inexpedient due to the overall situation that has developed, **then this defeat should be achieved through defensive means, forcing the enemy to break himself against an organized resistance and then, upon having achieved an advantage, to finish him off by an active attack, which, on the whole, crowns the defense.**

Our field manual puts forth this idea, saying that "Any engagement—offensive and defensive—has as its goal the infliction of a defeat upon the enemy" (*1936 Field Manual*, article 2), thus restoring the true meaning of the defense.

According to its final aims, the defense should therefore also pursue decisive goals, although achieved by other ways and methods.

The choice of these other ways and methods is chiefly due to the shortage of men and materiel along the given axis for an offensive, or by its overall inexpediency in the given situation.

However, the complete defeat of the enemy is achieved only by the offensive, as "only a decisive offensive along the main axis, concluded by an unbroken pursuit, leads to the complete destruction of the enemy's men and materiel" (*1936 Field Manual*, article 2).

3. A more complex problem is defining the content of the defense and the forms and ways of conducting it.

The defense, according to the forms and ways of conducting it, has an incomparably more complex, diverse and versatile content than the offensive. The offensive expresses itself in only one form—in the form of the attack. One can attack only by attacking; that is, only in the form of offensive actions. Even strategically this sometimes acquired its full expression. For example, in 1914 the Germans on the Western Front and the Russians on the Eastern Front carried out an offensive along the entire front of their strategic deployment.

The matter stands otherwise in the defense. One may defend using the most varied forms of combat activity, beginning with a stubborn passive resistance in place and concluding with the most decisive offensive.

4. In all cases, a single overall indicator unites the defense, which fundamentally distinguishes it from the offensive. Clausewitz defined this overall indicator as follows:

"What does the concept of defense consist of?—In the repulse of the attack. Thus, what is its indicator? The indicator of waiting and repelling is inherent in the concept of defense, as a whole."

However, the content of this indicator may be embodied in quite various forms.

One many foresee at least three main forms of defense:

a) A decisive defense.

Example: the German defense of East Prussia in 1914.

b) A holding resistance with a subsequent launching of an active blow.

Example: The French withdrawal to the Marne in 1914 and the launching of a counterblow against the Germans' right flank.

Our civil war of 1918–1921 offers many examples of such a defense. For example: the operation against Yudenich[5] and Denikin's[6] defeat around Orel.

c) A stubborn resistance along prepared terrain, with a subsequent counterblow, or without it.

These three typical forms of defense do in no way exhaust the entire abundance and variety in which the resolution of the defensive mission is generally possible. However, they embrace the two polarities in the resolution of this problem: from the decisive offensive to positional passive resistance.

5. One must distinguish resistance, consecutively and systematically moved back to prepared lines in the rear as a separate type of defensive form. This defense, known by the name of **mobile**, acquires its true expression only at the operational level, because any kind of tactical defense, when it is conducted, is stubborn in nature.

In operational mobile defense one must distinguish resistance, which upon being moved back, retains the goal of launching a counterblow against the enemy as an act that crowns, on the whole, the defense.

One must distinguish from this true form of operational mobile defense the defensive withdrawal maneuver, which sees its only mission as pulling its forces out from under an attack, in order to not accept the attack and avoid it altogether. Such a type of withdrawal maneuver were the actions of the already broken and defeated German armies along the French front, beginning in August 1918, when they first abandoned the Marne salient on August 4, falling behind the Aisne and Vesle rivers; then they abandoned the Amiens salient on August 30, and in September fell back to the Siegfried position and, finally, to the Hermann—Brunhilda position.

This was no longer defense from the point of view of inflicting a defeat upon the enemy, but rather a withdrawal, relying on previously prepared positions in the rear. If one calls this type of activities mobile defense, then the latter is in any case distinguished from defense in general by the fact that it does not set out to force the enemy to break himself against one's resistance and does not think about inflicting a defeat upon the enemy, while striving only to get out of the way of the attack, so as to avoid it altogether. Therefore, it is more correct to call this type of activity a defensive withdrawal maneuver.

6. A defense, conducted as an offensive or a consecutive holding resistance, according to its operational-tactical content, goes beyond the bounds of the defensive form of activities in the direct sense, for its content is either a direct offensive or an entire combination of maneuver activities, among which the direct defensive type of actions has only an insignificant share. For example, the French armies' withdrawal maneuver to the Marne in 1914.

In the modern understanding, the defense, as a problem having its own independent and special content, which essentially distinguishes it from other types of combat activity and which also requires special organization, consists of **stubborn resistance along previously occupied and prepared terrain.**

The defense acquires its most expressive nature and most complete content in organizing and conducting this resistance.

At the same time, this type of resistance and repulsing the attack imparts to the defense the significance of a special, independent type of combat activity, which requires a special organization and a special tactics; that is, it transforms it into a special problem with an independent organizational and operational-tactical content.

It is precisely this type of defense that puts forth special demands on modern military theory for its study and resolution. It is namely in this understanding that the defense represents a complex operational problem.

7. The tactical organization of the defense has a specific resolution. The possibilities and power of the tactical defense are beyond doubt. However, on the whole, the operational organization of the defense still leaves many unresolved questions.

What should the operational organization of the defense look like? In what operational forms should it be expressed? What should be the operational methods of conducting the defense? All of these questions are still seeking a definite answer.

At the same time, the necessity of imparting definite operational forms to the defense is beginning to be felt all the stronger. If defense in the recent past was chiefly viewed tactically, then now we are beginning to impart significantly greater importance to defense at the operational level. For the first time the new French manual of 1937 for employing major troop formations includes the chapter "The Army in the Defense," and attempts to define the fundamentals of organizing the army in the defense and the place and employment of army reserves as an organic component part of the defense as a whole.

This urgent requirement is understandable. One cannot count on resolving the entire problem of the defense within tactical bounds only; that is, through the tactical organization of the defense only. Not the fault, but the inevitable misfortune of the defense usually lies in the fact that it is intentionally conducted with small and insufficient forces. In these conditions, the tactical organization of the defense alone cannot resolve the defense's missions as a whole. The resolution of these missions should be looked for in the specific operational organization of the defense as a whole; that is, it is necessary to impart to the defense a definite operational organization and definite operational forms.

8. It is necessary to keep in mind that the organization and conduct of defense during the positional period of the World War was determined by special conditions. Then the defense, while embracing the entire strategic front for entire periods, was conducted by enormous amounts of men and materiel, thus supporting the concentration of all efforts in the immediate tactically organized resistance. Thus the organization of defense in the World War was, in general, resolved at the level of organizing tactical defense zones, each of which acquired completely independent significance as long as the defense was waged along it.

The fortification of the theaters of military activities during the World War was vastly developed into the depth. On the Russian front these fortifications had an overall depth of up to 150 kilometers to the north of the Poles'ye in 1917. However, the essence of the problem was in the fact that these fortifications consisted of individual defensive zones, each of which acquired significance only after the one lying ahead fell, remaining as dead capital until that time. On the whole, these fortifications did not represent a connected, organically united operational system in depth. They did not yet determine the operational organization of the defense as a whole and were only the sum of individual deeply-echeloned defensive zones of tactical significance.

Modern conditions present substantially different demands on the organization of the defense. However, in order to understand their prerequisites, it is necessary to at least examine in a cursory manner beforehand the evolution of the forms of defense during the World War and the results which it led to.

III. The Evolution of Defensive Forms in the World War

1. The armies entered the war of 1914 with the most primitive conceptions about defense as resistance conducted from one main position. When the thin line of positions immediately showed its poverty through the entire course of events, the forms of defense began to quickly adapt themselves to the new conditions and went through a complex evolution during the war years. This evolution unfolded under the banner of the development of two basic factors:

a) the development of the defense into the depth, and

b) a change in the very tactics of defensive activities.

The consistent development of these two basic defensive factors arose from the moment the front stabilized.

2. In 1915 the defensive system represented two positions, each of which consisted of two lines at distance of 100–150 meters from each other. The second position was located at a distance of up to two kilometers from the first.

Thus the entire depth of the defensive system in 1915 was barely 2½-3 kilometers.

Although at this time the controversy over the group system of positioning troops had arisen, in general, the positions were purely linear in character.

The experience of 1915 showed that the depth of the defensive was completely insufficient and that the first line, for the most part, could not be held at all and fell rapidly. The question immediately arose as to the necessity of developing the defensive system into the depth. The French proposed erecting three positions, each 4–5 kilometers from each other. At the beginning of 1916 the Russian command issued its "Instructions for Fortifying Positions," in which it was stated that "the defense must not represent thin lines, but a fortified area of such depth that it bring about the exhaustion of the enemy's materiel and moral forces."

At the same time, it was proposed to have two positions in each army—a forward and a rear one, separated by a distance of 12–15 and even 30 kilometers between them. Each position was supposed to consist of two zones at a distance of 3-5-8 kilometers from each other, thus requiring that the attacker completely shift his artillery each time. Each zone was to consist of three lines, with an overall depth of 1½ kilometers. The Russian command's instructions doubtlessly reflected very correct and progressive views on the development of the defense. However, their requirements were not fully realized.

3. In 1916 the defense was noticeably developed in depth. In general, views were

put forward that the defense should represent a "fortified area," broken up into zones, each 4–5 kilometers deep. The "group positioning" of troops, along with definite centers of resistance, lay at the heart of occupying the defense. In general, the defensive system consisted of two zones at a distance of 4–10 kilometers from each other and having an overall depth of up to 12 kilometers.

At the same time, as was the case in 1915, the defense's main forces were concentrated in the first zone.

At the same time, the combat experience of 1916 fully confirmed that the shifting of the resistance's center of gravity to the first defensive zone was not viable, because, for the most part, it fell quickly, causing enormous losses among the defender's personnel. The necessity of shifting the main efforts of the resistance to the second and subsequent defensive lines had become very obvious, in order to meet an already disordered and exhausted enemy with the most decisive resistance and a counterattack.

Therefore, alongside the development of the defense into the depth, the center of gravity of the resistance being offered was also moved to the depth.

4. In 1917 it was established that the first line should only be occupied by sentry posts in fortified strong points, consisting, on the whole, of the forward field or a covering zone up to 1,000 meters deep. Further on a combat zone for the main resistance would be organized to a depth of up to three kilometers and, finally, a rear zone would be fortified three kilometers from it. Before long, the distance between the main and rear zones rose to 6–8 kilometers. At the same time, defensive tactics changed radically and the tendency to shift the decisive resistance into the depth was expressed even more.

The instructions of 1917 pointed out that the essence of defensive tactics was "not the struggle in the first line, but the struggle for the first line and the area around it." Further on it was stated that "the engagement should not be conducted around fortified lines, but in the combat zones," and that "the defense should be active within the confines of the defensive zones, without binding the troops to the fortifications." At the same time, the possibility was admitted of abandoning individual defensive sectors, which, however, was not to lead to a withdrawal engagement. The instructions made the reservation that "the division commanders may make the decision to withdraw in an extreme case," and that "this cannot depend only on local conditions, because it is reflected in the neighboring sectors' situation."

Finally, in 1917 a new type of counterattack appeared—the so-called "counterattack from the depth," when significant infantry groups (sometimes several divisions) concentrate from the depth, at a distance of 8–10 kilometers from the front line and attack from the second zone, under the cover of artillery, after the enemy has overcome the first zone.

Thus in 1917 defensive tactics underwent very substantial changes.

5. In 1918 this tendency was developed further.

The Germans' forward zone reached more than 1,000 meters into the depth, finally reaching up to 4–6 kilometers. The entire defensive system took on the character of successively echeloned zones of resistance, consisting of at least three, with an overall depth of up to 15–20 kilometers.

At the same time, the garrisons of the covering zone operated as forward posts in the security area and were supposed to fall back at the first signs of an enemy offensive. On the whole, the idea that "the construction of positions should not bind the troops in place and should enable them to conduct a mobile and active engagement" was confirmed. At the center of the requirements were the freedom of movement and the principle of elastic and flexible defense, based upon the unexpected growth of resistance in the depth. The German instructions of 1918 even stated the following: "In case of doubts as to whether the garrison should fall back or attempt to hold its positions, it is better overall to prefer the withdrawal; in any event, this will be better than suffering useless losses due to stubbornly holding this or that sector of the position." These instructions already reflected the German army's complete loss of combat capability. It is, however, of interest from the point of view of the radical change of all defensive tactics. In general, the defensive system in 1918 already fully gave the definite impression of a deep zone of tactical resistance.

At this point the evolution of defensive forms and tactics in the World War came to an end.

6. The French, German and Russian defensive systems had their natural differences and shadings, although, as a whole, in viewing the entire process of the development of defensive forms in the World War, one may make the following conclusions:

a) the defensive system grew from a thin line into a deep tactical defensive zone of struggle;

b) the center of gravity of the resistance was shifted from the forward line into the depth; the resistance itself was supposed to remain hidden from the enemy and to [grow] {arise} unexpectedly for him in the depth;

c) the entire development of defensive forms did not extend beyond the confines of tactics and achieved only tactical expression;

d) in the final analysis, everything was reduced to only the fight to hold the tactical defense zone and within its confines; reserves were only assigned to resolve this task;

e) the very accomplishment of the defense's tasks was determined by the retention of the tactical zone; when this zone was penetrated and fell the defense was pulled back and organized a new tactical resistance;

f) all of the defense's operational possibilities were exhausted with the fall of the tactical defensive zone and a new concentration of reserves was required, which would stitch up the breakthrough already outside of the defensive system, as a completely new operational measure.

7. If we trace these points against the example of the Germans' March 1918 offensive, we arrive at the following conclusions.

Along an 80-kilometer front 62 German divisions, supported by an enormous mass of artillery, attacked the English forces that were two times smaller. The English had 31 divisions along the breakthrough front, of which 21 were in the first line, occupying the defensive front, with ten divisions in the army reserve in the depth. On the fifth day of the attack the English defense fell and all the army reserves, which had been used in the fight to hold the tactical defense zone, had been used up and a complete breakthrough up to 15 kilometers wide was formed. In this way all the possibilities of two English armies were exhausted and no operationally organized defense could any longer be erected. From this moment the organization of the defense had to be accomplished in a new strategic context, on the scale of a strategic regrouping along the theater of military activities as a whole. Two new French armies—the First and the Third— under the command of Gen. Fayolle,[7] were concentrated at the point of the breakthrough and stitched it up. The accomplishment of this task required the concentration of 45 infantry divisions and six cavalry divisions. Thus when the tactical possibilities were exhausted and there was no operational organization of the defense to be found, strategy immediately entered the picture. This was possible in the conditions of the World War, when positional war was waged not as a means for resolving the main task, but as the main task itself and the goal of the struggle along the entire theater of military activities. The entire front defended, while at the same time disposing of enormous amounts of men and materiel, which each time were able to flow into those sectors where the tactical defense was being penetrated.

8. Summing up, we arrive at quite an interesting conclusion on the nature of defense in the World War. This defense was conducted at the tactical level and then immediately at the strategic level. It did not have an operational organization, expressed in a definite army organization of the defense. This is why during the breakthrough of the tactical defense zone the command had to immediately take up strategic measures at the level of the entire theater of military activities. This leap from tactics to strategy was inevitable in the absence of any kind of organization of the defense at the operational level. To be sure, there were operational reserves, but they were employed only in the fight to hold the defensive front itself. And when this front fell this meant that all of the operational reserves were expended and all the operational possibilities at this level were exhausted. The army command could no longer undertake anything and strategy had to come immediately to the rescue. At the same time, if the 32 English divisions, which could still have been reinforced by the immediate operational reserves, had fought in March 1918 not for the tactical defense zone alone, but had had a definite operational organization and could have relied on an entire system of an operationally-organized army defensive area, then the overall strength of their operational resistance would undoubtedly have been expressed otherwise and would probably not have allowed such a major success by the Germans, who broke through, as is known, in this operation to a depth of up to 60 kilometers all the way to Amiens.

The necessity of imparting a definite operational organization and system to the defense arises from all of the defensive conclusions during the World War.

9. Of course, the problem of the operational organization of defense cannot be resolved in the modern wars being waged in Spain and China with relatively limited means, a lesser density of forces and an insignificant depth of echeloned forces. The system of defense being employed in Spain, both by the interventionists and the Republican army, has an obviously expressed linear nature. In the Republican army the entire main defensive zone consists of several lines of infantry trenches, echeloned to a depth of 2½–3 kilometers. A second zone, for the most part, is absent altogether. This, of course, cannot be a model, for a struggle of multi-million masses, supplied with all types of modern equipment, and requires most insistently the radical resolution of the problem of the operational organization of defense in a major future war.

IV. Prerequisites of the Modern Organization of the Defense

1. In studying the prerequisites of the modern defense, it is first of all necessary to keep in mind the fact that the correlation of the defense and offense in relation to the leading and decisive place has now radically changed.

Part One—IV. Prerequisites of the Modern Organization of the Defense

Given the weakness of suppression weapons in the first years of the World War and the enormous strength of fire in defense, the defense occupied the leading place and moved ahead of the offensive, according to its forms. The defense was the deciding factor and the methods of attack had to adapt themselves to the forms of defense.

This correlation has changed in modern conditions, although the tactical possibilities and power of the defense have undoubtedly grown.

All of military art following the World War developed under the banner of a search for new offensive methods and forms; that is, the radical solution of that problem which had not been resolved during the World War.

The new technical means of struggle have doubtlessly increased offensive possibilities and erected new foundations under the methods of attack, giving birth to the deep forms of attack as the simultaneous deep suppression of the enemy's entire combat formation throughout the entire operational depth.

This is now forcing us to construct forms of defense, depending upon the means of attack, which in modern times has become the leading and deciding factor.

2. It is known that the forms of the deep offensive are now being subjected to a major reevaluation. One must bear in mind that this primarily means the experience of the war in Spain.

It is not our task here to examine this problem. It is only necessary to point out the fact that one should not fully judge as to the strength of lack thereof of the deep operation's forms according to the events of the war in Spain, because no one has employed these forms there and because there has not been and there is not that mass of the technical means of struggle—artillery, tanks and aviation, without which the realization of the deep forms of struggle is impossible.

The deep operation is, first of all, an operation by an enormous mass of artillery, tanks and aviation, which support the infantry's deeply echeloned combat formations.

Only there, where these means will be represented in the necessary mass saturation, will the defeat of the entire operational depth become possible.

At the same time, the massed concentration of these means of struggle may make their employment on the basis of the deep strike objectively inevitable and logical.

It is difficult to presuppose what other employment the masses of aviation and tank formations may find in the breakthrough. While they are few, the employment of their efforts is realized only in the tactical depth of the defense. This, in general, is what is happening in Spain. But when the saturation of these means becomes massive, then one must expect, given the full development of a future war, that the spread of their effects throughout the entire operational depth of the defense will become inevitable, even despite all sorts of theories.

3. In regard to the employment of large masses of aviation, which isolate the breakthrough sector to a great depth and prevent the flow of new reserves to it, this cannot call forth any serious doubts. As regards independent tank formations (German armored divisions), in major breakthrough operations it is doubtful whether another place will be found for them other than in developing the breakthrough into the depth through a breach in the front. In any case, the fascist theory of total war is quite clear about this.

Gen. Guderian,[8] in describing the attack by a tank brigade, establishes its employment as follows:

> The first echelon passes directly through the tactical depth of the defense and attacks its reserves.

The second echelon attacks the enemy's artillery.

The third echelon attacks the infantry within the confines of the tactical depth of the defense.

In this case, in Guderian's opinion, the employment of armored divisions will be particularly effective, when the defense has already been cracked along a particular sector of the front and when the unexpected appearance of tanks at this moment will enable them to immediately penetrate beyond the defensive zone and achieve freedom of maneuver.

4. In any event, one of the most important prerequisites of the organization of the modern defense is the effect of offensive means throughout its entire operational depth and the extremely rapid increase of a crisis in the event of failure along some tactical sector of resistance.

During the World War the breakthrough to a depth of 5–6 kilometers did not yet bring about a crisis at the operational level, for this breakthrough could not as yet call forth immediate operational consequences, while the defensive reserves disposed of no fewer possibilities for plugging this breakthrough than the attacker had to take advantage of it.

Conditions are now developing otherwise.

Tank formations, upon appearing from the depth, may not only immediately take advantage of the accomplished breach, but may unexpectedly and more rapidly create one, if the defense is still maintaining its position with its last forces. In this way the crisis of the defense may mature much more rapidly than earlier and the operational depth of the defense may be much more rapidly subjected to an attack.

From this it is quite obvious that **if the operational depth of the defense should turn out to be unorganized for resistance and not have a definite operational organization, then the defense, as a whole, will be much more rapidly subjected to the threat of complete defeat than was the case earlier.**

Modern defense cannot be guaranteed against a rapid and unexpected breakthrough into its depth by large groups of tanks and must not be disorganized by this one fact alone.

This circumstance is one of the most important prerequisites for resolving the problem of organizing modern defense. Our 1936 field manual points this out, and says in article 225:

"In modern conditions the defense must oppose the enemy's superior forces attacking immediately throughout the entire depth."

5. From this it is evident that the organization of the modern defense cannot be resolved at the tactical level of defensive zones alone, even though they are developed in depth. Insofar as offensive forms consist of the deep strike, the defense must counter with a highly defined system of opposition throughout the entire operational depth, consisting in its operationally organized resistance of a single organic whole. This leads us directly to the concept of the **operational area of defense, with a highly defined organization of its construction.**

6. This tenet should in no way lessen our attention to the organization of the stability of the defensive front itself. The defensive front directly opposed to the attacker takes upon itself the greatest force of his attack, against which it is called upon to put up the chief resistance. The strength of this attack has increased all the more in modern conditions, because besides the enormous masses of artillery, tanks and infantry, aviation is also drawn in.

One of the chief conclusions from the war in Spain is that aviation takes part each time in the preparation of the breakthrough and in its conduct, operating on the battlefield, bringing about major results, and directly affecting the outcome of the offensive.

Overall, up to half of the entire number of combat sorties by both sides' aviation in Spain is directed against the ground troops and the positions occupied by them.

In certain operations, such as the Republicans' Madrid operation of July 1937, for example, the government bombers carried out 80% of their work over the battlefield.

As a general rule, aviation in Spain takes part alongside the artillery in suppressing the forward edge of the defense and in its depth.

The interventionists' bombers drop a large amount of fragmentation bombs along the axis of the main attack and along a limited sector of the first line of defense, and when the infantry attacks it concentrates its efforts against the defense's immediate reserves.

A strike against the defensive zone is often carried out in several echelons:

The first echelon bombs the enemy positions, with its main task to draw the fire of the defense's anti-aircraft batteries upon itself.

The second echelon, following behind at a distance of 5–8 kilometers, has as its primary task the suppression of the defense's anti-aircraft batteries.

The third echelon, consisting of the main air shock group, bombs infantry dispositions, artillery positions, command posts, and the immediate reserves.

The interventionists' bombers often fly over the Republicans' defensive lines for several hours and even for entire days, keeping the defenders in a stressful situation the entire time.

Finally, area bombing with incendiary bombs has been widely employed. There was such a case during the rebels' offensive on Bilbao in May 1937.

A large wooded area held by the Republicans was completed burned, which forced the defenders to abandon an entire defensive sector.

All of this subjects the main defensive zone, along which the main resistance is offered, to incomparably greater pressure than was the case earlier, while requiring special measures and greater resistance, particularly as regards anti-aircraft measures and defense against fires.

Thus the depth of the defensive front also places special demands for stability on itself.

V. The Fundamentals of the Operational Organization of the Modern Defense

1. The following principle requirements must lie at the basis of a developed system of the operational organization of modern defense:

a) The attacker should not be given the opportunity to approach unhindered the forward edge of the defense.

A deep forward resistance zone is necessary, forcing the enemy to expend significant efforts on overcoming it and leaving him until the very last in ignorance concerning the actual outline of the forward edge.

b) The immediate tactical resistance being opposed should represent a **deep zone** of continuous resistance, designed for the complete physical and moral exhaustion of the attacker.

This zone, which one could call the main one, should represent a complete and organically connected tactical system, allowing freedom of maneuver and active actions within its confines.

c) This main zone should be enclosed in depth by an obstacle zone, which has as its chief task the blocking tank units' access into the operational depth, in case they should break through the tactical depth of the defense.

d) The breakthrough of the main defensive zone by tank units should not lead to the smashing of the defense at the operational level, and the units that have broken through should not be allowed freedom of maneuver in the operational depth. For this, the operational defense zone, which lies behind the main zone, should be outfitted with a system of anti-tank areas and obstacles and, in its turn, be joined by a rear defensive line, which localizes the breakthrough in the operational depth and covers all of the army rear areas.

2. On the whole, the entire defensive system at the operational level should represent an area of defense as a unified operational system of resistance, which encounters the enemy's breakthrough in an organized manner in the entire operational depth, no matter where he has penetrated.

This area of defense should be a labyrinth for the attacker, in which he will encounter resistance and in which he will be bounded on all sides and localized. This area should thus be an operational trap, which is laid down for the enemy.

The very organization of this area should force the enemy in the depth to develop his attack not as he had planned it, but how it has been predetermined by the entire system of lines and positions created.

Just as rocky cliffs sticking out into the sea determine how storm waves will break against them; the modern operational defense area should determine how the attacker's combat formation will break. It is namely in this regard that the defense should and must completely impose its will on the attacker.

On the whole, the modern defense zone is calculated to completely exhaust the attacker's physical and moral forces; that is, his defeat and then the destruction of the enemy breakthrough by active counterblows, a combination of which in the most various ways must be secured by the entire system of the area's operational organization.

Therefore the defense should be operationally conducted to a great depth in the entire area.

3. **The army defense zone**, representing a single overall operational system, should consist of a series of echeloned zones (figure 1):

The forward defense zone (obstacle zone). Its purpose is to delay the enemy's approach to the defensive front and to hide from him the true location of the forward zone. The 1936 field manual suggest a norm for the depth of the obstacle zone at up to 12 kilometers (article 227). However, having important operational significance, the forward zone should be organized to a great depth—up to 20–25 kilometers—given the slightest opportunity. The most important axes in the forward zone should be crisscrossed by built-up areas and strong points echeloned in depth and defended by obstacles and contaminated areas.

The main defense zone. Its purpose is to put up stubborn defense against the attacker, to force him to break upon a deep system of fortifications, to exhaust his force and support the infliction of a defeat on him by an active offensive attack. The depth of this zone is up to 15–20 kilometers.

The main zone should basically consist of two defensive zones:

a) **The main defensive zone**, which is designed to put up the main resistance to the attacker, represents a closed and continuous fortified zone. Its depth is determined by the depth of the division's combat formation in the defense; that is, 5–6 kilometers.

b) **The second obstacle defensive zone** is called upon to localize the enemy's breakthrough of the main defensive zone and to bar his entry into the operational depth of the defense.

The obstacle defensive zone must bar entry into the operational depth of the defense to the enemy's tank units, which are developing the breakthrough through the piercing of some sector of the main defensive zone, which is why it is important to station it behind a natural anti-tank line. The 1936 field manual calls this the rear zone (article 227). However, this is not entirely correct on the operational scale of the entire defensive area, because the second defensive zone is still not the rear zone for the entire army defensive area as a whole. The second defensive zone should be located at a distance of 8–10-12 kilometers from the main defensive zone, depending on the terrain conditions and the nature of the defense, while requiring that the attacker shift of all of his artillery and regroup his forces. By its nature, the second defensive zone may represent a broken fortified zone, which stands astride with powerful centers of resistance the main axes leading into the operational depth of the defense. Given the slightest opportunity, the second defensive zone must be transformed into the same kind of continuous fortified zone as the main defensive zone.

The main and second defensive zones should not represent two independent defensive factors. On the whole, they represent a single and uninterrupted defensive zone, calculated to inflict the greatest amount of damage to the enemy within its confines, to localize his offensive and exhaust his offensive strength. Therefore the main and second defensive zones should be linked by an entire system of switch and auxiliary positions and obstacles, which constrict and outflank the enemy in the event of his breakthrough along individual sectors. Similar to compartments in a submarine, which wall off the spread of water, the switch positions must hinder the spread of the enemy in the main zone, if he has broken in, and draw him into a fire sack. At the same time, the main zone's switch positions should serve as the jumping-off lines for a counterblow. In the depth of the main zone the main axes leading to the second defensive zone must be covered by a system of obstacles (chiefly anti-tank), anti-tank areas and sectors prepared for contamination. Thus the depth of the main zone should present the enemy breakthrough with completely unexpected combinations, allowing the defense to maneuver with fire and movement. In conclusion, this is a single unified system.

The operational defensive zone. Its purpose is to localize the enemy's breakthrough in the operational depth of the defense, to hinder its spread there and to create favorable conditions for launching decisive counterblows from the depth.

The operational defensive zone should represent a system of anti-tank areas, switch positions and switch anti-tank lines, constricting the most likely axes for the development of the breakthrough by the enemy. It should be closed off by a rear army line, located 20–25 kilometers from the second defensive zone, depending upon the terrain conditions in the depth, and the nature of the defense and its tasks.

172   Three—Isserson's "The Fundamentals of the Defensive Operation" (1938)

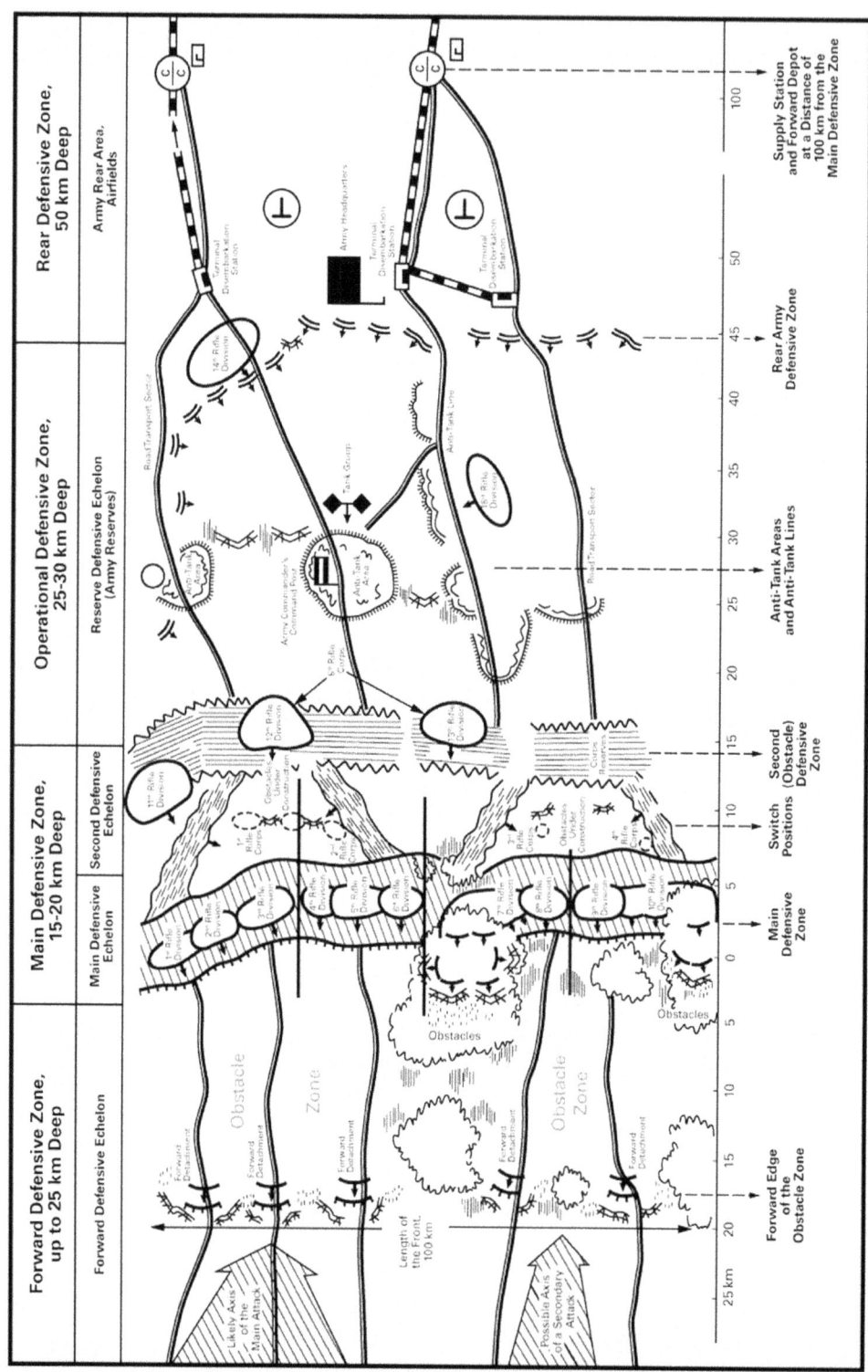

Figure 1. Outline of the Army Defensive Zone.

Thus the overall depth of the operational zone may be measured at 20–25 kilometers.

The rear army line will, for the most part, have the shape of a broken fortified line, which should be developed and improved all the time. Given time, men and materiel, we should try so that the army rear zone is fully developed and in readiness. The great importance of the army rear zone is that it covers the army rear, the forward depots, supply stations, airfields, and army headquarters.

Because the operational defensive zone is an operational one; that is, calculated in particular to take on the enemy's tank units that have broken through, it must be free of the presence in it of any kind of rear units, installations and airfields. Only landing strips for troop aviation and fighters are to be allowed. The operational zone is the area for the location of the army reserves and plays the role of a bridgehead for concentrating new reserves from the depth and the organization of counterblows.

Therefore, the overall operational depth of the army defensive zone, counting from the forward edge of the main defensive zone to the army rear line, can be measured up to 50 kilometers.

The rear defensive zone. Its purpose is to hold all of the army's rear organs, forward depots, supply stations, airfields, and the army headquarters. Because military operations may spread as far as the army rear line, then the supply stations should not be in its immediate vicinity and must be at least 50 kilometers from it, thus being, on the average, up to 100 kilometers from the defensive front and, in some cases, even further.

All of the rear zone's main targets, such as supply stations, the rear units' areas and auto transport, and airfields, etc., should be fortified and defended.

Thus the army defensive area represents a deep bridgehead, consisting of a series of zones with a definite purpose and with an overall depth, counting from the forward edge of the defense to the supply stations, of up to 100 kilometers and, counting from the forward edge of the forward zone, up to 125 and more kilometers.

VI. The Engineer Outfitting of the Army Area of Defense

1. It is quite obvious that the army defensive area represents a complex and powerful engineering system, erected upon definite general operational foundations.

If one calculates the length of the army's defensive front at approximately 100 kilometers, then the entire area from the forward edge of the main defensive zone to the army rear line occupies and areas of 5,000 square kilometers. The length of all the area's fortified sectors comprises about 350 kilometers.

It's quite obvious that their construction is a very complex task and requires enormous numbers of men and materiel.

If one believes that the main defensive zone should be at 100% readiness, the second defensive zone at 50%, and the rear army line at 10%, then it would take about 8–10 days in all to carry out the entire volume of work, on the condition that about 150 battalions of laborers were allotted for the job.

This enormous detail would require the daily diversion of up to 100 battalions of laborers from the troops, which comprises more than ten divisions of infantry. The remainder of the laborers could be made up for by levies from the local population to the tune of ten battalions. Besides, the mechanized equipment (graders, trench diggers, compressors, and excavators) could enormously compensate for the labor force and, given the modern army's authorized complement of this equipment, is approximately equivalent to 40 battalions of infantry. Nonetheless, even in these conditions, the required detail from the troops remains quite significant and will not always be feasible for the army constructing a defense. The outfitting of the army defensive area will thus be a prolonged process, continuing throughout the entire period for conducting the defensive operation. The sectors being fortified should be constantly developed and improved, thus achieving, in the end, a very high degree of readiness. Any break in combat activities, even at night, between the days of fighting, should be used for this purpose.

2. The requirement for engineering equipment necessary for strengthening the army defensive area is expressed to an even more significant degree.

History shows, for example that in order to strengthen the "Siegfried" position up to 300 boxcars of materiel was expended for every kilometer of fortified front.

According to modern calculations by foreign armies, up to 60 trains of construction material are needed for the stable fortification of a divisional defense zone.

In general, one should assume that the following are needed for one kilometer of fortified front:

- 165 tons of material for hurried fortifications;
- 500 tons of material for durable fortifications, and;
- 3,400 tons of material for concrete fortifications.

The army's minimal daily requirements for defensive work are calculated at one boxcar of material per kilometer of front, or per working battalion.

Given the daily work of 150 battalions, this amounts to 150 boxcars (four trains), with an overall weight of 2,400 tons (the weight of one boxcar is 16 tons). The dirt-road transport of this material from the supply station to the troops would require four auto battalions (ZIS)[9] daily. The cost of the material required for one day's work would be an enormous sum worth several million rubles.

From this it is clear that in this regard the idea of the defense as the cheaper means of struggle by no means corresponds to modern reality. It is also false that the delivery of supplies in the defense will be easier. Quite the opposite, in going over to the defensive, the delivery of supplies will inevitably have to work at a greater volume and with greater intensity.

VII. The Operational Organization of Forces in the Defense

The plan for the organization of the army defense area and the deep echeloning of its zones determine that the operational organization of forces in the defense should be deep and consist of several operational echelons with a definite task. This deep operational organization should include:

The forward defensive echelon is supposed to fight in the forward defensive zone.

It may consist of entire cavalry formations, if they are part of the army; or of individual forward detachments, removed from the forces occupying the main defensive zone.

The main defensive echelon is supposed to put up the main and stubborn resistance along the main defensive zone. It consists of the main mass of the defending forces and equipment, occupying a continuous and close defensive front.

The second defensive echelon is to bar the enemy's entry into the operational depth of the defense, support the fight in the main defensive zone, and be employed for active operations inside it. It is stationed along the second defensive zone, in individual groups along the most important axes.

The reserve defensive echelon, which comprises the army reserve, is supposed to repulse the enemy in the event of his breakthrough into the operational depth of the defense and launch decisive counterblows from the rear. It is stationed in the operational defensive zone, primarily in individual divisional groups.

The second and reserve defensive echelons, comprising the army's shock group, should be employed together for a decisive counterblow.

The rear defensive echelon consists of all the rear units and the army's rear installations and is located in the rear defensive zone.

VIII. The Composition of Men and Materiel in the Defense

1. The composition of the defense's men and materiel depends on its tasks and the length of the defensive front and may vary.

One should consider that the shift to the defensive, if it is carrying out at the front level the task of resolving the main mission along another direction, will never happen along too-narrow a front, for in this case it could not carry out any kind of remotely significant role for the point of view of economizing forces for the main axis and tying down significant enemy forces along a secondary axis. On the average, the length of the defensive front at the operational level can be calculated at 100 kilometers within the confines of the army. It will often be greater.

The base norm for calculating the defense's men and materiel is the average operational density norm. Our field manual defines the length of division front in the defense at 8–12 kilometers. However, it is necessary to bring up the strength of the defense's men and materiel to at least a 1:3 correlation with the attacker along the main axis and 1:5 along the secondary axis. Proceeding from these minimum requirements, it is necessary to keep in mind that along the main axis the attack front is calculated at 2–3 kilometers per division and at 6-8-10 kilometers for a three-division corps. In order to observe the correlation of 1:3 along the main axis, it is necessary to determine the defense's average operational density as at least ten kilometers per division. In many cases this front norm in the defense may prove to be insufficient. Thus we should consider that, overall, ten divisions will be required to defend a 100-kilometer front, if the entire front is open to attack and the given axis has, on the whole, important strategic significance.

If the given axis is not very important and it terrain conditions do not allow for an attack everywhere, then a 100-kilometer defensive front may be occupied by 7–8 divisions.

Therefore the composition of the main defensive echelon, given its conduct along a 100-kilometer front, should be calculated at 7–10 divisions.

On the average, it's necessary to hold one third of the main echelon's forces in the second defensive echelon; that is, 2–3 divisions.

Considering that within the army's confines the attacker's main attack sector occupies a front of approximately 30–40 kilometers, in this case we will manage to keep two divisions in the second defensive zone along the likely axis of the breakthrough's development.

The army's reserve echelon may vary in composition, depending upon the defensive tasks and its nature. On the average, its composition may be defined at one fifth of the main echelon's forces; that is, 1–2 divisions. Besides this, the composition of the reserve echelon should include all of the highly-mobile (mechanized, cavalry and motorized) forces that form a part of the army.

Therefore, **an army defending a defensive area along a 100-kilometer front should consist of, on the average, of 10–15 divisions, depending upon the defense's tasks, the importance of the direction and the nature of the terrain.** (7–10 kilometers in the main echelon, 2–3 divisions in the second echelon and 1–2 divisions in the reserve echelon).

2. One should more readily consider the army's calculated strength as small, rather than significant. It is necessary to keep in mind that a 100-kilometer defensive front, if it is everywhere open to attack, may be subjected to an attack by up to 30 divisions in the first echelon. Therefore, 15 divisions in the defense are opposed to 30 attack divisions, creating a twofold operational superiority for the attacker.

Actually, this correlation will always be less favorable for the defense in fighting along the main defensive zone. The offensive always puts forth the main mass of its forces in the first attack echelon; this can be up to 30 divisions along a 100-kilometer front. The defense cannot put forth the entire mass of its forces into the main echelon, while positioning them in echelon. An army defending along a 100-kilometer front with 15 divisions may put no more than ten divisions into the main echelon. Therefore, the attacker may normally enjoy a threefold superiority in relation to the main echelon. One must always take this into account. Aside from this, the defense will be forced, one way or another; that is, with this or that density, to always occupy and block the entire length of the front entrusted to it. The offensive may concentrate its forces along selected axes, achieving there an overwhelming superiority, while even leaving some sectors of the front weakly occupied, which will not always be immediately clear to the defense.

Therefore, the attacker has great opportunities for achieving superiority over the defense. These conditions are inevitable for the defense. They always include the defense's negative side. However, in the impossibility of immediately opposing all one's forces to the offensive lies the advantage of the defense, which supports its power of resistance in the depth, when the force of the attacker's blow is already fading.

3. The artillery means of defense will always be insufficient. However, as regards artillery saturation, it is necessary to observe a correlation of 1:3 along the main axis of the defense. In general, the abundant provisioning of our corps with artillery gives us the opportunity to maintain this correlation.

Our rifle corps has an authorized strength of 162 light guns and 126 heavy guns, for a total of 288 guns. If it, when consisting of three divisions, will be defending a 30-

kilometer front and if we assume that this front will be attacked by ten divisions, then the following correlation is created:

Polish division—48 guns × 10=480 guns
Plus three corps regiments—36 guns × 3=108 guns
Total—588 guns
That is; only a twofold artillery superiority for the attacker;
German division—89 guns × 10=890 guns
Plus three corps regiments—36 guns × 3=108 guns
Total—998 guns
That is; a 3.4 artillery superiority for the attacker.

Naturally, one must suppose that the attacker will have a large amount of artillery from the high command reserve (the Poles do not yet have it). Then, of course, the correlation will inevitably change in favor of the attacker.

However, the army in the defense should also count on reinforcement with high command reserve artillery. If a corps along the main sector of the defense were to receive one High Command Reserve howitzer regiment and one High Command Reserve cannon artillery regiment, then the number of its guns would increase, on the whole, to 366.

In any event, an army on the defensive may receive 2–3 artillery regiments from the High Command Artillery Reserve or, depending upon its task and the situation at the front, even more.

4. Tank formations in the army's defense are especially important. Without them the army cannot conduct an effective defense against enemy tanks that have broken through into the depth. The defending army should receive 1–2 tank brigades; while at the same time it is important that these be T-28 tanks,[10] as the most effective means of fighting other tanks.

5. The army in the defense should have its own aviation, which may play an enormous role. The attacker always presents an open organization of combat formations, concentrated in narrow and confined areas. And during the preparation for the breakthrough and during its development in depth, he offers abundant and rewarding targets. The army in defense may receive 4–5 air regiments, necessarily including a significant part of fighter aviation, in order to deprive the enemy of his freedom of movement on the battlefield. Of course, this aviation will always be insufficient. Therefore, the army in defense should be supported by front aviation.

6. Although in the defense the main part of the engineer work is carried out by the troops themselves, the requirement for specialized engineer troops is nonetheless quite significant. They are mainly necessary for organizing obstacles, anti-tank barriers, minefields, and other specialized barriers. A single engineer battalion may carry out this specialized work simultaneously along a front up to six kilometers in length. Thus up to 17 engineer battalions would be simultaneously required along a 100-kilometer front. Considering that this front will be occupied by four corps, consisting of ten divisions, the troops' authorized means will yield nine engineer battalions (four corps battalions and five battalions will yield ten divisional sapper companies). Therefore, the army should be additionally reinforced with 6–8 engineer battalions. Aside from this, the army in defense should receive the following:

a) 1–2 military construction directorates for work in the rear, where, for the most part, there will be a shortage of troop equipment;

b) 5–6 electrical companies for electrifying the wire network (one company per 15 kilometers of the wire network);

c) five companies of field water supply, not counting support for supply stations;

d) five masking companies, not counting support for supply stations;

e) 2–3 road-construction battalions;

f) 2–3 airfield-construction battalions, and;

g) 1–2 pontoon battalions, if there are significant water lines in the depth of the army defense area.

7. The employment of chemical equipment in the defense should naturally be widely practiced, which requires a significant amount of chemical troops. Individual sectors should be prepared for contamination not only in front of the defensive zone, but throughout the entire depth of the army area, thus reinforcing the defense and killing off entire areas of it, both in front of the front and in the depth. This will guarantee a great economy of force.

One chemical tank brigade, consisting of four chemical tank battalions may, given an average contamination density of 35 grams, contaminate up to 20 kilometers along the front (five kilometers per chemical tank battalion) with one charge. The depth of such a contamination is 500 meters. If one considers that about half of the length of a 100-kilometer defensive front will be subject to contamination, up to three chemical tank brigades (12 chemical tank battalions) should be attached to the army.

8. The army in defense should have a powerful anti-aircraft defense and is thus in need of a significant component of anti-aircraft artillery equipment. No less than 2–3 anti-aircraft battalions would be required to reinforce the troops' anti-aircraft defense in the main defense zone. Up to three battalions will also be required for covering the army reserves in the operational zone. Then up to five battalions will required for covering the army headquarters and supply stations. Therefore, on the whole, the army in defense should be reinforced by up to ten anti-aircraft artillery battalions, with a corresponding number of other anti-aircraft defense means.

9. As has been indicated, the intensity of supply delivery to the defense will inevitably be great, which is why the army should be supplied with a sufficient amount of auto-transport equipment. To be sure, the elongation of the dirt road delivery will, for the most part, be normal, because the supply stations will usually be no more than 100–120 kilometers distant. Auto transport runs, taking into account the troop delivery section, will thus be normal. However, the daily requirements of the army in defense are measured by significant tonnage.

	Weight of Daily Requirements (in tons)	Number of Trains	Amount of Auto Transport in ZIS Battalions
Engineer equipment	2,400	4	4
One combat load	6,000	10	10
Other types of supply	2,400	4	4
Total	10,800	18	18
While delivering ½ of a combat load	7,800	13	13

Because the army in defense should always have a certain auto transport reserve for maneuvering forces in the depth, one should calculate its amount of auto transport equipment at up to 15 battalions of ZIS.

Therefore, according to its composition, **the army in defense represents a large and powerful organism, consisting of all modern technical means of struggle and requiring a high degree of materiel support.**

We have examined only the skeleton of the army defense—its anatomy. Now we must examine how the army must operate in the deep defensive area and how it should conduct the defensive operation.

Therefore it is necessary to shift to the physiology of the defense.

This theme is laid out in the second part.

Part Two—The Fundamentals of Conducting the Defensive Operation

I. The Operational Conditions for the Assumption of the Defense

1. The defense belongs to that type of military activities that require the greatest amount of time for its preparation and the greatest volume of measures for its organization.

Behind the decision to assume the defense and to organize it are an enormous volume of measures necessary for realizing this decision.

These measures are incomparably more labor-intensive than for the offensive; they require an incomparably greater expenditure of efforts, while determining in the most decisive way the strength and power of that resistance, which it will be capable of offering.

The very conditions for assuming the defense and its organization represent a complex process and are wholly dependent on the operational situation in which this occurs.

2. If the occupation of the defense takes place in time, beyond direct contact with the enemy, or along terrain previously outfitted with permanent engineer fortifications, which have been erected along the borders of the majority of states, then, naturally, the very process of occupying and organizing the defense in the most normal conditions unfolds according to plan and in an organized manner.

3. If the shift to the defense takes place in the event of the onset of a forced pause during an offensive operation and the temporary stabilization of the front, when the enemy himself is not capable of the offensive—the process or organizing the defense is already significantly more complex, because it is necessary to construct the defense along non-prepared terrain. However, in this case the shift to the defense usually fixes that situation which had been achieved during the temporary stabilization of the front and is not linked to any kind of maneuver, requiring only the regrouping of men and materiel.

4. The process of occupying and organizing the defense is the most complex, when the necessity of assuming the defense arises given the impossibility of further continuing the offensive and when this offensive has died out, having encountered the enemy's superior forces and when its further continuance is quite inexpedient, according to the overall strategic situation.

In this complex situation:

a) All of the forces, which are supposed to go over to the defense, occupy an offensive combat formation, with its inevitable concentration of the greater part of them along a relatively narrow area.

This is completely contrary to the assumption of the defense and forces us to undertake a complex regrouping.

b) The front along which arises a situation that forces us to assume the offensive does, as a rule, not correspond the forward edge of the defense and must be, for the most part, changed by moving it back.

Of course, it would be incorrect to construct the defense along an accidental front that has been achieved at the given moment in the offensive.

c) The assumption of the defense is, for the most part, linked to a withdrawal maneuver and with the necessity of first occupying that area in which one has to defend.

This demands, for the most part, carrying out a withdrawal maneuver, which is linked to the necessity of breaking away from the enemy and covering oneself against him, sometimes in quite difficult conditions, when the enemy has gone over to active operations.

d) Finally, in these conditions one must construct the defense along completely unprepared terrain and transform it into an area of fortifications and obstacles difficult of access for the enemy.

5. All of these conditions transform the assumption of the defense in a situation, when the offensive operation has been broken off, when a further offensive along the given direction is impossible and, according to the overall strategic situation—inexpedient, into a complex process, which requires great organizational skill and great efforts from the command.

These conditions for assuming the defense are sharply distinct from the conditions of assuming the offensive against an enemy who has halted. We organize the operational formation for the breakthrough from the depth, while approaching the jumping-off area for the attack, and are organized for it, while being relatively unhindered by the enemy.

It is another matter when assuming the defense. Here the operational formation, from which one has to assume the defense, for the most part, in no way corresponds to it and calls for inevitable regroupings and a withdrawal. This circumstance takes on great significance as a major stage preceding the organization of the defense itself.

II. Covering the Occupation of the Defense

1. In such very complex conditions, the first task of the army command is to break one's main forces from the enemy and to quickly withdraw them to the area designated for defense. At the same time, it is necessary to observe complete surprise and unpredictability.

It is necessary to immediately organize the cover of the withdrawal by allotting covering units, which constitute the forward echelon. Because the forward echelon generally general carries out its operational task in the interests of the entire army, then its apportionment, organization and assignment of missions must be indicated by the army. Cavalry may be allotted to the forward echelon, if there is any in the army. A single cavalry division may cover a front of up to 15 kilometers. A cavalry corps, consisting of three cavalry divisions, may cover a front up to 50 kilometers. In this case, the remainder of the front should be covered by forward detachments, allotted from the troops. However, the army will not have cavalry in every case. Then the forward echelon will consist completely of troop forward detachments.

Forward detachments, detached from the rifle corps, should consist of, on the average, from one to three battalions, reinforced with artillery, tanks, sapper and chemical units. It is expedient to attach long-range artillery to the forward echelon.

It is expedient to leave tank formations in the depth of the forward zone, in order to launch brief counterblows.

Chemical and engineer units should be allotted in the necessary number for constructing obstacles and contaminating the terrain.

2. The danger for individual forward detachments is always that they may be destroyed in detail. Therefore the army should assign the general line of the main defense for the forward echelon, for the purpose of securing the forward detachments' cooperation along the front. Intervals should be kept under observation and obstructed. The forward detachments should not accept a decisive engagement, but operate by means of mobile defense.

The forward detachments are subordinated directly to the corps from which they were detached.

The time for the forward detachments' activities should also be determined by the army, depending upon the conditions of the situation and the time necessary for occupying and preparing the defense. The forward echelon's task is to slow down the enemy's arrival to the defensive zone and prevent him from arriving at it earlier than the defense is able to take on the attack.

Depending on the conditions indicated here, this time may be measured as 5–8 days.

The more time required for organizing the defense, the longer one will have to hold the forward zone and the stronger the forward echelon and the denser the obstacles.

The forward echelon should be supported by army aviation, which is how the latter should be employed during the enemy's approach to the defensive zone.

Reliable communications should be organized by wire, radio and aircraft. Gathering posts for reports should be pushed forward. Communications with the forward detachments should be laid down directly from the corps headquarters. If there is a cavalry corps in the forward detachment, it should be connection directly with the army headquarters.

III. The Organization of the Withdrawal

The rapidity of the break from the enemy and the withdrawal demand the immediate issuing of orders even earlier than the adoption of the entire decision to organize the defense.

This decision cannot be born that quickly and requires the evaluation of an entire series of data: it passes through quite a complex process.

A great deal of time would of course be wasted in organizing the withdrawal in expectation of when such a decision will be determined. Thus one should organize the withdrawal before the complete formulation of the decision. However, because it is necessary to pull back one's forces to that area, which has been selected for the defense, it is necessary to immediately, define the following:

a) the overall front of the main defensive zone;

b) the number of divisions designated for occupying the main defensive zone;

c) the overall disposition of the divisions along the main defensive zone (where to have the main group of forces);

d) what kind of forces to pull back behind the main defensive zone for stationing in the depth, while not yet predetermining their disposition and specific mission.

Once this has been decided, one can issue the order for the withdrawal.

The units will sometimes have to make 1–2 marches back and during this time the decision to defend will be completely drawn up. By the time the troops begin to approach the defensive front, they should receive the order to defend. At the same time, it will often be necessary to first transmit the mission by means of individual orders to those corps which are slated to occupy the most important defensive sectors.

Therefore, at the same time the troops begin their withdrawal; intensive work will begin in the army's headquarters for drawing up a defensive plan.

IV. The Fundamentals of the Decision to Defend

1. The decision to defend is very many-sided and varied according to its content. In this regard, it is incomparably more complex than the decision to attack, which is more of a piece and monolithic.

The decision to defend consists of the resolution of an entire series of completely independent, according to their content, tenets, as, for example, the organization of the forward zone, the organization of the main defensive zone and that of the main defense zone as a whole, the disposition of reserves, and the organization of counterblows, etc.

However, all of these questions are closely linked. Their resolution, on the whole, should result in a single operational defensive system and should therefore flow from a single overall operational **defensive plan** as the basis defining all of its component elements.

2. The adoption of the decision to defend goes through a complex process, which embraces the sequential resolution of a definite volume of questions.

Because this decision envisages the entire organization, construction and conduct of the operation; that is, the entire **defense plan**, then it is, according to its content, the **elaboration of a defense plan**.

An approximate outline of the elaboration of the army defense plan is presented in the following form:

I. The Evaluation of the Situation

1. The evaluation of the enemy.
2. Comprehending the defensive task.
3. Calculating the time for preparing the defense.

4. Calculating the defense's men and materiel and their possibilities.
5. Evaluating the terrain and noting defensive lines.

II. Determining the Defensive Plan

Determining the goal, methods and nature of the defense, as a whole
Given fully developed defensive forms:

1. Where, by what methods, and with what men and materiel do we offer a rebuff to the enemy, in order to crush and halt his offensive.

2. With what forces and along which main axes do we launch a counterblow against the enemy, in order to complete his defeat and restore the situation.

III. Elaborating a Plan for Constructing the Army Defense Area

1. Determining the depth and zones of the army defense zone.
2. Determining the main defensive sector, along the front and in depth.
3. Determining the system of defensive lines (the main defensive zone, the second defensive zone, switch positions, the rear army line, and the system of anti-tank areas and lines).

IV. Elaborating a Plan for Outfitting the Army Defense Area

1. The tasks of the engineer strengthening of the terrain.
2. The employment of chemical weapons (obstacle zones and areas).

V. Determining the Disposition of Men and Materiel

1. The forward echelon's units' mission and composition.
2. The missions and disposition of men and materiel along the main defensive zone.
3. The missions and disposition of the reserves along the second defensive zone and in the operational depth.

VI. Elaborating the Variations of the Counterblow

1. Determining the possible axes for the counterblows and the jumping-off areas for their launch.
2. Determining the variations of the group of forces for the counterblow and calculating the concentration to the jumping-off position.

VII. Planning the Defensive Operation

1. Determining the stages of the defensive operation.
2. Determining the special tasks by operational stages (counterpreparation, maneuver in the main defensive zone and in the operational depth, maneuvering artillery, the active employment of chemical weapons, and surprise and deception measures).

VIII. Aviation Employment

1. Aviation's general tasks and the distribution of its flight resources.
2. Aviation's local tasks by operational stages.
3. Aviation's airfield basing.

IX. Elaborating a Plan for the Organization of the Rear

1. Calculating materiel supply and delivery.
2. Determining a plan for the organization of the rear and delivery (supply stations, forward delivery stations and dirt road delivery sectors).
3. The distribution of auto transport.
4. The organization of medical evacuation.

X. Elaborating Measures for the Operational Support of the Defense

1. The organization of anti-aircraft defense.
2. The organization of intelligence.
3. The organization of command and control (location of headquarters and command posts).
4. The organization of communications.
5. A central point in elaborating a defensive plan is determining the defensive plan.

By the defensive plan we should understand the basis of the decision that determines the goal, method and nature of the defense, as a whole.

The defensive plan is fully complete when the decision as to the counterblow is resolved. However, one should keep in mind the fact that unfavorable conditions, the enemy's clear superiority and the shortage of one's own forces may in certain conditions force us to construct our defensive plan without calculating on the possibility of a counterblow and to renounce it, employing the army reserves for offering successive resistance.

The defensive plan will thus boil down to another method and nature of its conduct, resulting in a successive and stubborn resistance to the end.

Sometimes the counterblow's possibility and expediency will not be immediately and definitely realized and will remain hidden. Then the defensive plan will call for definite variations for conducting the defense, depending upon the conditions of the operational situation.

Therefore, determining the defensive plan imparts clarity to the goals, methods and nature of the defense, thus laying the foundation for resolving all individual problems of its organization and conduct.

Without determining the defensive plan, its organization and conduct cannot be exhaustively and purposefully resolved.

V. The Evaluation of the Situation

1. The determination of the idea of the defense depends upon an entire series of conditions, which should be studied beforehand, comprising the evaluation of the situation.

Five main factors comprise, on the whole, the volume of the evaluation of the situation:

a) the enemy,
b) the defensive task,
c) the time for preparing the defense,

    d) the defense's men and materiel (their possibilities),
    e) the terrain and its defensive possibilities (lines).
  All of these factors should be appropriately comprehended, so that the idea of the defense is correctly resolved.
  2. The evaluation of the enemy is of extreme importance.
  This is significantly more complex in the defense than in the offensive.
  In the offensive, in general, we have the entire enemy before us and in a non-mobile state. We can always determine the disposition of forces along the defensive front itself quite accurately. Only in the operational depth does the disposition and concentration of new reserves always present certain difficulties.
  The matter stands otherwise in the defense. The enemy appears before our defensive zone in the full strength of his combat formation only at the last minute, when there is no longer time to determine the disposition of his forces. The enemy is in motion before this, carrying out regroupings and observing all the conditions of secrecy. This requires very intensive and uninterrupted work from intelligence so as to uncover the strength and disposition of the enemy's concentrating forces. However, it is not enough for intelligence to fix the strength and disposition of the forces being concentrated for the offensive. It is necessary to determine the enemy's possible offensive plan, for without this the defensive cannot be correctly organized.
  An example of this is the Russian Southwestern Front in 1917, during the World War, when the incorrect evaluation of the enemy's attack plan at the strategic level led to a situation in which all of our positions were improperly erected according to their contour, facing southwest, when the offensive was launched directly from the west to the east. It is necessary to imagine the following:
    a) what are the enemy's goals in the offensive (an offensive with a decisive aim, or with a limited aim), and what task is he pursuing at the front level;
    b) where is the main attack most likely, and;
    c) along what axis is the development of the breakthrough most likely.
  Certainly without comprehending the latter question, it is basically impossible to in any way purposefully approach the organization of the defense in the operational depth and define the disposition of the army reserves.
  3. However, in one way the evaluation of the enemy's possible plan of action may always find a correct resolution in the defense. This involves determining the axis of the main attack.
  There are, of course, several axes for the main attack at the strategic level of the entire theater of military activities, and it will not always be possible to determine them beforehand.
  However, it should be mentioned that in February 1918 British intelligence along the French front quite accurately predicted that the Germans' next big breakthrough would take place in Picardy, near St. Quentin. At the same time, even the length of the main attack front and the number of divisions that were to be committed into the offensive were determined. As is known, this data from the great amount of intelligence was fully confirmed in March, although at the time neither the English or French commands attached the proper significance to them.
  However, at the army level of defense along a front of up to 100 kilometers, it is always possible to quite accurately determine the axis of the offensive's main attack, should one follow, for this is generally predetermined by the terrain conditions.

The terrain along a 100-kilometer front always offers quite definite conditions for launching the main attack.

Also, our western theater is forested to such a degree that along a particular front it usually offers only limited opportunities in this regard.

It is more difficult to determine the axis of the development of the breakthrough in the depth. Here several variations are always possible, which are resolved depending upon that strategic goal which the attacker is pursuing by the offensive. Only a profound operational-strategic analysis of the goals that the enemy is pursuing may lead us to the correct resolution of this problem.

Therefore, a great deal of operationally creative work, which assigns very important and crucial tasks to the personnel of the intelligence section, is required from intelligence in the defense. In this work a great deal of operational development is required, as the full expression of training in the General Staff. One should keep in mind that incorrect conclusions by intelligence are highly fraught with consequences for the organization of the entire defense.

This was true in the organization of the English defense in Picardy in the spring of 1918.

4. Of the remaining factors in evaluating the situation:

a) Comprehending the mission insistently demands that it be viewed not only at the level of the given army, but at the level of the entire front. For the most part, the defense is an instrument at the level of the entire strategic front, having the function of supporting the best achievement of the main task along another direction and in tying down during this time the greatest number of enemy forces.

It is necessary that the defense understand is place and role at the level of the entire front operation. Only this understanding imparts to the defense its true meaning and determines the essence of its idea.

At the same time, if the defense is being conducted alongside the main attack, it is particularly important to comprehend the demands being placed upon it in regard to securing the boundary and the methods of operations along it.

b) The terrain in the defensive area should be given a correct operational evaluation, bringing out its defensive possibilities and the system of defensive lines. Without going into excessive tactical detail, one should, however, involve oneself in the tactical evaluation of the lines. Terrain is a very concrete factor, which is determined precisely by tactical qualities. The task, however, is that the operational level immediately embraces all important components that offer a definite idea of the system of defensive lines suggested by the terrain itself.

The first section,[11] in highlighting these lines while evaluating the terrain, should immediately inscribe them on a map, giving a clear impression of their outline and nature.

c) Finally, the calculation of one's men and materiel and the time, according to the situation, for organizing the defense must decide what the defense can do, what tasks it can assign itself in regards to defeating the enemy through an active counter-blow, and at what depth can it put a halt to his offensive. The evaluation of these factors leads directly to the definition of the defensive plan, its goal, method, and character.

VI. Drawing up a Plan for the Construction of the Army Area of Defense

1. When at the conclusion of the evaluation of the situation and the assigned task the defensive plan will be defined and it becomes clear upon what basis, for what purpose, by what means, and by what employment of men and materiel it will be waged, it will be possible to immediately determine the plan for organizing the army defense area.

The following will be consecutively determined in accordance with the idea of the operation:

a) **The boundaries of the army defense area, as a whole**; that is, in which area the operational task should be resolved, as whole, and what depth this area should have. For the most part, this will be indicated by the higher (front) command. However, instances are not excluded when armies, operating along a separate direction, will determine this on their own.

b) **The zones of the army defense area**, their boundaries, depth and outline (the forward zone, the main zone, the operational zone, and the rear zone).

c) **The main defensive sector**, along the front and in the depth, in accordance with the revealed most likely axis of the enemy's main attack. Just as the offensive has a main attack axis, the defense must also determine its main sector, which it opposes to the attacker's main attack. It is quite clear that this main sector will be more densely occupied and more powerfully fortified.

d) The location and outline of the entire system of positions and the nature of their fortification (the main defensive zone, switch positions, anti-tank areas and lines in the depth, and the army rear line).

2. The main defensive zone, which as the first barrier immediately opposed to the enemy, is expected to put up stubborn resistance and inflict the greatest losses on the attacker, requires particularly careful selection.

Operational and tactical conditions influence the choice of the main operational zone.

The former require that the shape of the defensive front correspond to the army's assigned mission and the position of its neighbors. This is particularly important when assuming the defensive along particular direction of the front if; for example, the defense is pursuing the goal of pinning down the attacking enemy and setting him up for a flanking attack by a neighboring attacking army or group of armies.

In this case, the overall operational conditions and interaction with one's neighbor will influence the choice of the defensive front.

If in the theater of military activities one must hold an important area having great importance (strategic, political or economic), then its situation will always determine the choice of the defensive front. In this case, a sufficiently advanced position of the defensive front will be necessary, in order to secure the defended target against an immediate threat.

The tactical conditions that influence the choice of the defensive front come down entirely to terrain conditions, which demand the most careful evaluation.

The army command does not have the right to forego the tactical evaluation of the forward edge of the main defensive zone and is obliged to indicate its overall front.

Naturally, one must strive so that the forward edge along the main axes is inaccessible to tanks.

One should avoid the overly straight shape of the forward edge, as this excludes the possibility of organizing oblique fire.

Bends in the front and the flanking shape of its sectors will offer great advantages, putting the attacker in a difficult condition. It is precisely at the operational level that these advantages achieve their full importance.

However, because the winding shape of the defensive front inevitably increases its length and requires large forces, one should not seek to resolve this problem at the expense of the defense's stability.

At the same time, one should keep in mind that the narrowing of the length of the defensive front, while conserving our forces, creates the same kind of advantages for the attacker, enabling him to greatly concentrate his men and materiel for the attack.

This is how the situation developed, for example, for the Germans in 1918 along the French front, when falling back on their rear positions they shortened their front, while at the same time offering the Allies the opportunity to achieve a greater concentration of men and materiel for the attack.

Thus the very choice of the forward edge and the shape of its front put forth great operational demands on itself.

3. The entire fortification system of the defensive depth should be completely subordinated to operational considerations, which determine the idea of the defense.

If the main defensive zone must, in one way or another, cut across the entire defensive front, then the positions in the depth will not always equally cut across its entire length along the front. For the most part, there will not be enough men and materiel for this. Nor is there such an immediate necessity for this.

A decisive condition for organizing the entire defensive system in depth is the determination of the main defensive sector, along which the attacker's main attack is expected.

The main defensive sector should be crisscrossed in the depth of the main zone by a completely constructed second defensive zone. The second defensive zone may not immediately be continuous along the remainder of the length, but consist of individual large centers of resistance.

The most likely axis of the main attack's development should be crisscrossed in the operational defense zone by a series of anti-tank areas and be edged from the flanks by switch anti-tank lines and in the depth by the army rear line.

On the whole, this should lead to a situation in which the enemy's tank units, which have broken into the depth, should fall each time into an anti-tank sack, where their actions will be localized and defeated.

The army rear line will present along the remaining sectors in the depth a broken line of individual prepared positions. With the increase in engineer readiness, this line will gradually grow. However, for the most part, this cannot be immediately achieved, due to a shortage of labor. Thus one should very purposefully decide which positions in depth should first of all be fortified.

Finally, proceeding from the operational defensive plan, the entire system of fortifications in the depth should create convenient hidden and covered jumping-off positions for counterblows by the army reserves along the planned main axes. In this regard, switch positions in the depth should be built with the expectation of covering the army reserves' main concentration areas for a counterblow.

4. Therefore, the elaboration of an entire system for organizing the army defense area should proceed from the operational idea of the defense, as a whole.

The entire system of organizing the army defense area should specifically express the defensive plan, while determining its organization, character and methods of conduct.

A good deal of operational meaning and a great deal of purposefulness should be devoted to the organization of the army defense area.

It is quite clear that only after having determined the entire system of organizing the army defense area and the disposition of its fortifications; that is, having comprehended where and along which lines the operational mission will be resolved, on the whole, can one then intelligently and purposefully determine the disposition of one's men and materiel.

The very system of positions in the defensive area determines, to a significant degree, how the men and materiel should be grouped.

If, for example, it becomes clear that along certain sectors of the defensive area we will be able to employ the terrain's powerful defensive qualities or successfully construct obstacles and contaminated zones, then this will naturally be reflected in the disposition of men and materiel, in economizing them, and in the density of this or that defensive sector. However, this does not mean that the troops may be tied to their positions in the defense, for their overall disposition is determined to the same degree by the idea of the defense and how the system of organizing the army defensive area is determined by it. Besides, defense is not waged by positions, of course, but by its defenders. The most powerful positions fall easily if they are poorly or incorrectly defended. Thus the distribution of men and materiel is, in the final analysis, the basis for the decision to defend, concluding the execution of its organization.

VII. The Distribution of Men and Materiel in the Defense

A. The Main Echelon

1. The main part of the defense's men and materiel, no less than two-thirds of its strength, will, as a rule, occupy the main defensive zone, comprising the main defensive echelon.

The entire defensive front will never be occupied equally. The density of the defense will fully depend on the operational importance of the given sector of the defense. The main defensive sector will be occupied most strongly, while the secondary defensive sectors will be occupied more weakly, often having the character of a defense along a broad front.

Thus the army's main defensive zone will represent a combination of the most varied types of defense—of a narrow front, a normal front and a broad front, with a corresponding greater or lesser depth of their location.

2. The problem of the operational density of the modern defense has been, on the whole, little illuminated by theory, just as problems of the defense at the operational level, as a whole.

Only one thing is clear, and that is that the front norm in defense, which is determined by all modern manuals at 8-10-12 kilometers per division, is subject to increasing doubts, as too large and not capable of securing defensive stability. As a practical matter, the French are assigning six kilometers and, at a maximum, eight kilometers to a division in their lessons and exercises.

One should also bear in mind that according to the experience of the World War of 1914–1918, it was generally considered necessary to have 1,000 soldiers per each kilometer of defensive front.

Observing from this point of view, one of the French authors espies the reason for the Basques' lack of success at Bilbao[12] in the fact that there were barely 50,000 infantrymen along a 70-kilometer front, although the fortifications were generally recognized as excellent.

The front defensive norms during the World War show the following general density.

In the Western theater: a) the Germans had one division per 2.5–3-kilometers of front along the most important defensive sectors; b) the French norms coincided with those of the Germans.

In the Eastern theater the density of the defense was significantly lower and was distinguished by its extreme variety, depending upon the operational significance of the given defensive sector and the terrain conditions.

The Russians had one division per 9-10-15-kilometers of front. They had from 600 to 1,500 bayonets and seven machine guns per kilometer of front.

The Austro-Germans had a single division per 12-14-kilometers of front. They had from 350 to 1,000 bayonets and six machine guns per kilometer of front.

As regards artillery saturation, the Germans had on the average up to 20 light and ten heavy batteries per division in the defense, and 5–7 light batteries per kilometer for barrage fire, and 4–6 heavy batteries for other purposes. On the whole, the French had the same artillery saturation in the defense.

Along the Eastern Front this norm was significantly lower. Here the Germans had 18 batteries (of these, 12 were light and six were heavy) per division. The Russians had, on the average, no more than 8–10 batteries per division (of these, 2–4 were heavy). The Austrians had 12 batteries per division (of these, four were heavy).

In general, there was the following density per kilometer of defensive front: 20–27 light guns and 14 heavy guns on the Western Front; 3–5 light guns and 1–1.5 heavy guns on the Eastern Front.

3. One should recognize that given the defensive front norms called for by the field manuals, we are failing to achieve at present that artillery saturation which existed along the Western Front during the World War. As regards the artillery saturation on the Eastern Front, this is obviously insufficient in modern conditions.

Experience, in particular, that of the war in Spain shows that in the defense it is necessary to have a minimum of 4–5 anti-tank guns per kilometer of front. This norm alone forces us to assign the division a front of no more than eight kilometers.

However, in calculating the density of the defense, one should now proceed not only from artillery saturation norms, but also from anti-tank fire norms.

Because, however, this anti-tank fire norm is the minimal one, when actually taking into account the echeloning of anti-tank guns into the depth, a minimum of eight would be required per kilometer; then one should calculate a really securely held defensive front by a division at **less than eight kilometers** (approximately six kilometers).

Of course, such a high defensive density will not be required everywhere. Divisions will be able to occupy a greater defensive front along individual sectors. In this regard, one should have in mind three possible types of density in the defense—high, normal and low—calculated from the corps' authorized strength alone, and taking into account the reinforcement of the corps with High Command Reserve artillery, consisting of a howitzer regiment and a gun regiment.

With this, the defensive density will be expressed by the following norms:

TYPES OF OPERATIONAL DENSITY IN THE DEFENSE

	High Density	Normal Density	Low Density
Length of the division's front	less than 8 km	8–12 km	more than 12 km
Length of the corps' front (three divisions)	20 km	30 km	40 km and more
Artillery saturation (the authorized artillery of a 3-division corps)			
Total guns	288 / 126	288 / 126	288 / 126
Per kilometer of front	14 / 6	10 / 4	7 / 3
Artillery saturation (by attaching a howitzer regiment and an artillery regiment from the High Command Reserve to a 3-divisioncorps)			
Total guns	366 / 204	366 / 204	366 / 204
Per kilometer of front	18 / 10	12 / 7	9 / 5

Note: The total number of guns is shown in the numerator and the number of heavy guns in the denominator.

As can be seen from this table, even with a high density, the artillery saturation is hardly sufficient. However, it is necessary to keep in mind that the calculation of guns per kilometer of front in the defense is generally somewhat formal. The fire plan in the defense is based on its massing along definite approaches. If a 3-division corps is defending along a 20-kilometer front, then the fire of at least two-thirds of its artillery may be gathered along a front of 8–10 kilometers. This will yield, given a high density, up to 20 guns per kilometer of front. However, because fire may be concentrated along even narrower sectors within the confines of a 10-kilometer front, when conducting an artillery barrage along individual approaches, the density of fire, as a practical matter, can reach 25 guns per kilometer of front at certain periods.

When reinforcing the corps with artillery from the High Command Reserve, this norm may approach the artillery saturation norms along the Western Front during the World War.

4. The three approximate types of artillery density of the defense determine those combinations in which, one way or another, depending upon the importance of each individual defensive sector, the distribution of men and materiel is resolved in the main defensive zone.

One should strive to occupy the main defensive sector with a high density. Secondary axes may be occupied with a normal density. Often within the bounds of the army defensive front there will appear such sectors, along which an attack by more powerful forces is unlikely, due to terrain conditions. Such sectors allow for their occupation with a low density, on the basis of a broad defense.

5. It is extremely important to correctly distribute the sectors of the main defensive zone among the divisions and corps. It is necessary to place such a formation as the corps along such a sector that represents an operational whole, according to its operational significance, the axis which it intercepts, and according to the terrain conditions. Only this condition will secure the unification of all the corps' efforts for resolving a single organically connected local defensive task. Ignoring this condition may render the organized conduct of the engagement much more difficult, by putting the command and control and coordination of the means of struggle in very difficult conditions within the confines of the corps. The more the fighting takes place in the depth (which is inevitable in its development), the more this will tell, forcing the corps to engage along various axes differentiated by their importance and terrain conditions.

In the final analysis, this may lead to a situation in which the corps' activities prove to be completely disconnected.

Thus it is important, first of all, to determine the defensive sectors for the corps, to scrupulously study them from the point of view of the terrain conditions and the importance of the axes that they intercept.

Only as a result of such study will the defensive sectors and their length be determined.

Whether or not one has to unite two or three (and in rare conditions, four) divisions under a single corps in the main defensive zone will depend completely upon that length which defines the given sector as an operationally connected axis.

Thus corps in the defense may place under their control 2–4 divisions in the main defensive zone, depending upon the length and importance of their sector.

6. Reinforcement artillery, if it is attached to the army, should be subordinated to the corps occupying the main defensive sector. As regards independent tank formations, it is expedient to attach them to the corps along the main defensive sector only in that case where we plan to hold the main defensive sector at all costs and to concentrate all our efforts there. Only the absence of a clear superiority of the attacker's forces will allow us to count on this. Thus the resolution of this question depends completely upon the operational defensive plan.

The corps' reserves will, for the most part, consist of independent regiments, allotted from the main-echelon divisions and located along intermediate positions in the main defense zone, between the main and the second defensive zones.

In special cases the corps may have along the main defense sector a second-echelon division in reserve, located in the second defensive zone.

In this case, the corps fights throughout the entire depth of the main defensive zone and includes within its combat formation two echelons: a main one and a second one.

The corps' headquarters should be located in anti-tank areas, under the cover of switch positions, or in the second defensive zone, although outside of those axes which we plan to use for maneuvering within the main defensive zone. The boundaries between the corps should be extended as far forward as the forward edge of the forward zone, and in depth as far as the second defensive zone, inclusively.

If one plans to conduct a defense by successively pulling it back into the depth as far as the army rear line, then the boundaries between the corps should immediately extended throughout the entire depth of the army defensive zone. This also depends on the operational idea of the defense. According to their shape, the boundaries should reflect the idea of the defense as regards maneuver in the depth of that group of forces, which the main echelon must take on in falling back, so as the pull the attacker into a sack and, while relying on switch positions in the operational defensive zone, to secure a favorable jumping-off position for the counterblow.

B. The Second Defensive Echelon

1. The second defensive echelon, although it closes the main defense zone with its group of forces, is, however, a factor of operational significance, because it is called upon in modern conditions to struggle simultaneously with the main echelon against the enemy's tanks that have broke through, barring their entry into the operational depth of the defense. This, however, is only one of the tasks of the second echelon, which independently or along with the reserve echelon (the army reserve), comprises the shock group for launching counterblows while fighting in the main defense zone.

As has been indicated, the second echelon consists of 2–3 divisions, located in individual groups in the second defensive zone.

The second-echelon divisions may be subordinated to the corps occupying the main defensive zone, or remain subordinated to the army.

2. If a second-echelon division along a given axis may be employed only within its confines and, in this regard, may be given a definite assignment, then it is expedient to subordinate it to the corps defending the given axis.

If a second-echelon division cuts across the most important axis of the likely development of the breakthrough into the depth and, in this regard, assumes operational significance at the army level, then it is necessary to leave it subordinated to the army, because the army commander must directly unify the fighting against the success development echelon, while in no way distracting the corps commanders from carrying out their direct task in the main defensive zone.

If a second-echelon division is located behind the boundaries of two corps in anticipation of being employed along the sector of one or the other of them and cannot immediately be of a definite operational significance, it is also expedient to subordinate it to the army.

The second-echelon divisions, located along a definite sector in the second defense zone, may with equal operational intent, be united under corps control. In this case, the second echelon may contain not individual divisions, but rather a corps, which unites the fighting in the depth of the main defense zone in the second defensive zone. This is possible, given the timely organization of powerful resistance along the most likely axis, for the development of the breakthrough. In this case, one should not, however, put in reserve the second-echelon corps' corps artillery, but reinforce the main defense echelon with this artillery.

3. Flank boundaries are not assigned to the divisions and second-echelon corps under army control. The sphere of their responsibility for the strengthening of a definite sector of the second defensive zone is indicated by the length of this sector along the front and the assignment of the task, as a whole.

The employment of the second-echelon divisions in the fighting in the main zone should call for the following:

a) the launching of a counterblow; in this case, the possible axes and jumping-off areas are indicated, or

b) the occupation of the important switch positions, in order to localize the enemy's breakthrough within the confines of the main defensive zone.

C. The Reserve Defensive Echelon

1. The reserve defensive echelon comprises the army reserve, which is designated for fighting in the operational defense zone, should the enemy manage to break into it, and for launching decisive counterblows from the depth. In this regard it comprises the army's shock group, for the most part, together with the second defensive echelon. By this time it will often be reinforced by front reserves.

The army's reserve echelon consists of individual divisions (up to 2–3), all of the army's highly mobile formations (tank and cavalry), a mobile artillery reserve (particularly anti-tank), and chemical and engineer units.

2. The army's reserve echelon is stationed under the cover of the second defensive zone in the operational defense zone, beyond the reach of the enemy's ground forces, which are attacking the main defensive zone—at a distance of 25–50 kilometers from the forward edge of the defense.

The depth of the location and disposition of the reserve echelon's forces depends entirely on the operational idea of the defense.

If one can count on localizing the enemy's breakthrough within the confines of the main defensive zone and the very rapid appearance of a situation for launching a counterblow, one should station units of the reserve echelon closer. If, on the other hand, the enemy disposes of a great superiority in men and materiel and all data indicate a stubborn and prolonged battle, with the possible shift of activities into the depth, the reserve echelon's units should be stationed further away, so that they are not drawn into the fighting during the course of the battle earlier than they can be employed for those purposes for which they are designated.

3. In issuing tasks to the reserve echelon's units, one should keep in mind four required instructions:

a) which positions should be strengthened in the operational defense zone;

b) in which area should the hidden intermediate position be occupied—upon completing the engineer work, or upon the start of the enemy attack;

c) which positions should be occupied during the breakthrough by the enemy's tank units into the operational defensive zone, and how this breakthrough is to be localized;

At the same time, the task of the reserve echelon's tank units is to be particularly active in attacking the enemy's tanks that have broken through.

d) along which axes do we plan to launch a counterblow, and in which areas do we occupy a jumping-off position for this.

VIII. The Counterblow in the Defensive Operation

1. The problem of the counterblow is the most complex one in the defensive operation. Its resolution depends on a great many conditions and matures during the process

of the operation's development, as its active conclusion. Only the counterblow imparts developed forms to the defense and concludes it with an act organically inherent in its nature.

Clausewitz stated the following about the role of the counterblow in the defense: "The transition to a retaliatory attack should be conceived of as a tendency of the defense, and thus as its important component part...." "When the defender has achieved important advantages, the defense has carried out its task and must, by employing the advantages gained, pay back, blow for blow, if it doesn't want to meet its inevitable doom...." "The rapid and mighty transition to the offensive is the shining sword of revenge and comprises the defense's most brilliant moment."

Foreseeing and carrying out this brilliant defensive moment should be the particular concern of the army command, requiring from it very profound operational foresight and a clear picture of the entire perspective of the forthcoming defensive operation.

2. Counterblows in the defense may be conducted in the most varied conditions of the situation, and for different purposes.

One should distinguish the following three typical cases of the counterblow:

a) When the enemy has broken the line along an individual sector which, on the whole, should be held—a counterblow is necessary in order to destroy the enemy who has broken through and to retain one's position.

In this case, the counterblow is of a local nature and is **conducted to hold one's position**.

Often such a counterblow will be carried out by the corps reserves alone (particularly if the corps has a second-echelon division). However, this does not exclude cases when the army reserves will be committed to carry it out and when the army itself takes command of the counterblow.

b) When the enemy has achieved a major success and threatens the defense's position, as a whole, a counterblow is necessary in order to repel the looming threat and to restore the situation. This counterblow is already carrying out its mission in the interests of the entire army and is conducted by the army reserves under direct army control. However, it is not yet linked to the complete shift of the initiative to the defense and is only a forced retaliatory attack against an enemy threat.

c) Finally, when the enemy is exhausted and when his offensive has been halted—the counterblow will be the decisive conclusion of his defeat. Such a counterblow is conducted by all available army reserves, pursues the decisive goal of defeating the enemy and delivers the initiative in the fighting into the hands of the defense.

This counterblow expresses its true calling in the defense.

3. The counterblow in the defense, undertaken in the interests of the entire army, should not be an isolated manifestation of activity, but rather a decisive attack, by which the enemy, already exhausted and, in one degree or another, weakened by his losses, has a final defeat inflicted upon him. The operational activity of the defense is expressed in the realization of this act. One should draw a definite line between operational activity and tactical activity.

One may accompany the entire conduct of the defense with an entire system of local counterattacks, while manifesting in this regard the most fervent tactical activity, but nonetheless not making the defense operationally active, if one does not conclude it with a decisive counterblow for the purpose of inflicting a final defeat on the enemy.

One may, on the other hand, conduct the defense on the basis of the most stubborn and consistently rendered tactically passive resistance, but nevertheless make the defense operationally active, if one concludes it with a decisive counterblow, which crowns the defeat of an enemy who has already exhausted his forces.

The counterblow in the defensive operation is a major and decisive operational undertaking.

It is necessary to weigh whether or not the situation has matured to carry it out; but once one has decided upon this, to carry it out with all firmness to the end.

In this regard, the operational counterblow takes on incomparably more decisive forms, pursues more decisive goals and brings about more decisive consequences than counterattacks of tactical significance.

The latter always pursue local, one-time goals; they are conducted during a short span of time and cannot have such decisive consequences, for even when they fail they still leave many other combat possibilities in the hands of the commanders.

A counterblow, carried out at the army level, itself becomes a local operation. All available reserves are brought in to carry it out: it occupies an entire period of time and pursues decisive goals, upon the realization of which depends the very fate of the defensive operation, as a whole. In throwing all of one's reserves into the counterblow, the army exhausts all of its possibilities and no longer disposes of any kind of means for waging the struggle if the counterblow does not enjoy success.

This is why the decision to launch a counterblow is one of the most crucial moments for the army command, requiring from it a clear understanding of the maturing situation and a great deal of resolution.

4. The formulation of the question about the counterblow in the defense often contains a great deal of romanticism and bravado, when defensive activity is understood as the mandatory encounter of the enemy's breakthrough force with an immediate offensive attack.

Given this line of conduct in the defense, one may very quickly expend all of one's reserves and, having achieved nothing, lose all opportunities for continuing the fight.

The defense's first and main task is to crush and halt the enemy's offensive. This is basically achieved by stubborn and consecutive resistance, which exhausts the enemy's physical and moral forces. As a rule, it is only after the enemy's offensive has been exhausted and halted that the situation for launching a counterblow will mature. While the enemy is continuing to attack and has not yet exhausted his reserves, the division in the defense, having occupied a position prepared in depth (a switch position or in the second defensive zone), and while repelling attacks with its fire, will play an incomparably greater role than by going over to a counterattack, which will inevitably quickly die out, having failed to achieved any kind of substantial results.

The assertion by the French author Bouchacourt[13] makes undoubted sense, when he says:

"Counterattacks during the World War often resembled a tub of water poured into the flames of a fire. Only a momentary flash occurred, the water evaporated in an instant, and the fire continued to burn."

The evaluation of the situation that has matured for the counterblow is thus one of the most crucial responsibilities of the army staff during the development of the defensive operation. It is necessary to correctly determine when the moment has matured for the counterblow.

To hurry with the counterblow, when the situation has not yet matured for it, often means to prematurely exhaust all of one's opportunities for continuing the struggle, when it is still going on and has not reached its culmination point. This may often put the entire defense under the threat of a complete defeat. To miss the moment for the counterblow and be late in launching it means, in the best case, to doom oneself to a passive exhaustion and sometimes to complete defeat.

5. The situation for a counterblow may mature in different conditions, depending on the correlation of forces and the overall course of the operation.

The weaker the enemy and the sooner his offensive will be halted, the more rapidly the moment for the counterblow will mature. In these conditions it will often be possible as early as in the confines of the main defensive zone.

If the enemy enjoys a heavy superiority and commits new reserves into the attack, the moment for the counterblow will inevitably be postponed.

If one is unable to hold off the enemy offensive within the confines of the main zone and the defense is pushed back into the operational depth, he will have to organize the counterblow in the operational zone while employing its switch lines.

In this case it is not the loss of territory that determines the outcome of the defense, but the exhaustion of the offensive at a great depth and the retention of forces for a final attack. The deep army defensive zone makes it entirely possible to accomplish this task, while demonstrating its superiority over the linear defense.

Cases are possible when a counterblow will become possible in conditions in which the enemy's offensive has not yet been halted and when the planned withdrawal to some defensive sector or another will create favorable conditions for an immediate and decisive flank attack.

Thus the conditions for a counterblow may vary greatly, depending completely on the course of the operation and the developing situation.

These conditions cannot be foreseen in their entirety. On the whole, however, their appearance should be foreseen by the defensive plan and the entire system of organizing the army defensive area. It is namely in the depth that the entire defensive system should predetermine how and along what axes the enemy's offensive will break and in what condition it will be placed in order that the counterblow be supported by the most favorable conditions.

In the struggle for the realization of these conditions lies the art of conducting the defensive operation.

6. What kind of counterblow variations should be foreseen will proceed from the operational defensive plan. We can never reduce the problem of the counterblow beforehand to one sort of decision. Variations will be inevitable. However, this does not mean that one may count on their realization, as had been assumed. Certain deviations will always be inevitable. However, the significance of elaborated variations lies in the fact that they foresee the most typical situations for a counterblow, being determined by the idea of the defense and the system of its organization. On the basis of these variations, all individual deviations that have arisen, according to the situation, will be easily resolved. Finally, in determining the disposition of forces for the counterblow and the order of their concentration in the jumping-off position, elaborated variations create more solid grounds for resolving these problems, depending on the situation at hand.

The variations of the counterblow should foresee the following:
a) the goal and direction of the counterblow;
b) the composition of the shock group of forces;

c) the jumping-off position for the counterblow;

d) the order of concentration in the jumping-off point and calculating the time required for this;

e) the counterblow's interaction with the main defensive echelon.

In all cases it is necessary to strive so that our counterblows are conducted against the flank of the enemy's breakthrough. Switch lines in the depth, which outflank the enemy who has broken through, must fully ensure this. However, the maneuver of troops falling back into the depth should also be completely subordinated to the interests of the counterblow, while placing the enemy in a situation to be outflanked and setting him up for a flank attack. In this regard, the counterblow's close interaction with the forces waging the defense in the main zone will always be required.

7. One should bear in mind that it is late to organize a counterblow when the situation has already matured for this. In anticipation of the maturing of this situation, the army reserves should be moved to the jumping-off areas beforehand for the counterblow so that they will already be concentrated there by that moment when the situation will have matured for the counterblow. Otherwise, the favorable moment will, for the most part, be missed.

Thus the army headquarters should comprehend the line of development of events and their immediate prospects. It is necessary during the course of the operation to precisely imagine what kind of situation might arise in order to resolve the question of the counterblow in a timely manner.

The army command must personally take the organization and conduct of the counterblow upon itself. This stage of the defensive operation is the most crucial and decisive for it.

Of course, this does not exclude the active conduct of the defense within the confines of the divisions and corps and their organization, on their own initiative, of a series of counterattacks, according to the situation.

{8. Given a favorable situation, the exhaustion of the enemy and the successful development of the counterblow, the latter may develop into a general counteroffensive along the entire front, signifying the seizure of the strategic initiative and the passage from the defense to the offensive. For the most part, the concentration of new front reserves will be required for this.}

IX. The Employment of Aviation in the Defense

1. Aviation in the defense has a broad field of activity. The concentration of the enemy's men and materiel opposite the defensive front and the accumulation of his dense combat formations before the start of the offensive offer numerous and inviting targets.

However, aviation's opportunities in the defense will always be limited, because it will usually be weak in strength and the enemy, for the most part, will be stronger, dominating the sky.

Thus the employment of aviation in the defense must be very purposeful and one should avoid any kind of dispersal of efforts. Missions should be limited to the most important targets along the most important axes and during the most decisive stages of the development of the operation.

On the whole, all of army aviation's efforts should be concentrated against the attacker's personnel, his reserves, mechanized forces, and artillery.

Front aviation should take upon itself the fight against the enemy's aviation on its airfields, attacking the attacker's deep rear areas and his railroad supply system.

2. Army aviation's main tasks in the defense are as follows:

a) to defend from the air the main defensive sector, the army reserves and the forward depots;

b) to delay the enemy's arrival at the defensive front;

c) to operate against the enemy's deployment along the axis of his main attack;

d) to combat the enemy's artillery along the main axis;

e) to cut off the enemy's reserves arriving to develop the offensive;

f) to attack the enemy's mechanized units in their jumping-off positions and at the moment they pass into the breach;

g) to attack the enemy's units, which have broken through into the operational depth of the defense;

h) to cover the retreat of the main echelon's units;

i) to support the counterblow from the air;

j) to combat the enemy's fighters over the battlefield, for the purpose of securing the defending forces' freedom of action at specific and decisive moments in the operation, and;

k) to reconnoiter the enemy group of forces and the arrival of new reserves.

The multiplicity and variety of aviation missions in the defense require that the army command purposefully determine the most important of these, depending upon the specific conditions of the situation that have arisen during the development of the operation.

3. The depth of army aviation's combat activities in the defense should be insignificant.

From the moment the enemy approaches the forward edge of the main defensive zone, army aviation it will operate in an approximately 20–25 kilometer strip in front of it; that is, the depth of the operational formation for the breakthrough. Upon the beginning of the enemy's offensive, this depth will constantly shrink, in the end limiting itself to the depth of the battlefield. However, the missions of front aviation must change accordingly.

The depth of the army's air reconnaissance should extend to the enemy's railheads; that is, approximately 50–60 kilometers from the forward edge of the main defensive zone.

4. According to the defensive plan, the army command should determine beforehand at what stages of the defensive operation the greatest efforts will be required of aviation and to accordingly assign tasks and distribute aerial resources. At the same time, in order to carry out the most important tasks at the decisive stages of the operation's development; for example, all available aviation, both fighter and troop aviation, should be brought in for repulsing a breakthrough by the enemy's tank units. At these moments front aviation should also appear over the battlefield at the summons of the army command.

5. All of army aviation's airfields should be removed to the rear zone, be covered by the army rear line and be dispersed and hidden.

There may only be landing strips, used by troop aviation and fighters, within the confines of the operational zone. They should be covered by anti-aircraft areas and switch positions.

Reserve airfields should be foreseen to a sufficient degree, giving army aviation under enemy air pressure the opportunity to change its basing unhindered.

X. The Development of the Defensive Operation

1. The army command should, on the whole, have a clear idea of the possible development of the defensive operation. Only by proceeding from such a perspective can we determine the intensity of the forthcoming engagement, the materiel means required and the specific tasks at each separate stage of the operation's development.

The planning of the operation should, on the whole, reveal these stages—their content, the elements of struggle that will be employed at a given stage and the method (character) of activities. The development of the defensive operation in its complete forms may be imagined in the form of the following five successive stages.

Each of these stages of the defensive operation makes its own special demands on the work of the army command.

2. Although the struggle in the forward zone is conducted, for the most part, by forward detachments, detached from the main echelon's rifle corps, the army command should, on the whole, control this fighting. The forward echelon must, in the event of necessity, be reinforced by equipment for making obstacles and contamination.

Large tank formations may be pushed ahead into the forward zone for carrying out specific tasks for attacking the arriving enemy columns. Finally, army aviation against those same columns should be purposefully controlled by the army command.

The main task consists of ensuring that the enemy does not get the opportunity of approaching the main defensive zone before the entire defensive system has been prepared. The army headquarters is thus obliged to strictly keep an eye on the course of events in the forward zone. Besides this, at this stage intensive work by all types of intelligence will be required, in order to receive in a timely manner information about the strength and disposition of the approaching enemy units.

3. The stage of the struggle in front of the forward edge of the main defensive zone is now acquiring particularly great significance in the broader formulation of the question of a major counterpreparation. As early as the World War on the Western Front the artillery counterpreparation yielded great results for the defense in a number of cases.

Our artillery field manual (part II) states that "the counterartillery preparation has the task of foiling the attack being prepared by the enemy" (article 243). In and of itself, this is a major task. In modern conditions, it may be carried out while bringing in other means of struggle (aviation and tanks) at a broader operational level. A concentrated artillery attack and an air strike may completely suppress along a given sector of the enemy's combat formation, which has already deployed for the attack. In certain conditions, this artillery and air attack may be concluded with a raid—an attack by a large tank formation through the forward edge of the defense, which may lead to the complete disruption of the enemy's combat formation. For this at least the temporary neutralization of the enemy's artillery along a given sector is required, which may be carried out by aviation. Thus the counterpreparation may grow into a major operational undertaking during the fighting in front of the forward edge of the defense and have major results.

THE DEVELOPMENT OF THE DEFENSIVE OPERATION

Operational Stage	Content of the Stages	Elements of Struggle	Nature of Activities
First Stage	Fighting in the forward defense zone.	Aviation. Forward detachments. Tank and cavalry formations. Obstacles.	Holding resistance and active attacks against the enemy's approaching columns.
Second Stage	Fighting in front of the forward edge of the main defensive zone.	Aviation. Artillery. Tank formations.	Counterpreparation. An artillery attack and strike against the enemy's deployment.
Third Stage	The struggle in the main defensive zone.	The main echelon's defensive system. The second defensive echelon's troops. The possible involvement of the reserve echelon's forces in shifting the decisive fighting to the main zone. Aviation.	Stubborn defense. A counterblow while localizing the breakthrough in the main zone.
Fourth Stage	The struggle in the operational defensive zone.	The reserve echelon's forces, in conjunction with the main echelon. Aviation.	The tank formations' struggle against the enemy's tanks that have broken through. A combination of stubborn defense in the rear army defensive zone with counterblows from behind the switch lines.
Fifth Stage	A counterblow for the purpose of destroying the enemy who has broken through and restoring the situation.	A shock group. Aviation.	A decisive offensive.

The counterpreparation, based upon the interaction of aviation, artillery and tanks, requires special organization and control. As an undertaking on the operational scale, carried out along the main defensive sector in the interests of the entire army, it should be directly controlled by the army command.

4. Control of the struggle in the main defensive zone is mainly the responsibility of the corps commanders. However, when the matter reaches the second defensive zone the army command must directly enter the fray. For the main echelon's forces the question is only one of stubbornly defending and maintaining their positions to the end. However, the main echelon's forces may often be subjected to such destructive artillery fire along a certain sector that their continued occupation of the main defensive zone may threaten them with complete physical annihilation. This, of course, should not be

permitted and the army command is obliged in such cases to authorize a withdrawal into the depth of the main defensive zone.

Maneuver in the main zone and the interaction of the first-echelon corps among themselves and with the second-echelon divisions should be the subject of constant control on the part of the army command. If, according to the defensive plan, all of the fight's efforts are transferred to the main defensive zone, then this will often require the second-echelon divisions' arrival there to occupy the switch positions, to localize the enemy breakthrough, and to prepare the counterblow. All of this should take place on the army command's orders.

5. However, while employing the army reserves for the fighting in the main defensive zone, one should remember that these reserves' main task is not the fight for the main defensive zone, as was the case during the World War, but the creation of a new fire resistance to the enemy in the depth until his forces are exhausted.

The defense's success in conditions of the struggle in the deep army defensive area should in no way be measured only by the retention of space. It should be measured by the size of the losses suffered by the enemy; the exhaustion of his offensive strength and the maintenance of one's own forces. The army command should bear all of this in mind, while controlling the course of the fighting in the main defensive zone. In conforming to the conditions of the situation, it should try to resolve the question of employing the reserves during the fighting in the main defensive zone.

6. The struggle against the enemy's breakthrough development echelon should be entirely led by the army command. The main echelon's corps commanders should not be distracted by anything in the event of a breakthrough by major tank units through the main defensive zone. The army command should organize the interception of the breakthrough tank units in the second defensive zone in a timely manner and to directly control their repulse, employing the second-echelon divisions and tank formations from its reserve.

Given the presence of a second-echelon corps the accomplishment of this task will be entrusted directly to it.

In the event of a breakthrough by the enemy's tank formations into the operational defensive zone, the army command must immediately organize their annihilation. While employing anti-tank areas, switch anti-tank lines and the army rear line, one should first of all localize the area occupied by the breakthrough tanks and then, while thus holding them in an anti-tank sack, to organize their systematic annihilation by aviation, tank formations, the mobile anti-tank artillery reserve, and, during the night, by the detachment of special infantry destruction detachments.

7. The shifting of the struggle to the operational depth of the defense will require energetic and feverish activity from the army command and an entire series of organizational measures. One should bear in mind that with the withdrawal of forces to the second defensive zone and the commitment of the second-echelon divisions into the front line the organizational composition of the corps will inevitably change. One should strive to constantly create a new army reserve for oneself. If some sector of the main defensive zone is passive, part of the main echelon should be removed from it and put into the reserve. All of this calls forth significant regroupings and sometimes the inevitable mixing of forces. The army headquarters is obliged to precisely resolve all the frictions that arise at this time and in all conditions try to create the necessary order in the troop's disposition. Thus, as the defensive operation develops and the fighting is

shifted into the depth, the work of the army command becomes more complex and difficult, facing it with particularly high organizational requirements.

8. Finally, as has been shown, the organization and conduct of the counterblow should be completely realized by the army command, thus raising the art of conducting the defensive operation to a high level.

Depending on the operational idea of the defense and the development of the operation, the situation for the counterblow will either mature within the confines of the main defensive zone, or in the operational defensive zone.

In all cases, the army command should first of all strive to localize the enemy breakthrough.

In the main zone this will be mainly achieved by means of enveloping the enemy breakthrough from the flanks by switch positions and in the depth of the second defensive zone. Thus the enemy will be trapped in a sack and a counterblow will become possible against one of his flanks, depending on the situation.

For the most part, such a counterblow in the main zone will have been foreseen by elaborated variations and will enable the defender to concentrate a shock group in a previously-foreseen strength of 3–4 divisions and tank units, given the presence of about five divisions in the second echelon and army reserve.

The counterblow will development in incomparably more complex conditions in the operational defense zone; that is, in the event of the fall of the second defensive zone. The system of organizing the army defense area should be foreseen ahead of time, and how the enemy's breakthrough will be localized in the operational zone. The system of anti-tank areas, switch positions and the army rear line must predetermine this. However, one can never determine beforehand in what kind of grouping and in what strength the troops will occupy these positions, for at this moment part of the army reserves may be already employed and, in the opposite case, a number of main-echelon divisions may have been pulled back into the reserve. Although definite variations of the counterblow should also be elaborated in shifting the fighting into the operational zone, however, the army command will have to manifest incomparably greater initiative and creativity in putting them into practice. The first task will again consist of localizing the breakthrough and drawing the enemy into the sack. At the same time, the army command should immediately create a reserve out of a specific group of divisions and concentrate it in that area from which, depending on the situation, it will be most favorable to cut off the entire sack along its base and against one of the flanks of the enemy's breakthrough group.

One should keep in mind that at this time the fighting may take on very complex forms, for individual divisions along the flank sectors of the main defense zone will still be holding out in the second defensive zone and possibly in the main defensive zone along a secondary axis.

The tasks for these forces should be issued in definite interaction with the counterblow being organized in the depth, which at this moment takes on an active character.

In certain conditions of the situation, when the enemy breakthrough into the operational zone has been completely localized, these salients, which have remained in the second and main defensive zone, may be employed as jumping-off bridgeheads for the counterblow, if it leads directly to the flank of the main enemy group of forces.

A number of the army command's measures in organizing the counterblow consist, besides this, of the regrouping of reinforcements (artillery and tanks) and the organization of interaction with the aviation.

9. The successful development of the counterblow, which has as its immediate task inflicting a final defeat on the enemy breakthrough and fully restoring the defense's position, may grow into a general counteroffensive, when the defeat of the enemy will enable all of the army's forces to advance beyond the confines of the main defensive zone. However, as a rule, for this it will be necessary to reinforce the army with new men and materiel from the front reserves.

XI. The Rear in the Defensive Operation

1. The army's materiel supply in the defensive operation is measured in modern conditions by no fewer requirements than in an offensive operation.

As has already been shown, the single daily requirement of engineer equipment necessary for preparing the army defense area along a front of up to 100 kilometers is measured at 2,400 tons. The requirement for munitions in the defense will be enormous, because the entire defensive system is based on fire, particularly on the most omnivorous type—the artillery barrage.

During the most intensive days the demands for munitions may be expressed in 2–2.5 combat loads of gun artillery and no less than three combat loads for plunging and heavy artillery, thus requiring the preliminary accumulation of munitions.

The demand for chemical equipment for its employment in the forward zone and in front of the forward edge also reaches 1,000–1,500 tons.

As a result, the daily requirement for an army of 12 rifle divisions in a defensive operation (given the daily delivery of ½ of a combat load) reaches 8,000 tons. At the same time, one should bear in mind that during the operation's preparatory period, when it will be necessary to accumulate munitions beforehand, this norm will increase.

The intensive work of the rear and the entire delivery system in these conditions is inevitable in the defensive operation.

2. Railroad supply delivery in the defensive operation may be carried out as far as the army rear line; that is, at a distance of up to 50 kilometers from the forward edge of the main defensive zone.

However, it is necessary to bear in mind that the fighting in the defense may spread throughout the entire depth of the army defensive area, reaching right up to the army rear line and not excluding the possibility of its penetration by individual enemy tank units.

In these conditions, the location of the forward depots along the line of the army rear line does not secure their safety, either in the air or on the land.

Therefore, the deployment of the supply stations and forward depots should be moved to a distance of up to 50 kilometers from the army rear line in the defensive operation; that is, at a distance of up to 100 kilometers from the main defensive zone.

While the fighting is going on within the confines of the main defense zone, the railroad delivery from the supply station to the army rear line may be carried out by individual mobile units.

Thus the plan for the organization of the rear in the defensive operation should, besides supply stations located at a depth of up to 100 kilometers, foresee railhead stations, moved up to the line of the army rear line. This feature of organizing railroad supply in the defensive operation proceeds from the conditions of conducting the defense in the deep army area. However, dirt road sectors should be deployed from the main supply stations and passing through the railhead stations. At the same time, the front should afford the army a broader basing.

3. Dirt road supply delivery in the defensive operation works in normal conditions as long as the fighting is being waged in the main defensive zone. Upon the breakthrough of the enemy's tank units into the operational defensive zone, the delivery of supplies along the main routes of the dirt road sectors may be disrupted. Thus auxiliary routes of the dirt road sectors, leading in a roundabout way to the main echelon's forces, should be foreseen along the axis of the likely development of the enemy's breakthrough. They should be covered by the operational defensive zone's switch lines and assure the safety of delivery in the event of a breakthrough by the enemy's tank units through the second defensive zone.

During the defensive operation's preparatory period, munitions on a scale of one to two combat loads, depending on the expected intensity of fire, should be accumulated and dispersed in the depth of the second defensive zone in anti-tank areas and under the cover of the main defensive zone's switch positions. They should in no way be concentrated along those sectors of the second defensive zone, through which the likely axis of the development of the enemy's breakthrough lie.

4. Special protection and security measures should be adopted in the rear defensive zone. One should always take into account the possibility of a breakthrough by the enemy's tank units through the army rear line and the possibility of an airborne landing and diversionary attacks. The protection of the rear zone should be foreseen by special measures, which consist of organizing the immediate defense of important targets and in the formation of individual motorized detachments for destroying diversionary bands. These measures should be chiefly carried out by the rear units' forces.

In the event of a special threat, individual units from the reserve echelon's forces will have to be detached to the rear zone.

This will be particularly necessary in the case of an enemy airborne landing. One should bear in mind that, as a rule, the conditions of organizing the operational defensive zone, all built up with an entire system of fortifications and occupied by the army reserves, exclude the possibility of an airborne landing within its confines. Should the latter be employed, then the enemy's landing will be carried out, for the most part, within the confines of the rear defensive zone. This will require the organization in the latter of a special observation and warning system and the maintenance of a special mobile reserve, ready for decisive and active operations.

XII. Anti-Aircraft Defense

1. The anti-aircraft defense of the army defensive area is acquiring particularly important significance. It has already been shown that aviation plays an extremely large role in the suppression of the defense.

Keeping in mind the limited amount of anti-aircraft artillery, it is necessary to avoid their dispersal. All anti-aircraft weapons should be concentrated for covering specific and very important sites, the determination of which comprises a special task of the army command.

Individual anti-aircraft centers, guaranteeing the secure and reliable air cover, should be organized in the army defensive area.

These centers should embrace the following:

a) the most important centers of resistance in the main defensive zone;

b) the positional area of large artillery groups;

c) an important sector of the second defensive zone, through which lies the likely axis for the development of the enemy's breakthrough;

d) major anti-tank areas in the operational defensive zone;

e) the location of the army reserves and the army headquarters;

f) the jumping-off areas for the counterblow, and;

g) supply stations.

Because there may not be enough available anti-aircraft artillery equipment for covering all of these sites, the army command should purposefully chose the most important of them, depending on its idea of the defense.

2. The organization of the army defensive area's anti-aircraft defense should not present an unvarying plan. With the defensive operation's development and the shifting of the fighting into the depth, it should change, coming to cover new and important sites in the changing situation.

Thus the maneuver of anti-aircraft weapons should be foreseen. This will become particularly important upon the beginning of the counterblow's preparation, when the main body of anti-aircraft artillery should be concentrated for covering the army reserves' jumping-off position.

3. An entire series of other measures, particularly measures for the fire protection of wooded areas occupied in the defense, important inhabited locales, crossings, and depots, should be foreseen in connection with the organization of anti-aircraft defense. This is acquiring special significance due to incendiary weapons employed by the attacker's aviation.

XIII. *The Organization of Command and Control*

1. The army headquarters in the defense is located at an approximate distance of 40–50 kilometers from the forward edge of the main defensive zone, under the cover of the army rear line and near a convenient road junction, facilitating travel to the most important sectors of the defensive front along the shortest routes. The location area of the army headquarters should be covered by an anti-tank area or an outfitted sector of the army rear line.

In order to directly observe the course of the battle in the main zone and the commitment of the army reserves into the counterblow, the army should have command posts prepared for itself behind the second defensive zone, chiefly in those areas, which have been identified as jumping-off positions for the counterblow.

Two to three such command posts may be chosen along various sectors; they should not be located along avenues of the possible development of the enemy's breakthrough and should be covered by switch positions or anti-tank areas.

With the shift of the fighting into the operational defensive zone and to the army rear line, a reserve command post, located in the depth of the rear zone, should be prepared for the army headquarters.

2. The initial decision on the defense is announced in the general operational order.

As a rule, the defensive plan, along with a motivated exposition of its entire organization and the methods of waging it, is not written down. This may take place only with the timely organization of the defense, along previously prepared terrain, and beyond immediate contact with the enemy.

However, in individual cases a particularly complex defensive situation may force us to issue, as a supplement to the order, an operational instruction, which defines the special requirements for the organization of the defense, its special tasks and methods of operation in different variations.

Such an operational instruction as a supplement to the order may often be required for the reserve echelon's forces, establishing in a detailed manner the order of concentration for the counterblow and its organization in different variations.

Operational instructions are signed by the chief of staff and confirmed by the army command.

Given the hurried assumption of the defense, so as not to delay its organization, the overall operational order should be anticipated by the issuance of individual orders, laying out individual tasks to each of the formations.

During the defensive operation all of the command's decisions are transmitted in the form of individual orders. A general order may be required only for organizing a major counterblow as a result of the stabilization of the situation achieved.

During the breakthrough of the enemy's tank units into the operational defensive zone, command and control must be especially flexible and mobile, requiring the dispatch of army headquarters commanders to the scene for the personal transmission of orders and the organization of their execution.

3. The organization of communications in the defensive operation should call for the special durability of wire communications means, which is achieved by means of broadly employing permanent lines and the broad organization of bypass lines, sometimes through neighboring armies, as well as the creation of a developed network of lateral lines and communications centers in places where the wire lines cross. Besides this, reserve centers should be prepared in the event of a shift of the army's headquarters.

As a rule, radio communications in the defensive operation works only to receive. However, specially elaborated and prearranged message code and a radio signals table should guarantee the complete opportunity to transmit a number of orders over the radio: to open artillery fire, to commit the tanks and the reserves, to fall back, and to occupy specific lines, etc. Such tables should be drawn up each time by the army headquarters, in accordance with the requirements of the situation.

In the event that some defensive sector is encircled or isolated, communications with it should be maintained by radio and aircraft.

All of the army's radio sets should observe the enemy's radio network and its own radio sets.

XIV. Political Support for the Defensive Operation

1. The conduct of the defensive operation places high demands on the troops' tenacity, endurance and staying power. A great exertion of force will be required of the troops for fortifying their positions, which should be constantly developed and improved. The troops will often have to continue their work and restore destroyed fortifications at night, following combat operations. This requires special concern for the troops, the organization of their normal feeding and offering the necessary rest, according to the situation.

A high degree of tenacity in defending positions, the great exertion of force and a willingness to sacrifice oneself all present special demands on the commissar and the political and command element for maintaining the troop's political-moral condition at a high level. When the course of the operation reaches its highest development and the moment matures for the counterblow, a particularly high combat élan should be guaranteed. Each soldier's understanding of the situation and his task will be particularly important at this time.

2. Severe measures should be adopted throughout the entire army defensive area for combating espionage and diversions. It should be borne in mind that while defending in enemy territory spies and diversionary forces will always be left behind in the army's rear. In this regard, special measures for maintaining military secrecy, vigilance and combating enemy agents will be required from the command at all levels.

3. The large area embraced by the army's defense requires appropriate political organization and the area's entire civilian population should be touched by political influence. Keeping in mind that the defense, as a component part of the front operation, will be, as a rule, conducted on the enemy's territory, this acquires particularly important significance. The entire civilian population should be evacuated from the main defensive sector to the rear defensive zone. The operational zone along the most important sectors and axes should also be freed up. The population should be enveloped by broad political agitation and should be organized for working to construction fortifications in the rear. This will be the task of the army's and corps' political sections.

Conclusions

The organization of the army's defense, as laid out above, has chiefly in mind the conditions of our western theater of military activities.

It is quite clear that the nature of the defense and the methods of conducting it are most directly dependent on the conditions of the given theater of military activities, its strategic significance and the terrain. A mountainous theater, while offering great advantages to the organization of the defense, for example, does not require such a developed army defensive area in the depth.

One should, besides this, bear in mind that when the defense is conducted in previously prepared fortified areas, relying on a system of permanent defensive structures, its nature changes materially.

The durability of the fortifications and the presence within them of their own equipment impart to the defense a significantly greater stability and ability to resist, enabling them at the same time to significantly broaden the length of the front of the field forces occupying the main defensive zone (from 3–5 kilometers per battalion).

Finally, the very strategic task being carried out by the defense is reflected in its nature in the most substantial way. If for example, one is forced to defend and hold an important area (political, economic) adjacent to the front of struggle, the question of shifting it to the depth may completely fall by the wayside and all efforts will be concentrated along the main defensive zone, for the retention of which all reserves will be employed. For example, Madrid is now in such a situation.[14]

Thus all of the special conditions of the defense change its character, organization and methods of conduct.

In general, all of the above-listed fundamentals of the organization and conduct of the defensive operation had in mind the defense that arises [in a front operation on the enemy's territory] along a secondary direction, when the further continuation of the offensive along it is impossible and, according to overall strategic conditions, is inexpedient. In this case, the defense, while economizing forces for the main direction, will always be waged with purposely small forces and in intense conditions. A counterblow in such a defense will not always find its full and decisive expression without being materially reinforced with front reserves.

However, in all cases the defense should be "impregnable for the enemy, no matter how strong he is along the given axis" (*1936 Field Manual*). The advantage of the deep army defensive area consists namely in that it secures the resolution of this task and must, on the whole, render the breakthrough of the defense impossible for the enemy and must, on the contrary, lead to his complete exhaustion and destruction.

Should a large group of forces, well supplied with equipment, be forced to temporarily halt is offensive and assume the defensive for the purpose of accumulating new men and materiel for continuing the attack {along the main direction}, which is always the origin of a positional war, then the defense, which in general is built along the same foundations, will take on a substantially different nature according to its forces and power. It will be conducted by enormous amounts of men and materiel and conclude with a very decisive counterblow, which, in the final analysis, grows into a general offensive.[15]

Aviation will participate in such a defense in enormous masses, often playing a decisive role.

In these conditions, the growth of the defensive operation into an offensive one will always be logical, thus resolving in full that goal, which is pursued by the very essence of the defense.

In a decisive offensive operation conducted at the front level, the defense will always be only a means for supporting the concentration of overwhelming numbers of men and materiel for the main attack.

The conduct of decisive offensive operations will often be impossible without combining with the defense.

The defense will be inevitable each time, when along secondary {and sometimes main} directions and given a shortage of forces, we will have to hold the enemy's offensive, and when for the purpose of winning time, regrouping and occupying a favorable position, we will have to temporarily halt our offensive movement along individual sectors of the front.

In all of these cases, the defense should be impregnable for the enemy, while expressing the entire power and might of the resistance.

However, the defense must, at the first available opportunity, go over to the offensive and destroy the enemy with a decisive attack.

Four

"The Fundamentals of Conducting Operations" (1939)

Brigade Commander Isserson
(An Outline)
Installment II: The Fundamentals of the Operation
Educational Section of the RKKA General Staff Academy, 1939

Table of Contents

Introduction	212
Chapter I. Fundamental Concepts of the Operation	213
Chapter II. Operational Formations	218
Chapter III. Fundamentals of the Offensive Operation	221
Chapter IV. The Front Offensive Operation	227
Chapter V. The Army Offensive Operation	231

Introduction

The first installment of the outline, *Fundamentals of Conducting Operations*, studies the character of our war and the fundamentals of our operational art. It places a theoretical base under the applied resolution of the problems of organizing and conducting operations by the RKKA's unified armed forces.

At the heart of the resolution of these problems lies the doctrine of **conducting combat operations to destruction and through a decisive offensive on the ground and in the air to destroy the attacking enemy on his own territory.**

"**Should the enemy force a war upon us, the Worker's-Peasant's Red Army will be the most offensive of all attacking armies that ever existed**," is how in words full of deep meaning comrade **VOROSHILOV** defines the entire essence of our doctrine, placing before our operational art the task of conducting the most decisive offensive operations, directed toward the complete destruction of the enemy's men and materiel.

Offensive operations have been conducted in all recent wars and by the majority of armies, although, for the most part, they did not yield decisive results and did not resolve the task of truly defeating the enemy.

We are faced with the task of realizing such of the offensive operation's forms and methods that lead to the consecutive defeat of the enemy throughout the entire depth and the actual achievement of a decisive victory.

The nature of our war, as the most just war of all wars in human history; the new man of [the Stalinist][1] {our} age and the abundant supply of our army with the most modern and refined means of struggle—offer us everything necessary for such a resolution of these tasks in the offensive operation.

At the same time, one should keep in mind that the enemy will also be powerful and armed with all the modern means of struggle.

All subsequent installments of the outline, *Fundamentals of Conducting Operations*, have as their task the elucidation of the forms and methods of conducting a decisive offensive operation.

However, they are not a theoretical study of these forms and methods and do not offer a full-blown theoretical basis for their resolution.

All subsequent installments of the outline, beginning with this second installment, formulate the resolution of the crucial problems of conducting operations, in the form of brief theses, and pursue the goal of an applied exposition of fundamental operational tenets.

It is quite clear that such an exposition of the outline is based on a large amount of preliminary work in studying the character of modern operations and the conditions of their conduct and is a conclusion from this work.

A brief applied exposition of operational art's tenets is always in danger of creating a certain template and mold.

It is necessary to absolutely guard oneself against this.

The tenets laid out here are not a template. They only put forward the principle forms of conducting operations, which only in a specific and real situation, and only depending it, can acquire this or that specific realization.

Besides this, one should take into account the fact that the tenets laid out here primarily concern our western theater of war and the conditions of conducting operations in it. At the same time, the outline's main tenets are primarily concerned **with those**

operations that will develop along the main directions and with a large concentration of men and materiel. It is quite clear that along a number of secondary directions the operations' deep forms will not be so fully developed.

Subsequent installments of the outline consist of three separate parts.[2]

The second installment elucidates the fundamentals of the operation.

—Brigade Commander Isserson

Chapter I. Fundamental Concepts of the Operation

1. The determination of the tasks of military operations corresponding to the political goals of the war; the employment of the armed forces for achieving these goals; the waging of war as a whole—on land, air and sea, and the organization of the country's resources for feeding war—comprise the field of **military strategy as the continuation of politics**.

The employment of the armed forces for the resolution of tasks assigned by strategy comprises the **field of operational art**.

Operational art consists of the organization, support and conduct of military operations. **Joint actions by the armed forces, directed toward resolving the tasks set by strategy, are called military operations.**

2. Military operations are divided into ground, air, naval, and combined types.

Ground operations are conducted by all the combat arms of the ground armed forces together with aviation and, with river flotillas along major river lines.

Air operations are conducted by combat aviation against the enemy's ground forces, his air force and his lines of communications, his naval forces and naval bases, and against his important political and economic centers in the country's rear.

Combat actions in the air may be conducted as follows:

a) in **tactical cooperation** with the ground forces for directly supporting their combat success and striking the enemy's men and materiel on the battlefield and in the immediate rear;

b) in **operational cooperation** with the ground forces, beyond tactical contact with them, but in the direct interests of their operation (against the enemy's operational reserves and their concentration along a given direction, against the enemy's aviation along a given direction and above the area of the operation, against the enemy's supply routes, and his supply stations);

c) in **strategic cooperation** with operations being conducted, lacking direct contact with them, but in their overall interest and in the interests of the war as a whole (against the enemy's strategic reserves and transportation; against his aviation for the achievement of air superiority; against the enemy's military, political and economic centers in the depth of his country, and; against the enemy's naval forces and naval bases).

Air operations conducted in strategic cooperation with ground operations and in the interests of the war as a whole **have a completely independent nature** and are **independent air operations**.

The chief task of aviation's combat activities is the direct support of the combat success of ground operations and the defeat of the enemy's men and materiel along the main directions.

Naval operations are conducted at sea by naval forces together with aviation against the enemy's naval forces, against his coastline, for blockading his ports, for disrupting the enemy's maritime communications, and for defending one's own coastline.

Combined operations are conducted jointly by ground, air and by naval forces along the seacoast and mainly consist of operations by the ground forces, supported from the sea; in a naval landing operation and in repelling the enemy's landing operation.

3. Operations are conducted in theaters of military activities.

The theater of military activities is that territory in which a single overall strategic task is resolved.

The boundaries of the theater of military activities, along the front and in depth, are determined by the war's goals, the strategic tasks, the enemy's geographic situation, and the geographic nature of the terrain.

Proceeding from these conditions, in warring with several countries, or with a country, the frontier territory of which is divided by a major geographical obstacle—operations may be conducted in several theaters of military activities.

The territory in which a single strategic task against several small and contiguous countries is being resolved may comprise a single theater of military activities. Naval theaters of military activities are limited by the confines of each naval basin, taken separately.

4. According to their scale, ground operations are divided into front and army operations.

An operation conducted at the level of an entire given theater of military activities, or in its greater part, by a major part of the armed forces, for the resolution of a strategic task is a front operation and is led by the front command.

An operation conducted by a limited number of the armed forces for the achievement of a partial goal along one of the directions in the theater of military activities, either as part of a front or separately, is an army operation and is controlled by the army command.

In special cases, along individual directions, limited in length, the operation may be conducted by an independent troop formation (an independent corps).

5. A ground operation consists of joint actions of the armed forces, purposefully and consistently aimed at the defeat of a specific enemy group of forces, or in opposing it.

The modern operation is characterized by the depth of its attack and is a complex system of employing heterogeneous combat efforts in a single interaction along the front and in depth, on the ground and in the air.

The art of conducting the modern operation consists of unifying these heterogeneous combat efforts, not directly tactically connected, along the front and in the depth, on the ground and in the air, into such a [service] {connected} system that they, in a single purposeful effort, lead to the task of defeating a specific enemy group throughout its entire depth.

6. Aviation taking part in the ground operation, as its component organic element, is employed for the following:

for defeating the enemy's combat formation on the battlefield in direct tactical connection with the ground activities, and;

in operational connection with ground activities for striking targets in the enemy's operational depth in the direct interests of the ongoing operation and for maintaining air superiority over the area of the operation.

7. Being a single general strike force throughout the entire depth of the enemy's group of forces, the operation consists, in its indissoluble development, of a number of separate stages, one directly growing into another. Movements, the engagement and the battle constitute the operation's stages.

Movements may be carried out by railroad, auto transport and by marching. They are carried out for the purpose of concentrating men and materiel for conducting the operation, to close with the enemy, changing their disposition, and to achieve the most favorable position for conducting the engagement and the battle.

The engagement is an act of direct pressure on the enemy, which conditions the employment of all armed efforts in tactical and fire liaison. The engagement thus plays out along a limited terrain sector, allowing for the direct tactical interaction of the combat arms.

The battle represents a combination of engagements, broken up along the front and in the depth, unified in space and time by a unity of goal and directed at resolving the overall task.

8. Having resolved the task of defeating a specific enemy group, the operation does not yet achieve his overall defeat, because the enemy, for the most part, is in a condition to oppose it with new forces.

Bringing the struggle to final victory requires a **series of consecutive operations**, developing one after the other and united by the single strategic goal of the overall defeat of the enemy.

A series of consecutive operations may develop throughout the entire depth of the theater of military activities.

For the most part, it is linked to temporary halts and pauses, necessary for the preparation of a new operation, for restoring and regrouping one's men and materiel, for concentrating new reserves, and for bringing up the rear and accumulating materiel supplies.

The development of consecutive operations throughout the entire depth of the theater of military activities requires the unbroken feeding of the attacking forces with reserves and materiel supplies, as well as their reinforcement, in case of necessity, with technical suppression means.

9. The **goal and plan** (idea) must form the basis of every operation.

The operation's tasks may vary (capturing an important area, the seizure of a favorable line, the isolation of a specific enemy, etc.).

However, the goal of activities in the operation should always be a specific enemy group slated for defeat.

The main thing in the operation is the destruction of the enemy's men and materiel. Any task is resolved upon the achievement of this goal.

The **operational plan** determines the method of activities leading to the achievement of the goal. **The drive to encircle and destroy must always lie at the basis of the operational plan**. However, the paths toward achieving this idea are quite varied. Only the scrupulous and profound study of all of the conditions of the given situation may determine them.

The goal and idea of the operation should be determined specifically and clearly. If this requirement is not observed, the operation will inevitably lead to non-purposeful actions and disunited efforts.

10. The goal and idea of the operation determine the direction of its development into the depth.

The direction, along which, according to the operational plan, the goal of defeating the enemy should be achieved, is called the **operational direction**.

The operational direction is the specific terrain area leading to important targets in the enemy's territory and allowing, according to its geographical conditions, the unification of separate efforts in a single connected and purposeful operation.

An operational direction, along which are concentrated the main forces for defeating the enemy and where the main task is resolved, is called the **main** operational direction.

11. Operations are basically divided into **two types**, according to their task: offensive operations and defensive operations. During the course of their development these two types of operations may grow from one into the other. This should always be kept in mind, so that in the development of events we take into account changes in the situation in a timely manner, determine the turning point in the operation and adopt a new decision, striving in all cases to preserve the offensive initiative.

12. **The offensive operation**, which is directed at the defeat of a specific enemy group, is, according to its content, a combination of various types of offensive actions, broken up in depth. The chief ones are:

the **approach-march**, consisting of the advance of the entire mass of men and materiel in the direction of a specific enemy group, in order to come to grips with and destroy it;

the **meeting battle**, which develops directly from the approach against the enemy attacking toward us;

the **breakthrough**, which is conducted against the enemy's occupied and fortified front;

the **development of the breakthrough**, which is conducted immediately following the breaking of the front through the open breach into the enemy's operational depth, for his complete destruction, and;

the **pursuit**, which is conducted for completing the defeat of the enemy, who is retreating or wishes to avoid a battle.

The sequence of these actions and battles, which grow from one into the other during the course of the offensive operation, **is quite varied and depends** in each separate case **on the situation and the enemy's actions**.

The offensive operation may begin both from the approach and the meeting battle and from the breakthrough and its development, which then grows into the pursuit or into a meeting battle, or a new breakthrough.

13. The **defensive operation**, which is directed at holding a particular area of terrain and resisting the enemy, when the resolution of one's task through an immediate offensive is impossible, due to the situation, or inexpedient, consists of stubborn resistance along prepared lines and a decisive counterblow, which concludes the enemy's defeat.

It may, according to its content, have various forms.

The basic ones are:

the **rigid defense** has as its goal putting up steady resistance to the enemy in one selected terrain area, while stubbornly holding on to it;

the **mobile defense** has as its goal putting up resistance to the enemy at a specific depth, with the successive backward movement of organized resistance into the depth.

I. Fundamental Concepts of the Operation

The task of the mobile defense is to preserve one's forces without meeting the enemy's full-blown attack along intermediate lines, and launching a decisive attack against him upon achieving favorable conditions for this.

the **withdrawal** has as its goal pulling out one's men and materiel from under the enemy's attack, in order to regroup them and occupy a favorable position.

In all cases, the defensive operation pursues the goals of economizing men and materiel and should create an invulnerable resistance to the enemy, no matter how strong he is. Any favorable opportunity during the course of the defensive operation should be employed for going over to the offensive for the purpose of inflicting a decisive defeat on the enemy.

14. The rapid and unexpected concentration of one's men and materiel must lie at the heart of each operation, so as to place them in the most favorable position for conducting combat activities.

Troop movements carried out for this purpose are called **maneuver. Maneuver is a means, not an end, and does not yield a result without the subsequent destruction of the enemy**.

Maneuver consists of:

the creation of groups of forces and in the regrouping of one's men and materiel;

the concentration of one's men and materiel at the enemy's most vulnerable spot;

the interaction of various groups of troops along an axis for the purpose of inflicting the most decisive defeat on the enemy;

changing the direction of the attack during the development of the operation, and;

creating a superiority in men and materiel along the decisive axes and at the decisive moment.

Rapidity, secrecy, surprise, and flexibility are the foundation of maneuver.

15. The basic and most decisive form of maneuver is the wheeling into the flank and rear for the purpose of completely encircling and destroying the enemy.

The enemy's encirclement is achieved by:

turning both his flanks;

turning one of the flanks, which may also lead to a complete encirclement, particularly if the other enemy flank is anchored on an impassable or inaccessible terrain obstacle, and;

penetrating through the broken front into the enemy's operational depth, so as to attack him from the rear.

In all cases, maneuver should be accompanied by a simultaneous attack from the front, so as to tie down the enemy's forces being encircled.

The encirclement of the enemy should also be supported by his isolation from the air, in order to cut him off from his so that he can't feed himself. The encirclement may also be combined with an airborne landing in the enemy rear. The maneuver should fool the enemy and for this purpose should be accompanied by demonstration actions that conceal the true direction of the main attack.

16. If the enemy has no open flanks for developing the maneuver, the **flanks must be created by breaking through the front**. The breakthrough should already lead to a maneuver in the forms of its attack, and for this purpose it is expedient to:

break through the front along two sectors along converging axes, or

while breaking through along one sector, to develop the attack toward one or both flanks.

Following the breaking of the front, the breakthrough should be developed into the operational depth, where the most decisive maneuver into the flank and rear should lead to the enemy's complete encirclement and destruction.

17. Unexpected and stunning actions yield the greatest result in the fighting. In organizing and conducting the operation, it is thus necessary to try and manifest **complete surprise.**

Surprise is achieved by:
the rapidity and secrecy of the troops' movement;
the employment of new means of struggle *en masse*;
the observation of all preparations for the operation in complete secrecy, and;
carrying out false activities and demonstrations.

18. The modern operation requires the expenditure of enormous materiel supplies and the organization of their unbroken delivery to the front.

Any operation may be undertaken and justified only in if the necessary conditions have been strictly taken into account and observed.

The calculation and accumulation of the necessary materiel supplies and the organization of their deliver are therefore one of the most important tasks in organizing the operation.

Chapter II. Operational Formations

19. The armed forces, which are designated for waging war, consist of the ground forces, air force and navy.

In order to wage military operations, the ground forces deploy in a particular group of forces, forming front amalgamations (fronts) and army formations (armies) and, in special cases, independent groups.

20. The **front** unifies the armed forces in a given theater of military activities, or the greater part of it, and embraces several operational directions, along which, according to their geographical shape, the resolution of a single overall strategic task is achieved.

The front is a strategic instance and resolves strategic tasks assigned by the High Command.

In special cases, the armed forces carrying out an independent strategic task along a separate operational direction or in a small isolated theater of military activities may be united into an **independent army.**

21. The **army** unites the troop formations of various combat arms along a single operational direction and is the **chief operational formation.**

In uniting the individual tactical efforts for achieving the defeat of a specific enemy group along a given direction, the army resolves operational tasks set by the front command.

An army, which consists chiefly of combined-arms formations and aviation, is called a **field** army. A field army which attacks along the main operational direction and which is especially supplied with reinforcements is called a **shock army.**

An army, which consists primarily of cavalry and tank formations, is called a cavalry-mechanized group (KMG).[3]

An independent group or an independent corps, directly subordinated to the front command, may operate along a separate operational direction that does not require significant amounts of men and materiel.

22. Depending on its importance and tasks, a front may consist of several field armies, a KMG, aviation formations from the various aviation arms, airborne units, various types of reinforcements (tank, artillery, chemical, and engineer, etc.), and front reserves from various formations, a number of specially-designated auxiliary weapons and services, and materiel supply installations constituting the front rear.

23. The front air force, depending on its composition, may form several aviation groups.

Front aviation is employed for directly facilitating the combat success of the front's armies and for resolving independent tasks in the operational depth of the theater of military activities (combating the enemy's aviation, disrupting the enemy's railroad and auto shipments, suppressing his reserves, the enemy's materiel exhaustion, and the destruction of important targets in his rear).

During decisive periods of the front operation, all of front aviation concentrates its main efforts on defeating the enemy's main group of forces along the main direction.

Besides this, the front may be supported by the High Command's aviation, which usually consists of an air army with an independent strategic mission.

The depth of front aviation's activities is determined by the High Command, in accordance with the front's tasks, and should support the concentration of all air efforts in that depth in which the outcome of the front operation is being decided at a given stage. This will usually be 200–300 kilometers.

24. The army (field) is the basic executor of the ground operation and is equally suitable for all types of maneuver and battles. Being the chief bearer of the shock force's penetrating power, it may independently resolve any task in the operation.

The army (field) may consist of a group of 3–5 rifle corps; cavalry and tank formations; various reinforcement means (artillery, engineer, chemical, and others), aviation formations, an airborne unit, and a number of specially-designated auxiliary means.

Depending on its purpose and task, the field army's composition and the degree of its technical outfitting may vary.

The inclusion of cavalry formations in the army is not required and is expedient particularly in those cases when the army is operating along the flank and enjoys definite maneuver opportunities.

25. The shock army, according to its composition, should dispose of:

the capability of independently conducting an operation throughout the entire depth of its assigned mission;

penetrating power for overcoming the enemy's resistance of any kind and character, and;

the capability of deeply striking the enemy throughout the entire depth of his operational position.

Proceeding from these requirements, the shock army may consist of 4–5 three-division corps (12–15 rifle divisions), a cavalry corps (2–3 cavalry divisions), 3–4 tank and 1–2 motorized rifle brigades for independent actions; 4–5 tank brigades for reinforcing the rifle formations; 10–12 High Command Reserve artillery regiments for quantitative and qualitative reinforcement, 4–5 aviation brigades of bomber, assault and fighter aviation, airborne units, and special-designation units.

Such a shock army may attack along an average front of 50–80 kilometers, depending on its task and the strength and the nature of the terrain in the offensive sector. Such an army's complete daily requirements can be measured at 15,000 tons. The supply of such an army requires:

about 36 trains along the railroads (including trains necessary for restoring the railroads), and;

up to 20–24 auto transport battalions of 2.5-ton trucks along the dirt roads.

26. The army's aviation will form an army aviation group (AGA), which will mainly consist of short-range bomber, assault, fighter, and reconnaissance aviation.

The AGA operates in direct tactical and operational connection with the army's forces and has as its chief task facilitating their combat success.

The AGA's chief targets are:

the enemy's combat formation on the battlefield;

the enemy's reserves;

forward depots, supply stations and dirt roads for supplying the enemy, and;

his troop aviation's forward airfields.

The AGA's depth is determined by the front and is measured by the depth of the army's activities at a given stage. This will usually be from 60 to no more than 100 kilometers.

The more the ground efforts directly oppose each other, the more the depth of the aviation's activities should diminish. During the beginning of the breakthrough the AGA's combat activities may comprise no more than 20–25 kilometers. Accordingly, the depth of front aviation's activities must shrink, which during decisive periods in the operation may be employed at the same depth as army aviation.

As a rule, the AGA is employed in a centralized fashion in the hands of the army command.

In special cases, the AGA's formations may be attached during separate stages of the operation to the troop formations operating along separate axes. As a rule, the AGA's units should always be attached to highly-mobile formations (cavalry and tank) for accompanying their maneuver and supporting their attack.

27. The KMG is a formation of strategic significance and the main strategic means of rapid maneuver, attack and capture.

Through the rapidity and unexpectedness of its activities, the KMG may play a decisive role in the development and outcome of the front operation. Disposing of great shock power, it is capable of independently conducting a battle with a decisive goal and consolidating its success. However, at the first opportunity it should be freed from the necessity of holding space, in order to retain it as a means of maneuver.

The KMG's composition may vary. It may consist of the following: 6–8 cavalry divisions, individual or unified into cavalry corps of 2–3 divisions; several tank and motorized rifle brigades, separate or unified into tank groups; several brigades of combat aviation and various reinforcement and support means (artillery, chemical, engineer, and transport, etc.).

28. The KMG is employed in the front operation:

given the absence of contact between the sides, ahead of the front for attacking and destroying a specific enemy group and creating a favorable flanking situation for developing the offensive operation;

in the maneuver development of the operation, along the front's turning wing for operating against the enemy's flank and rear, and;

in the breakthrough, for developing the breakthrough through the breach in the front into the enemy's operational depth for the purpose of completely defeating him.

The diversity of the situation must dictate the employment of the KMG in each case that offers the opportunity for decisive maneuver activities. At the same time, we should strive to direct the KMG's activities into the enemy's flank and rear for the decisive purpose of encircling and destroying him.

In modern conditions the KMG will often be forced to operate frontally against the enemy's mobile or hastily organized defense.

The KMG must therefore be ready to independently overcome oncoming frontal resistance.

Upon the establishment of the enemy's continuous immobile front, the KMG should be relieved in a timely manner by the field army and pulled back into the immediate rear, while waiting out a favorable situation for developing the success and renewing its decisive maneuver into the enemy's operational depth.

In this situation, the KMG is a decisive means in the hands of the front command for developing and concluding the front operation.

While employing the KMG, one must bear in mind that after 3–4 days of activity it needs a brief rest and restoration.

In these conditions, the depth of its penetration and loss of contact with the overall front is measured as far as the first halt by a distance of up to 100–200 kilometers.

29. The armies conducting the operation are not, in their composition, permanent operational formations. For the most part, with the development of the operation into the depth a new situation arises, which requires the strengthening of some armies and the weakening of others.

Finally, the situation may bring about the necessity of creating new armies. All of these measures, which come down to the regrouping of men and materiel, are carried out at the front level and require of the front command a great deal of flexibility in managing the operation.

The number, composition and disposition of the armies should not be standardized and must always correspond to the given specific situation in the theater of military activities.

Chapter III. Fundamentals of the Offensive Operation

30. The offensive operation is the RKKA's main and decisive type of conducting military operations and pursues the goal of destroying the enemy on his own territory.

It is conducted by pushing forward on land and in the air all available men and materiel, organized and directed for launching crushing and deep attacks, following one after the other, up to the achievement of the final goal.

These attacks are carried out:

a) by the striking power of aviation throughout the entire depth of the enemy's operational position;

b) by an attack of all the ground combat arms from the front and into the flank and rear;

c) by a breakthrough of cavalry and tank formations into the enemy's operational depth.

31. The chief goal of the offensive operation is the destruction of the enemy's men and materiel. Any task is resolved through the achievement of this goal, for example: the occupation of important geographical areas, the occupation of the enemy's territory and the capture of his political and economic centers; depriving the enemy of bases necessary for him to wage the struggle, and, finally, compelling him to cease the struggle. Thus the enemy's group of forces is the main object of the offensive operation.

Offensive operations will often have as their main task the occupation of important areas and the capture of favorable lines. However, the chief means of resolving this task will always be the destruction of a specific enemy group standing athwart the path toward a given target.

32. In the joint employment of troop formations of all the arms, the chief ones of which are combined-arms (infantry) formations and, supported by aviation, cavalry and tank formations, airborne units and chemical weapons, the destruction of the enemy in the offensive operation should be achieved by a deep attack throughout the entire depth of his group of forces. In this manner, the withdrawal of the main body of the enemy's forces should not be permitted and all of his group of forces should be suppressed and destroyed throughout the entire depth.

The deep defeat of the enemy retains its significance in all types of offensive operations, whether this is the approach march, the meeting battle or the breakthrough, changing only its forms.

The offensive operation must therefore be calculated throughout the entire depth and must be ready to overcome the entire depth.

At its conclusion it should play out as a struggle conducted on the ground and in the air at several levels of the overall depth and take on the appearance of a multi-tiered battle that destroys the enemy in the entire depth.

33. During the approach march, given the availability of open distance between the sides, the essence of the deep strike consists of the long-range attack against the enemy's personnel. One should take advantage of any opportunity in order to push one's operational efforts forward and to suppress a specific enemy group before he has time to get closer and present an organized front of struggle. In this regard, the capabilities of aviation and highly mobile cavalry and tank formations should be employed in all possible ways. Any retention of the highly mobile formations in the same line as the attacking front or, even more so, in the reserve, deprives them of their chief quality of speed and mobility as the capability of pushing their shock power forward.

Thus, **in corresponding conditions of the situation and with a corresponding correlation of forces**, one should, given the presence of a spatial gap between the sides, push the highly mobile formations forward for long-range pressure on the enemy.

At this stage, the deep strike means that the forward enemy group of forces is destroyed before it has time to organize its own attack or to turn into a continuous front of fire, requiring a breakthrough. This task is accomplished by pushing up the **forward echelon** of highly mobile cavalry, tank and motorized formations, which in direct cooperation with aviation, attack a specific group of the advancing enemy and, destroying it, in this way disrupt the integrity of the enemy front and immediately cause it to waver.

The range of the forward echelon's movement is always determined by the possibility of supporting it with the main forces at the decisive moment and should never place it in danger of a local defeat. Depending on the situation, this range may be measured at a distance of up to 40–50 kilometers and, only in certain cases, greater.

The attack by the newly-arrived main forces, which immediately develop the forward echelon's success, and the further penetration of the highly mobile troops into the enemy's depth, should ensure its overall defeat in the mobile battle. At the same time, the airborne landing in the enemy's rear may play a major role. One should bear in mind that the enemy may also dispose of a powerful group of highly mobile formations and push them forward. In this case, the forward echelon's task consists in first of all routing the enemy's highly mobile forces and achieving superiority in the operational forward outpost area.

The forward echelon should always be supported by powerful combat aviation.

At the front level, the KMG may carry out the task of the forward echelon.

34. Upon the establishment of the front of struggle, which requires a frontal attack and a breakthrough, the essence of the deep strike consists of the following:
- that the tactical breaking of the front immediately grow into the operational development of the breakthrough into the depth;
- that the enemy's entire depth be struck by a deep attack, and;
- that the entire enemy front being broken through be isolated from the air from his rear so that it cannot feed himself.

At this stage the deep strikes consists of developing the breakthrough through the breach in the front directly into the depth and, through a deep attack, isolate the enemy from his rear, in order to completely encircle and destroy him.

This task is carried out:
- through activities by long-range bomber aviation against the enemy's deep rear, for the purpose of isolating the entire breakthrough front;
- through activities by short-range bomber and assault air aviation against the enemy's defensive position on the battlefield, and;
- by carrying out the breakthrough in two echelons:
the attack echelon, which breaks the front, and;
the breakthrough development echelon (ERP), made up of highly mobile cavalry and tank formations, which are thrown forward through the breach into the depth, in order to rout, together with aviation and an airborne landing, the enemy's reserves and, in conjunction with the attack echelon, to destroy the defending enemy from the rear.

35. A mandatory requirement for organizing the deep strike consists of such an interaction of attacks along the front and into the depth that they, in a single, operationally unified effort, purposefully lead to the joint defeat of the enemy.

Pushing forth the deep strike to such a distance, which as a result of its great distance isolates the attack in the depth and makes any kind of cooperation with the attack along the front impossible, leads only to the dispersal of forces into the depth and cannot yield a result, and is therefore **inadmissible**.

The range of the deep strike must be defined in each separate case by the operational depth of that enemy group, the destruction of which is the goal of the operation.

The maximum line of this depth should, as a rule, be determined by the location of the main railroad supply stations on which a given enemy group is based. On the average, this depth may be measured at a distance of up to 40–60 kilometers and only in special cases may it be greater. Any extension of the deep attack beyond the confines of these distances must inevitably place it beyond the confines of the battle area and to **its dropping out of the overall sum of efforts directed at the defeat of the opposing enemy**.

In all cases, the significance of the deep attack must, in the final analysis, manifest itself in the fact that it is turned directly into the rear of that enemy group which is simultaneously being attacked from the front.

36. An important task of the offensive operation is the preliminary suppression of the main body of the enemy's combat aviation. Only the achievement of air superiority can secure freedom of development for the offensive operation.

The offensive operation should thus be prepared by a powerful attack against the enemy's aviation.

This task is accomplished, as a rule, by front aviation. However, army aviation may be brought in to resolve it, in case of particular necessity, if the air enemy's active operations created a particular threat.

However, one should not count on the rapid resolution of the task of suppressing the enemy's aviation. Usually the struggle for air superiority will require a long time. Thus independent of the struggle with the enemy's aviation, the forces conducting the offensive operation should be securely covered from the air.

37. The goal of the offensive operation is resolved along selected operational directions, which in each individual case are recognized as those most corresponding to the achievement of defeating the enemy. It is impossible to achieve the offensive operation's resolution along all axes. The striving to attack everywhere actually leads to an inevitable dispersal of forces and the weakness of each attack in isolation; it cannot lead to a decisive result. The offensive operation must therefore be conducted through the concentration of an overwhelming superiority of men and materiel along the selected axis of the main attack, in relation to which the remaining axes will be for pinning down the enemy. Depending on the scale of the operation along the front and the availability of men and materiel, there may be one main attack axis or several. However, in all cases **decisive significance** must be accorded to one of the axes: it must have a clearly expressed superiority of men and materiel on the ground and in the air.

38. One of the most important tasks in the offensive operation is the establishment and maintenance of such cooperation between the main and pinning directions, so that having tied down as much of the enemy's men and materiel along a broad front as possible, it can thus secure the offensive's decisive development along the main attack axis.

The pinning axis should, as a rule, be assigned active missions, but calibrated to the amount of men and materiel operating here and with those of the opposing enemy.

The forces operating along the pinning axis should not be put in danger of a local defeat that could, in the conditions of an enemy breakthrough along the boundary with the main direction, lead to the disruption of the entire offensive operation.

Securing the boundary between the main and pinning axes therefore comprises a subject of particular and important concern. When faced with the threat of powerful enemy pressure along the pinning axis, one should not hesitate to go over to the defensive and even the withdrawal to such a depth that does not disrupt the integrity of the entire situation at the front.

At the same time, we should not hesitate to leave behind along the pinning front weak rearguard units, reinforced with chemical and engineer equipment, in order to, having created here a powerful zone of chemical obstacles; to regroup the main body of forces to other directions.

In this case the enemy, having been pulled along the pinning axis, will all the more easily be outflanked from the main attack axis and, as a result, encircled and destroyed. In these cases, the enemy may have a decisive defeat inflicted on him before he can reach the final line of the offensive operation.

However, at the same time one should take into account that activities along the pinning axis should in no way threaten the operation of the main forces and should reliably secure the boundary with them.

In any conditions, the activities along the pinning axis should in all regards be subordinated to the interests of the main attack.

39. The axis of the main attack determines the form of the offensive operation.

This form is possible in the following types:
- the turning attack with all forces against one of the enemy's flanks;
- an attack against both flanks along converging directions, putting the enemy in pincers, and;

—a frontal attack from the center for the purpose of splitting the enemy's front.

The offensive operation's forms should always strive to strike the enemy's entire depth and to encircle and destroy him.

40. The definition of the offensive operation's goal and idea, the choice of the main attack direction and the appointment of men and material for the shock group comprise, on the whole, the plan of the offensive operation.

The selection of the main attack axis has a decisive significance in the offensive operation's plan.

In the final analysis, to adopt a decision to attack means to decide with what goal, where, with what amount of men and materiel, and how, the main attack is to be launched.

41. Proceeding from the task imposed by strategy, the offensive operation should foresee the overall prospects of its development all the way to the achievement of the final goal.

The planning of the operation should foresee this prospect in the overall calculation of time and space, for without this the operation cannot be supported by the necessary supply of materiel and the adoption of timely measures for feeding it.

However, **the operational plan may be determined only up to the defeat of the immediate enemy**. The outcome of such a battle gives rise to a new situation and requires a new decision.

Much of what we had in mind before becomes incapable of accomplishment and, on the other hand, much that could not have been considered earlier becomes possible.

42. In its development, the offensive operation cannot be confined to a single unchanging main attack axis from beginning to end.

The conduct of the operation along unchanging axes, without regard to the newly dawning situation, results in a unilinear and indiscriminate advance into nothing and inevitably loses sight of the main goal of defeating the enemy's forces.

Depending on the newly arisen situation, the main attack axis in the operation may thus change, requiring, in one case, turns and flanking movements and, in another, a regrouping of men and materiel. Flexibility in the conduct of the offensive operation consists of making sure that these turning points in the course of the operation should be comprehended **in time** and that each time a new situation arises new decisions are made immediately and purposefully.

However, without new data on the changing situation the operational plan adopted should not change and should be taken to its final decisive outcome with all firmness and inexorability, even under the burden of a difficult situation.

43. In modern conditions a threat to the flank and rear may bring about local complications in the development of the offensive operation.

This will be most likely along open flanks. However, even given continuous flanks the fluctuations of the front and the breakthrough by the enemy's motor-mechanized and cavalry units along boundaries may create an unexpected threat to the flank and rear.

Long-range intelligence along the threatened directions should warn of a developing threat in time.

The main thing is to establish the threat from the very beginning, when the enemy is still only concentrating his forces for an attack against the flank and rear. The enemy's maneuver must be nipped in the bud.

The employment of a massed attack from the air may, in specific cases, secure the flank against a developing threat.

On the ground resistance measures may be passive or active, depending upon the situation.

If the offensive operation is close to its successful resolution and should not be broken off, the threatened flank is secured by moving forward a screen, which bars the enemy's path along favorable axes by taking up the defense and constructing powerful engineer and chemical obstacles. In these cases, contaminating the terrain will be particularly profitable. The further forward the screen's front is moved for securing the flank, the more freely can the operation continue along the main attack axis.

If the developing threat to the flank and rear will make it impossible to continue the operation, it is necessary, depending on the situation, to first gain freedom of action and turn against the new enemy group with all one's forces.

In this case, an immediate regrouping and decisive meeting battle along a new axis must pursue the goal of quickly routing the enemy who has appeared along the flank, for the purpose of, upon completing this task, returning to the previous axis and continuing the interrupted operation.

Any half-way decision and desire to resolve the task through simultaneous active operations along the former attack axis and against the new enemy group usually lead to subordinating oneself to his will and may result in a serious defeat and should not be allowed.

44. Depending on the nature of the enemy's operation and his forces, the offensive operation may, in its development, encounter the enemy in a different sequence of his attack, having gone over to the defensive, falling back, and going over to a counteroffensive.

The offensive operation will therefore have to attempt to overcome the entire depth, up to the achievement of the final goal, in an unbroken series of combat efforts, growing in a different sequence from the approach march into a meeting battle, into a breakthrough, the development of the breakthrough, the encirclement, the pursuit, and the repulse of a counterblow, etc.

At the same time, one should expect the greatest intensity and the operation's crisis at the end, when the enemy, facing the threat of a complete defeat, will be forced to manifest the entire strength of his resistance in order to save the situation. The offensive operation must therefore bring the troops to the accomplishment of its final goal, fully equipped with the men and materiel necessary to completely defeat the enemy.

For this the deep echeloning of men and materiel and the presence of powerful reserves is required. Only in this condition will the force of the attack be able to grow along with the development of the offensive and achieve its maximum strength at the final concluding stage of the operation, when all available men and materiel must be thrown into the fighting along the decisive axis for achieving the complete defeat of the enemy.

Chapter IV. The Front Offensive Operation

45. The overall defeat of the enemy's main forces in the theater of military activities and carrying the struggle to final victory is resolved at the front level by the front offensive operation, which embraces the entire given theater of military activities, or a large part of it.

Local goals of defeating a separate enemy group and occupying a specific area of his territory may be resolved at the army level through an independent army offensive operation, conducted along a separate direction.

46. The depth of the front offensive operation is determined by the strategic mission, as well as by the distance of important strategic lines, the capture of which gives us access to the enemy's major centers, deprives the enemy of a specific operational base and forces him to shift the fighting back or renounce it altogether in a given theater. Such lines can be: lines of fortified areas, a major river line, a mountainous or wooded and swampy area, an important lateral communications line, a line of major railroad junctions, and political and economic centers.

Depending on the conditions in a given theater of military activities, such lines may lie at a depth of about 200 kilometers.

A single front offensive operation is conducted throughout the entire depth, upon the accomplishment of which, depending upon the developing situation, a new operation begins, which grows out of the previous one. The front operation, in favorable conditions of its development, may immediately be continued further. However, for the most part, overcoming a depth of up to 200 kilometers, and sometimes less, will require the organization of a new operation and the regrouping of men and materiel.

The entire depth of the front operation is broken up into a series of consecutive stages, which are determined by the capture of intermediate lines. At the same time, the front proceeds from its operation's overall strategic mission and its final goal in a given theater of military activities.

47. The front conducts the offensive operation with a deployed echelon of army formations, which occupy, on the average, a depth of up to 100 kilometers.

The KMG, as a rule, occupies a place along the front's outflanking wing and, given the presence of an open distance between the sides, pushes its efforts ahead of the attacking front in the capacity of its forward echelon.

The front reserves are located in several groups in areas of major railroad and dirt road communications lines and along the directions where they are likely to be employed, on the average, at a depth of 100–200 kilometers. They are supported by operational maneuver means along the railroads and by auto transport, designed to commit them into the operation in a timely manner.

In a case where the front reserves are slated for directly developing the first echelon's attack, or lengthening its flank during the course of the operation, they should be employed along the planned axis at a distance of a 3–4 days' march, thus forming the front's second echelon.

Front aviation groups, based on a network of permanent airfields and landing strips, are located in the front's depth at a distance of 200–400 kilometers.

The front's base depots and all the means for feeding the front operation are usually located at this same depth. The front's men and materiel are therefore deployed at an average depth of up to 400 kilometers.

48. The front's deployed echelon of army formations is grouped, in accordance with the operational plan, along operational directions, forming shock and pinning groups of forces. Depending upon the scale of the given theater of military activities, the available men and materiel and operational plan, the shock groups may be as follows: one along the front's outflanking wing, or in the center, or along two of the front's enveloping wings.

The front's shock group should contain a penetrating strength capable of breaking any resistance encountered and overcoming the entire depth of the offensive operation.

Depending upon the front's composition, the significance of the theater of military activities and the strategic mission being resolved there, the shock group may consist of 1–3 shock field armies and the KMG. The pinning group of forces may consist of 1–2 pinning armies. Each field army has its own operational direction and attack zone, with a width depending upon its strength:

50–80 kilometers for the shock army;

80–100 kilometers for the pinning army.

The KMG is assigned a separate direction, not limited from the open flanks.

49. The front, in controlling the course of the offensive operation and organizing an attack throughout the entire depth, does the following:

• employs its aviation against the enemy's rear and personnel to a depth of 200–300 kilometers, in the interests of the ground fighting;

• systematically suppresses the enemy's aviation and maintains air superiority throughout the entire operation;

• assigns tasks and indicates directions to the armies, organizes the interaction of the shock and pinning groups of forces and their armies by regulating movement along directions and changing their directions, depending on the situation;

• directly controls the KMG as the chief means of front maneuver and for developing the attack into the enemy's depth;

• dispatches an operationally-significant airborne landing, if it is employed, for operating in the enemy rear, in connection with the front's overall maneuver;

• directly organizes the augmentation of the attack and the feeding of the operation from the depth with the forces of the front reserves, committing them along the main directions for achieving decisive results, and;

• organizes the uninterrupted delivery of materiel-technical support means along the railroads and automobile roads, as far as the forward depots.

The front directly influences the development of the front operation; it organizes the deep defeat of the enemy in the theater of military activities and, upon the penetration of highly mobile troops into the enemy rear, controls its general defeat in depth.

At the same time, the front is obliged to directly control the shock group's offensive and, given its makeup of several shock armies, secure their interaction and direct support by aviation.

50. The flanking maneuver by the front's shock group should be conducted at high speed, so that the enveloping shock wing reaches the enemy's flank and rear before he can slip out from under the blow. At the same time, the scope of the turning movement should be proportionate in such a way as to immediately create a threatening situation for the enemy. For this, the turning movement should not be directed at the emptiness of the deep rear, although it should not be directed against one of the flank's extremities.

The KMG ahead of the enveloping wing of the front's shock group, as its forward echelon, has a decisive significance in the turning movement and should appear first in the enemy's rear, setting about its encirclement in coordination with the airborne landing here.

51. One should bear in mind that in modern conditions one should not count on the free development of large turning movements to a significant depth. The opportunities for the enemy's operational maneuver may before long oppose a new group of forces to such a movement, while his aviation and obstacles may create significant barriers. Therefore, in the final analysis, the shock group directed toward outflanking the enemy will encounter a newly organized front.

Front aviation should establish the concentration of the enemy's new group of forces in time and suppress it with an air strike. The front's forward echelon should complete the air attack with its own impulsive attack on the ground and disrupt the integrity of the enemy's group of forces before it has time to organize its own front of struggle.

The subsequent attack by the shock group's main forces should inflict a decisive defeat upon the enemy.

However, if it has not been possible to do this fully and the enemy manages to organize a defensive front, the shock group must be ready to directly transform the flanking maneuver from the march to a frontal attack, with its development into the enemy's depth, with all its available highly mechanized weapons.

In modern conditions the formation of a continuous front during the course of the operation, or the enemy's timely assumption of the defense, will most often be a **logical phenomenon**.

Even an insignificant interruption in the operation always causes the formation of a defensive front, which is often based upon previously developed lines or permanent fortifications already created in peacetime.

These conditions, in embryonic form, always lead to the appearance of positional forms of war. In all of these cases, the **breakthrough of the defensive front** is necessary, requiring the concentration of the greatest number of men and materiel for carrying the operation to a decisive outcome.

52. The front is obliged to organize the breakthrough of the enemy's defense from the march throughout the entire depth and to pull back the highly mobile formations into the second echelon in time, while employing the KMG as the front ERP for penetrating the broken enemy front into his depth.

However, in breaking through heavily fortified zones, the operation will require special preparation, the concentration of reserves, the regrouping of men and materiel for deploying the shock group, and for organizing the rear. Depending on the situation, this may occupy several days in time, and sometimes a lengthier period.

All the preparations for the breakthrough should be carried out under the cover of the beginning of our aviation's combat activities, which carries out the **aviation preparation of the breakthrough** by suppressing the enemy's combat formation in the defensive zone, and of his aviation, reserves, and rear.

The breakthrough by the front's shock group is accomplished, depending on the nature of the theater of military activities, by a closed formation of shock armies, or along separate adjacent directions. The pinning group is brought in to the breakthrough along the boundary with the shock army, thus supporting its flank.

The front ERP's KMG may penetrate through the broken enemy front in separate groups along different axes, or be thrown forward as a whole through the breakthrough breach, which has been broadly punched through along the boundary of two shock armies. As a rule, the front ERP is committed into the breach when it has already reached a depth of 15–20 kilometers, with the mandatory capture of the enemy's second defensive zone, and when the width of the breakthrough has reached approximately 20–25 kilometers.

Combat aviation and the shock armies, along the boundary of which the front ERP is committed, are obliged to adopt all measures so that the enemy's reserves cannot reach the new breakthrough front and close it sooner than the KMG's forces can pass through it. The KMG, which is preceded by the airborne landing and supported by combat aviation, breaks through into the enemy rear throughout the entire depth of his operational formation, on an average of up to 100 kilometers.

At this depth the KMG destroys the enemy's operational reserves, paralyzes his rear and resistance capabilities and, in conjunction with the shock armies attacking from the front, concludes the operation by the encirclement and destruction of the main enemy group of forces.

Front aviation blocks the enemy's deep reserves' access to the breakthrough front, isolates him from the deep rear, paralyzes the delivery of supplies, and assists the front breakthrough development echelon.

Thus the front breakthrough operation should lead to the enemy's complete rout.

One should bear in mind that it will not always be possible to conduct the front breakthrough operation with a front ERP. The development of the breakthrough may be resolved at the level of the shock armies, by means of army ERPs.

However, in this case the establishment of coordination between the adjacent shock armies' ERPs will be required, when the fighting shifts to the enemy's depth.

53. The pace of the front offensive operation's development along the main directions depends on the force of the offensive blow, on the enemy's powers of resistance, the nature of the theater of military activities, its engineer preparation, available transportation means, and the pace of restoring the railroads.

Depending on these conditions, the average daily rate of advance by the front's shock group may vary. The march speed should, on the average, be as follows:

20–25 kilometers for combined-arms formations, and;

up to 50 kilometers for the forward echelon's units and the breakthrough development echelon.

The pace of the front operation's development throughout the entire depth may reach up to 12–15 kilometers per day.

The restoration of the railroads at an average pace of 8–10 kilometers per day, with a capacity of up to 18 pairs of trains for a single-track railroad and up to 36 pairs for a double-track railroad, will be capable of supporting such a pace of advance for the front operation in conditions when each shock army has a single railroad. However, one should keep in mind that the pace of the operation is the product of many givens and must be specially calculated each time, proceeding from the conditions of the given specific operation, and its character and content. This comprises a crucial task for the front in elaborating and planning the front operation. **In this case, the enemy's resistance and stability are most important**. It is also necessary to keep in mind that in modern conditions more time will often be required for preparing the operation than in conducting it. This touches, in particular, upon the preparation of major breakthroughs against the enemy's powerfully fortified zones. Therefore, on the whole, the average pace of advance in the front operation may prove to be significantly lower, depending on the time spent on its preparation.

54. The front operation, which unfolds in the entire theater of military activities, is a complex phenomenon and is suffused with a great variety of the most varied army operations, which unfold simultaneously along a large front and to a significant depth.

Each army, operating within the confines of the front, will, for the most part, be placed in conditions of a specific operational situation and carry out its own special army operation, different from the others, but comprising together with them a single whole, united in the front operation. The front's different armies may simultaneously break through the front, defend and even fall back. During the front operation the front is obliged to unify the course of various army operations for accomplishing the overall goal of defeating the enemy and in this manifest its art of control. It is impossible to count on achieving a final decisive victory in a single unified front battle. For the most part, the geographical character of the theater of military activities will determine how the fighting unfolds around specific areas and along individual directions. In these conditions, the front will reach its final and decisive victory through a series of independent decisive battles. On the outcome of the latter at the army level depends, in the final analysis, the achievement of the overall goal and the outcome of the front operation as a whole.

Thus to each army in the front operation belongs an independent place, which presents its own, special and independent demands to the conduct of the army operation.

Chapter V. The Army Offensive Operation

55. An army, which operates within the confines of a front, is the executor of the front operation and independently carries out an army offensive operation along its assigned operational direction, in conjunction with the other armies.

An independent army and an army that is attacking along an independent direction, although as part of a front but lacking direct contact with the other armies, operates on the overall basis of the army operation, disposing of, however, more freedom in the choice of means and goals for resolving its task.

In favorable conditions, an army offensive operation may develop to the depth of the front operation (an average of about 200 kilometers). In this case, the front operation's stages and intermediate lines determine the army's immediate and subsequent tasks. However, in conditions of the front operation's intensive development and given the participation of major forces on both sides, each stage of the front operation will usually be a particular operation for the army, with its independent content. In this case, the depth of the army operation is determined by the depth of the front operation's stage (this may be from 60–100 kilometers, depending on the situation).

The shock army should be ready to independently resolve its assigned task throughout the entire depth.

56. The army conducts the offensive operation with a deployed front of combat formations, which form its **main echelon**.

The army's highly mobile cavalry and tank formations are pushed forward, given the corresponding conditions of the situation and the presence of free distance between the sides, forming the army's forward echelon.

Given the absence of an opportunity for free maneuver, they are pulled back into the second echelon and, as the army ERP, are designated for developing the attack into the depth through the enemy's broken front.

The army reserves, which form the army's **reserve echelon**, follow along the axis of the main attack, at a distance of 1–2 days' march from the main echelon.

The army's aviation group (AGA), is based on landing strips and is located in the depth at a distance of 60–100 kilometers.

The army rear organs, forward depots and supply stations are located at this depth.

Only the shock army may have such a deployed formation. The weaker pinning armies may not have the means for creating a forward echelon and an ERP.

The reserve echelon in these armies may be represented only by a weak force, or be absent altogether.

57. The army's main echelon will form shock and pinning groups in accordance with the operational plan adopted.

Depending on the army's composition, the width of the attack front and the operational plan, the shock groups may be as follows: one along a single shock flank or in the center, or two along both shock flanks. Thus, there may be one pinning group along a single pinning flank, or two along both pinning flanks.

Depending on the army's composition, the shock group may consist of 2–4 reinforced rifle corps. The pinning group usually consists of one rifle corps.

Each rifle corps is assigned the following attack zone width: 8–12 kilometers for the shock group's reinforced corps, and 12–18 kilometers for the pinning corps.

The army's forward echelon should consist of all the army's highly mobile cavalry, tank and motorized infantry formations and be supported by the AGA. It may also be reinforced by rifle troops delivered by auto transport.

The forward echelon is assigned an axis in accordance with the operational plan. It should, as a rule, operate in conjunction with the shock group and may be assigned an axis along its flank, attacking echeloned in front. In this case, it should receive along the flank a free zone of corresponding width, in anticipation of joining up with the main echelon's overall front and its withdrawal into the second echelon upon the appearance of conditions for a frontal battle. The reserve echelon will usually be formed from individual divisions. It is expedient to support it with auto transport for a rapid movement.

58. The army's offensive operation, while developing throughout its assigned zone, may not embrace the entire length along the front with its combat activities, if the interests of concentrating the shock group along the main axis will prevent the pinning group from covering the remaining sector of the army's front. In this case, individual secondary axes may be only weakly observed by small detachments, or blocked by obstacles. This will be most often employed in the pinning army's zone of activities, which should in no way strive to engage along the entire length of its front, always concentrating its shock group along a narrow sector along the boundary with the shock army.

Besides this, individual inaccessible sectors in the offensive zone, such as wooded and swampy areas, lake areas and others, often shorten the front of combat activities.

In all cases, the army group of forces must be strictly subordinated to the requirements of concentrating the greatest amount of men and materiel along the axis of the main attack.

59. The army commander directly controls the course of the offensive operation and personally organizes the deep attack by his highly mobile forces.

For this, he:
- employs his aviation, in direct coordination with the ground operation, against the enemy's personnel and operational rear to a depth of up to 60 and not more than 100 kilometers;
- indicates the axes and assigns immediate and subsequent tasks to the main echelon's corps, organizing their interaction by regulating their movement according to lines and by changing their axes according to the situation. At the same time, he directly controls the shock group's attack;
- directly controls the highly mobile group as the main means for the deep strikes against the enemy and for developing the attack into his rear to a depth of 40–60 kilometers;
- launches an airborne landing, if it is employed, for operating in the enemy's operational depth, in direct coordination with the army maneuver;
- directly organizes the augmentation of the attack and the feeding of the operation from the depth with the forces of the reserve echelon, committing it along the main axis, for achieving a decisive result, and;
- organizes the railroad basing of the army and the uninterrupted delivery of materiel-technical supplies along the dirt roads from the supply stations to the troops.

The army commander is thus the direct organizer of the army operation and personally directs the defeat of the enemy throughout the entire depth.

60. The front and the depth of the army offensive operation may be suffused with combat phenomena of the most varied content.

Troop formations may be simultaneously placed in conditions of carrying out flanking movements, conducting the meeting engagement and carrying out the breakthrough, and, in special cases, even going over to the defensive and withdrawing. The latter may have a place along the pinning group's sector and, depending upon the plan and form of the operation, should be subordinated to the development of the main attack and used in its interests. The nature of the army battle is determined, in accordance with the content of the actions developing along the axis of the main attack.

On the whole, one should always strive so that the army's shock group represents a single, closed ramming blow, and that the offensive along the main axis is unified by the unity of goal, axis and action.

The art of conducting the army offensive operation consists in unifying various combat efforts along the front and in the depth and their direction toward achieving the single goal of defeating the enemy.

Five

The New Forms of Struggle (1940)

(The Experience of Studying Modern Wars)
Installment 1
Moscow. Voennoe Izdatel'stvo, 1940

This book is an experiment in studying the new forms of struggle, which, in the author's opinion, have revealed themselves in the war in Spain and in the German-Polish war.

The author puts forth a series of problems of operational art for discussion.

This book is designed for the Red Army's command element.

—G.S. Isserson

TABLE OF CONTENTS

Foreword	236
Introduction	238

Part I. The War in Spain

Part II. The German-Polish War

1.	Introduction	250
2.	Entering the War	252
3.	The Polish Command's Mistakes	253
4.	The Polish Strategic Deployment Plan	254
5.	The German Deployment	257
6.	The First Stage	260
7.	The Second Stage	264
8.	Why the Poles Could Not Create a Front	266
9.	The Third Stage (The End of the War)	271
10.	The New Forms of Struggle in Action	276

Foreword

The history of military art illustrates the unbroken replacement of some forms and methods of waging war by others. Comrade Stalin writes that "The methods of waging war and the forms of war are not always the same. They change, depending on conditions of development, primarily depending upon the development of production."[1]

Each time when the development of productive forces creates new technical means, when social relations and social conditions change, and when politics puts forth new goals of struggle—the forms and methods of waging war change.

In major historical eras, when enormous popular masses are drawn into the struggle and the struggle has great historical significance, the replacement of some forms by others takes on a particularly stormy and radical character. However, this replacement does not take place spontaneously, does not take place in and of itself and not in a smooth process of the development of events. More often, it is accompanied by harsh trials and is born in a cruel struggle.

In the final analysis, the historical process of development leads to the victory of that which is emerging and developing. The new always comes to replace the old. However, the paths that lead to the confirmation of the new are varied. When historical conditions have matured for this, the new forms of struggle arrive at their realization either by their conscious adoption on the basis of profound theoretical study of the new conditions, or they themselves spontaneously break through into life in the historical course of events.

In the first instance, the influence of brilliant commanders manifests itself in the fact that they, by taking cognizance of the new conditions of their time, "adapt the nature of struggle to new weaponry and new soldiers" (Engels) and thus consciously direct the struggle along a progressive course, which has been laid out by history.

In the second instance, backward military theory, which is plodding along in the tail of history, ends up unexpectedly for itself before the facts of the new manifestations pushing their way into life.

In this case, the new forms of armed struggle are realized in excruciating childbirth, having passed through difficult and prolonged trials. Finally, they achieve their belated recognition at the cost of cruel and bloody losses sacrificed to the old and out-of-date forms of struggle that no longer correspond to the new conditions.

The events of a series of wars show that the new forms of struggle usually came to their realization by both paths.

This is once again being confirmed in our era.

The enormous changes in all conditions of waging modern war have by no means been comprehended. The recognition of the new forms of struggle occurred too late in some places. At the same time, it was clear that a new war should look substantially different than the war of 1914–1918. All of military affairs have undergone too many major and unprecedented changes since the time of the first imperialist war. Military equipment has never before made such a major strides forward. Never before have armies been subjected to such a radical reconstruction. All of military literature after 1918 has been devoted to the study and prognostication of the nature of a future war. For a long time strategic thought has been wandering in search of new solutions. Many theories have been put forward. In the end, a path was found and it led from the antiquated forms of the linear strategy to the still unexplored bases of the deep strategy.

When following the difficult heritage of positional war and the stagnation of military art, a new and refreshing thought produced the abstract outline of the deep operation as the simultaneous deep striking of the enemy's entire operational base, promising to resurrect crushing attacks and brilliant maneuvers, there were more skeptics than supporters of this theory. It was labeled a fantasy and poetry.

But this theory has cloaked itself in real forms. Historical reality has set a limit on doubts. Facts have arrived to replace them.

The modern wars of the XX century's third and fourth decades, which have unfolded in a complex political situation and on a new and still untested materiel basis, have opened a new page in the history of military art. Of course, people knew about the new forms and methods of struggle earlier. They spoke and wrote about them. But few believed in them. Their enormous and efficacious strength was not understood everywhere.

Now the events that have unfolded on the fields of Europe have revealed them in action. Historical reality always speaks for itself. It is only necessary to take it as it is.

The new forms of military art, which have been brought forth by the enormous changes of the new age, have ceased being a historical problem.

They have now moved from the sphere of theory to the field of practice. One may already speak of them not in the sense of a hypothesis or theoretical forecast, but in the form of a military-historical description of the new **forms of struggle in action**.

Such a military-historical description may pursue two aims: the first—to render a descriptive exposition of facts in their most complete content, without at the same time setting oneself the tasks of a special theoretical study; the second—to render a theoretical study of military-historical events from the point of view of their significance for the development of military art, without at the same time setting oneself the special tasks of a complete exposition of the events in all their details.

In the first case, the factual exposition of events is an end in itself; in the second case, the entire sum of facts is only the material for theoretical study and conclusions in the field of military art.

We are chiefly pursuing the second goal in our work. It was not the reproduction of the factual side of events in their full volume and content that was in the center of our attention. We are faced with only a single task: **to study the new forms of struggle in action**.

Although in many instances we considered it necessary to adhere to the chronological sequence of the course of events, laying out in basic terms their factual content, this was done only to study the new forms of struggle in the historical process of their realization; that is, not in the static form of ready-made theoretical conclusions, but in the dynamics of their appearance and development. This historical exposition of events showed the internal causal connections and the regularity of the development of the new forms of struggle, which in modern conditions have undergone major and fundamental changes.

At the same time, we are not laying down here general historical justifications for the new forms of struggle, for we assume that this was done to a certain degree in our work *The Evolution of Operational Art*.

The present first installment contains two parts, examining the war in Spain and the German-Polish war.

The approach to examining these wars cannot be identical. The events of the war in Spain, which were generally positional in character, are not, in and of themselves, of particular interest. According to their operational-tactical content, they do not represent anything new in comparison to the war of 1914–1918.

Of much greater importance is the evaluation of those conclusions that the experience of the war in Spain should have led to. Thus its study lacks the character of a military-historical description of facts and is only a critical-strategic critique of the war's nature.

Quite the opposite, the interest the German-Polish war presents consists namely in the essence of those events that unfolded in the course of military operations. Their description in chronological sequence thus comprised the main canvas for examining this war.

The work's third and main part, according to its significance, examines the war in Western Europe, which is still in full swing. This part could thus not be completed and will follow in a separate second installment.

Introduction

The forms and methods of waging war are always the product of political, economic, geographical, technical and other conditions in which the war arises and is conducted.

These conditions are extremely varied and diverse, each time giving rise to special operational-strategic forms of armed struggle.

Lenin wrote that "An era ... embraces the sum of varied phenomena and wars, both typical and atypical, both big and small, and characteristic of both developed and backward countries."[2]

Our era, which is distinguished by the extreme complexity of political interweavings and abundant in the extreme diversity of the phenomena of war, fully confirms this.

Each of the modern wars, which are taking place between different countries and in different territory, possess their special character and unfold in a special situation.

Of course, one should not mechanically generalize the experience of these wars independent of the conditions in which they were waged and are being waged. To the same degree, one should not ignore the experience of these wars, alluding to the fact that the given war is only an individual case.

The wars that take place in one era and are generally waged by one and the same means of struggle always reveal some kind of general conditions and phenomena, which to one degree or another are characteristic of the given era's wars. Of course, only a major war, which immediately embraces enormous masses and a broad territory, may acquire, on the whole, a character typical of its era. Clausewitz wrote that such a major war "represents a separate era in the history of military art."

However, even in individual so-called "small" wars, the characteristic features of the age acquire this or that manifestation, while lifting the curtain on the nature of modern war.

Our age is abundant with wars of the most varied scope and nature, from the "smallest" wars to a major one in the full sense of modern war, which is what the second imperialist war in Western Europe is.[3]

These wars offer extremely rich material for theoretical study.

The task of such a study always consists, having determined the special conditions and special nature of each modern war, of establishing that which is common and inherent to the wars of a given era in general and which is typical and logical for them.

In this regard, three modern wars, which have unfolded in three different parts of Europe in the space of four years—from 1936 through 1940—attract particular attention: the war in Spain, the German-Polish war and the war in Western Europe. Two of these wars have already ended. The third has concluded a major stage of completely independent significance and is shifting to the next phase of its development.[4]

We do not touch on modern wars that have unfolded on other continents. We omit the Italian-Abyssinian war as a colonial war, which was waged in a sharply different theater and with such a qualitative and quantitative correlation of forces which cannot be of significance for a major modern war.

We also omit the war in China, although it has very great significance for conclusions in the field of the nature of modern war. This war is generally unfolding as a war of maneuver in conditions of a large territory. It has, however, acquired an extended character and has taken on all the signs of a war of attrition, and in this regard is not much different from a positional war, according to the strategic paths of its development. The difference is only in that its front has not taken on the aspect of a continuous and immobile defensive line, which has anchored itself in the ground; given the ever increasing activity of the Chinese army, it is thus to a significantly greater degree subject to fluctuations not linked each time to the breakthrough of a fortified zone.

The war in China, according to its operational-strategic forms, is somewhat similar to the war in Spain, although it promises a completely different outcome. The great upsurge in the national consciousness of the Chinese popular masses is obviously tilting the scale in favor of China, having created the necessary conditions for its victory.

We halt on the war in Spain, the German-Polish war and the war in Western Europe not only because they have most vividly concentrated within themselves that new element which is characteristic of a major modern war, although it is possible that they are far from having fully revealed all of the possible paths of its development.

We halt on them because they, following one after the other over the course of four years, have comprised, in the operational-strategic sense, a certain common and ascending link of events, in which the new forms and methods of waging war have gradually forced their way into life and have finally found their historical realization.

From the point of view of realizing the new forms of military art, the war in Spain could be called the prologue to the drama, while the German-Polish war is the opening and the war in Western Europe is its development. It is precisely in this understanding that the three modern wars in Europe are acquiring their great historical significance. The finale of the entire drama is still hidden in the historical future.

However, the finale's possible nature, from the point of view of the forms and content of the military art of the future, are undoubtedly revealing themselves already in the modern wars in Europe; this is why their study has enormous significance for determining the forms and content of armed struggle for the immediate future.

The war in Spain may also be numbered among the so-called small wars,[5] which in the field of strategy yield relatively little experience. This war, of course, yielded an incomplete picture of an armed struggle between large modern armies.

The German-Polish war unfolded in special conditions against a state, the entire internal bankruptcy of which predetermined its destruction at the first serious military test.

Despite this, it yielded a very definite picture of the possible forms of a separate campaign in a modern war.

The war in Western Europe, which, following an extended waiting period, was marked by an armed struggle of enormous scope, has displayed a true picture of a major European war of the imperialist type.

According to its operational-strategic forms, which are derived from an entire series of conditions, these three wars in Europe represent three different strategic types, in which the new forms of military art have found their gradual realization.

The war in Spain began as war of maneuver, turned into a war of position and ended with the overcoming of the positional front.

The German-Polish war began, unfolded and ended as a war of maneuver.

The war in Western Europe began as a war of position and, following a prolonged period of positional stagnation, turned into a maneuver war of enormous force. This is how it concluded in the first round of its development.

If one speaks of possible operational-strategic forms of modern war, then these three types, as derivative forms, essentially exhaust all the possibilities, because a war on land can only be either one of maneuver or position, or a combination of both.

The task of military-theoretical research consists in explaining why these wars, according to their character, became that which they were and are, and what conditions led to this, and what possibilities they reveal for waging war.

This research must reveal what the essence of the new military art of the new age consists of.

Part I—The War In Spain

The civil war in Spain was from the very beginning a war of improvisations, according to the political conditions of its origins. It was conducted without previously organized armies and without their normal deployment, as one usually views these things, without a previously prepared theater of military activities and without outfitted fortified lines. The armed forces, the armies' deployment and fortified lines—all of this arose and developed during the war. Finally, this war was waged with relatively limited forces, the amount of equipment of which cannot in any way be compared with the scope of a large modern war.

In these conditions, it seemed, were all the prerequisites for a war of maneuver, because the correlation between the front of strategic deployment that could be occupied by the troops and the geographic extent of the front that could contain them would inevitably create conditions for the formation of significant empty areas and open spaces.

The war in Spain did begin as a war of maneuver. Limited forces were concentrated along individual important directions, along which combat activities unfolded around definite strong holds and important political centers.

However, the period of maneuver with which the war in Spain began proved to be of extremely short duration and did not lead to any kind of decision. Of the two infantry masses that opposed each other in the beginning of the war, the more powerful one had the opportunity to advance, while the weaker was forced to retreat.

However, this quite natural process, in which the war's maneuver period found its expression, changed the position of the sides only in space. It did not to any degree bring closer the war's goal for the attackers; it did not in any way weaken the Republicans. The reason for this was in the old linear forms of waging war, in which the attacker only follows behind the retreating party.

Missing were highly mobile weapons, which could have overtaken the retreating Republicans, cut off their route of retreat and preempt them near the country's important centers. Missing also was a powerful air force (there were about 200 aircraft on each side during the first period of the war), which could have, in general, delayed an unimpeded withdrawal. As a result, no where could the depth be affected and the paths of retreat remained everywhere free and a decisive result could not be achieved.

The Republican army's detachments were able to fall back unhindered, gather their forces and organize a resistance.

Franco's[6] attack was beaten off and halted at the walls of Madrid along the insignificant Manzanares River. At first this was achieved by completely insignificant forces numbering 1,400 rifles, eight machine guns and one artillery piece, which amounted to a correlation between the Republicans and their enemy of 1:20.

Later, in the fall of 1936 the by-now organized Republican forces put a final limit to Franco's offensive maneuver along the Toledo—Madrid axis. At this point the short and rapid movement period of the war in Spain was over. Madrid was transformed into a Spanish Verdun[7] and remained as such until the end of the war. The Spanish Marne occurred along the Manzanares River and this, just as in 1914 along the Marne, was a turning point of the war in Spain.

To be sure, the offensive maneuver once again tried to revive and did not immediately die. From January through March 1937 there were three major attempts to resume it.

These were the rebels' three operations:
in January near Valdemorillo;
in February along the Jarama River, and;
in March near Guadalajara.

If Franco had had the strength to carry out these three operations simultaneously as a concentric maneuver to outflank Madrid from the north and south, it is possible that the war's maneuver period might still have resumed. However, these operations, each of which were conducted separately and with significant breaks in time, encountered the Republicans' organized resistance and suffered a defeat, one after the other.

If the maneuver period of the war in Spain failed to yield a decision due to the absence of the necessary highly mobile means for maneuver, then these operations failed because of the absence of the necessary breakthrough strength for the attack.

The rebels attacked with the following forces per kilometer of front:

At Valdemorillo (along a 23-kilometer front)—500 men, five tanks, 15 guns, and 25 planes.

Along the Jarama River (along a 15-kilometer front)—2,500 men, ten tanks, 12 guns, and ten planes.

Near Guadalajara (along a 40-kilometer front)—1,500 men, 7.5 tanks, 19 guns, and five planes.

In these conditions the Republicans' defense, which was strong in their staunchness and willingness to fight, but weaker in the technical and fire regard than the resistance modern organized armies can offer, shows the power possessed by the nature of the defense.

No matter how weak the defense is, once it has put up an organized front of fire, then the breakthrough strength of the attack, suffused with a definite norm of suppression weapons, is required. In any event, this is not 15 guns and ten tanks per kilometer of front.

Here, near Guadalajara, the enormous influence of the air force on the ground forces' attack was confirmed. One foreign observer wrote of the attack by the Republican air force against Italian motorized forces around Guadalajara that "during a raid on an auto column, every two automobiles out of six were burned, while the others were delayed, a large part of the drivers were wounded or killed, and the vehicles destroyed or damaged."

The Republican air force's activities showed that a ground offensive without an offensive in the air and without securing air superiority threatens heavy consequences in modern conditions and is, in general, hardly possible.

Thus following three unsuccessful attempts which, by the way, were already unfolding in conditions of an emerging positional struggle, any prospects for a war of maneuver had been lost and the offensive maneuver had been finally halted.

The further general course of events in Spain created a striking analogy with the course of the first imperialistic war's development in France. It was as if the picture had been repeated in full.

As in 1914 along the French front, following a brief maneuver period that did not yield any kind of decision, a positional war ensued in Spain. As in 1915–1916 in the French theater, in 1937 in Spain the opposition of fronts ensued, with a series of unsuccessful attempts to break through. And, finally, as in France in 1918, a period ensued in Spain, with its clashes at the turn of 1938–1939, which yielded a decision to the side that was much superior in men and materiel.

All of this amazingly reminds one of the course of the 1914–1918 war's development along the French front. And the duration of these wars occupied almost the same amount of time. And although they unfolded in a completely different situation and were of a deeply different nature and completely different scope, evidently certain overall conditions determined one and the same logic of their development.

These general conditions, which resulted in the transformation of a war of maneuver into one of position, involved the helplessness of the means of maneuver there where it was possible by the conditions of space, and in the absence of the attack's breakthrough strength there where it was required to renew maneuver.

From the beginning of 1937 a front was established along the enormous length of the Iberian peninsula, from the Cantabrian Mountains to Malaga. It was not yet continuous or positional; but it stabilized the situation and, in this regard, began to play the role of the First World War's positional front.

This phenomenon immediately attracted broad attention.

The front in Spain arose without any kind of intentional considerations and without any kind of preparation. There were no prerequisites for this in the way of any kinds of previously existing fortified lines or strong points, although the geographical conditions of the mountainous terrain were undoubtedly favorable to this.

The front in Spain immediately stretched to 2,000 kilometers. When the Bilbao[8] area fell in the beginning of the summer of 1937, the front was established from the

Pyrenees Mountains to the southern shore of the peninsula, with an overall length of 1,500 kilometers. This was exactly twice the length of the front in France at the beginning of 1918.

The Western Front's 750-kilometer length in 1918 held an Allied army numbering 4,000,000 men. Within 20 years the Republican front in Spain, 1,500 kilometers in

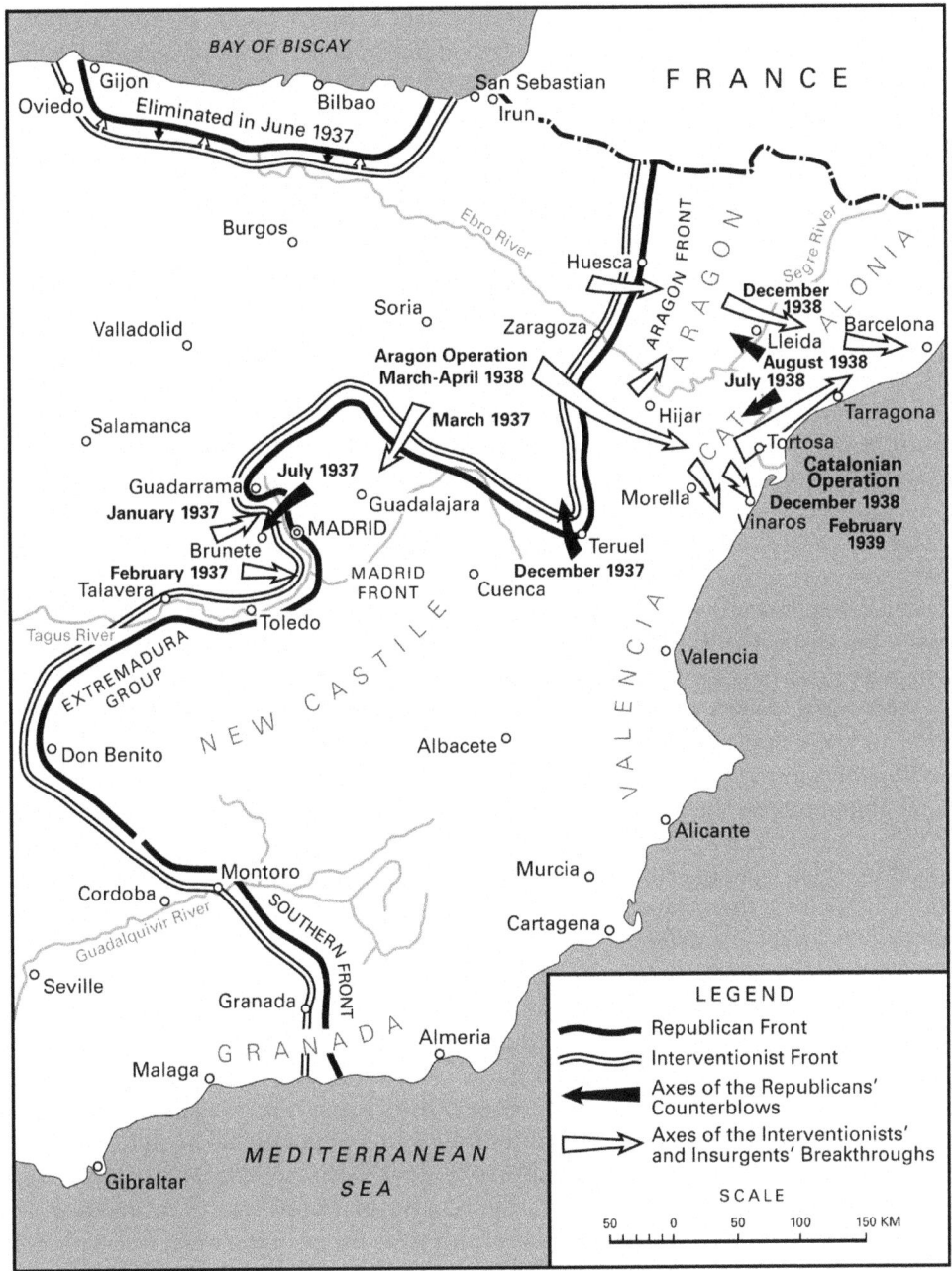

Map 1. The Situation at the Front by March 1937 and the Chief Directions of the Interventionists' Consecutive Breakthroughs and the Republicans' Counterblows during 1937–1939.

length, held an army of 500,000–600,000 men; that is, approximately one-eighth, or 12% of the Allied army in 1918. Thus, given a front twice the length of the one in France in 1918, and an army eight times smaller than that of the Allies in 1918, a front was never the less established in Spain. Naturally, it should have its own special character.

At least 1,500,000 soldiers (based on a calculation of 1,000 men per kilometer of front) are required to hold a 1,500-kilometer front. The Republican army had less than 50% of this number. In these conditions, the front in Spain could not have the nature of an unbroken line and was rather a screen along an entire series of sectors. In this it was distinguished from the positional front of the 1914–1918 war. Along individual, secondary axes only the important areas were fortified as strong points. Often, particularly in the mountains, only observation posts were left behind. On the whole, the front was weakly covered. However, on the whole, this did not alter its significance. This was a front that had stabilized the situation, divided both sides by a wall of fire and which imparted a positional nature to the war.

People hurried to draw from this phenomenon the conclusion that a war of position is not a passing phenomenon of the First World War of 1914–1918 and that if in Spain we were, in general, unable to avoid the establishment of a continuous front, this means that it is inevitable in any modern war.

It seemed, undoubtedly, that all the objective conditions favored the maneuver nature of the war in Spain. After all, the bigger the front and the smaller the size of the army, then, it would seem, the greater the conditions for a war of maneuver.

However, it was precisely the events in Spain that once again showed that the real reason for the establishment of a front consists not in the simple correlation of space to the numbers of the belligerent armies, but in the absence of highly mobile weapons for developing maneuver there where it is possible, according to objective conditions, and in the absence of the attack's penetrating power there where the opportunity for maneuver must be acquired at the price of overcoming the enemy's frontal resistance.

Of course, there were special reasons for the establishment of the front in Spain. Concern lest a single province or piece of land fall into the hands of the enemy forced the Republicans to spread themselves broadly along the front.

Limited forces, the absence of prepared reserves and the impossibility of making up their losses forced the Republicans to strictly economize their men and materiel and prevented them from placing everything on a decisive operation along a selected direction. This inclined them more to a defensive strategy and inevitably led to their dispersal along the front for the defense of their territory.

Thus the war in Spain became one of position.

The first attempts to overcome the positional front suffered defeat and showed, in general, the same tendency; thanks to the enemy's superiority in men and materiel, the attacker was at first able to achieve a local success and penetrate into the defensive position. However, as a result of the absence of deep suppression weapons, the tactical intrusion was nowhere able to transform itself into an operational breakthrough and each time the defense's reserves would freely flow to the breakthrough sector. Auto transport assisted them by supporting the transfer of troops with great rapidity. For example, during the fighting near Brunete both sides threw in their reserves within 2–3 hours from areas up to 100 kilometers from the front.

With the arrival of the defense's reserves the offensive would halt and the lost situation would be quickly restored.

This is the way the first attempts at a breakthrough unfolded in Spain in 1937.

Finally, the Republicans' enemies were able, thanks to assistance from abroad, to concentrate relatively large forces, thus achieving a decisive superiority on the ground and in the air.

At the same time, the Republicans, blocked from all sides, were deprived of this assistance and were left to their own devices.

For them the war was transformed more into a **war of materiel helplessness**; in 1938 the final, difficult stage of the struggle ensued for them.

The enemy was now already conducting decisive operations, consecutively cutting the major areas of Republican Spain from each other.

In Aragon as early as March-April 1938 24 infantry divisions (250,000 men) attacked along a 90-kilometer front, supported by 1,800 guns, 250 tanks, and 700 planes. A density of 60–70 guns and 15 tanks per kilometer of front was achieved along the axis of the main attack.

This was still twice as small as in 1918 (then there were 120 guns, 30 tanks and one division per 1.5–2 kilometers of front).

However, this already guaranteed the necessary breakthrough strength against the Republicans' technically weak defense.

During the war's final stage, at the turn of 1938–1939, this suffusion grew even more in the Catalonian operation.

The Republicans' opponents achieved an enormous, crushing superiority: two to one in infantry, and ten times in artillery, tanks and aviation.

In February 1939 for every 50 tanks and 50 planes possessed by the Republicans their enemies had 500 tanks and 800 planes. The correlation even reached 1:50 for certain types of equipment.

At the same time, the Republican army was straining its last sinews and did not even have enough weapons for its infantry.

The Republicans' great heroism and just cause could not make up for this materiel weakness and in these conditions the outcome of the struggle was preordained. When the people were betrayed and the high command turned its coat, the tragic end inexorably ensued. Thus one may less seriously speak of the significance of the experience of the final stage of the war in Spain, because the result was determined most of all by the simple correlation of forces.

However, the course of events that led the enemy to the final goal is worthy of even greater attention.

Despite their enormous superiority, the interventionists[9] were unable to achieve a decisive goal during the final period of the war by a single deep and overwhelming attack along the entire front and in its depth.

Their offensive did not result in the general and simultaneous collapse of the Republicans' front and the unhindered development of the breakthrough into the depth all the way up to a final result. Such an operational form was still an unknown phenomenon for the linear strategy employed in Spain.[10]

Thus independent motor-mechanized formations directly developing the breakthrough into the depth were not employed, although their employment could have imparted a completely different turn to events.

The interventionists' offensive was conducted according to the old linear tenets. It unfolded in a series of individual operations with a limited goal, which each time developed along a single selected axis and soon reached a definite limit in its development. Following this, the attacker was forced to turn against another sector of the front and, following a significant break which was necessary for regrouping and a new deployment, began its operations along another axis.

This method of consecutive breakthroughs with a limited goal, when the attacker throws himself first against one and then another sector of the front, and each time drives a wedge in, is well known from the experience of war in 1918 in the French theater. It demanded a lot of time. As a result, the fighting in Spain took on a prolonged character and all the signs of an attrition struggle. In March-April 1938 the interventionists began the Aragon operation and in a series of individual consecutive attacks, first to the south from the Ebro River and then to the north of it, and finally, again to the south, reached the seacoast, having separated Catalonia from Valencia.

This was an operation of individual dashes, which nowhere were able to congeal into a single general squall of a decisive offensive developed to the end.

Then, delayed in the summer of 1938 by the Republicans' counterblows along the Ebro River and the Segre River, the interventionists were only able in December 1938 to undertake their next major operation in Catalonia, which finally fell in February 1939, following heroic resistance.

This was already the final tragic stage of the struggle for the Republicans, when their forces and opportunities were exhausted.

It required an entire year (from the beginning of 1938 to the beginning of 1939) in Spain for the method of consecutive operations with a limited aim to finally achieve a decisive strategic result.

At the same time, the central Madrid sector of the Republican front maintained its stability to the end of the war and remained unbroken; it was simply opened to the enemy by Franco's agents. If the interventionists had had to continue the war following the fall of Catalonia, then they evidently would have needed a good deal of time to break through the most powerful central front.

Such were the results of the strategy of consecutive operations in Spain.

And all this was against the background of the attacker's enormous and overwhelming superiority in men and materiel.

As to whether this method will yield the same outcome as in Spain, given a powerfully organized defense and significant reserves, remains in question. In any event, this would require significantly more time than in Spain. And 1918 in the French theater showed that the Germans did not achieve a decision by this method and that the attacker's forces were subject to attrition, and not the defense's reserves.

If the defense disposes of reserves and opportunities for continued resistance, then each break between the attacker's operations will be employed by it in order to meet the next offensive with restored forces.

The Republicans had neither the opportunities nor the reserves for this. The losses and attrition following each breakthrough could neither be made up by them nor restored. Thus the conditions in Spain were substantially different and in this manner were distinguished from the Allies' condition in the French theater in 1918.

The enemy's consecutive breakthroughs led to a dispersal of the Republicans' already weak reserves and thus enjoyed success.

When Catalonia was isolated from central Spain the opportunity of maneuvering reserves at the strategic level fell by the wayside altogether. As a result, reserves maintained only a local significance, were quickly drawn into the fighting along their axis and, finally, began to be exhausted altogether.

After this, the intrusion into the defensive zone and its tactical breakthrough did not require, essentially, its operational development; because even without this the breakthrough would be subsequently transformed into a relatively free offensive advance that no longer encountered new defensive forces in the depth.

In these conditions the striking of the entire defensive depth simply lost its significance, because to a significant degree there was no depth. The breakthrough was resolved through the single tactical overcoming of the defense. They did not attempt to carry out in Spain the deep forms of struggle, such as the general suppression and striking of the defense's entire depth.

The necessary means for this, mainly major tank formations for developing the attack into the depth, were not available. Also, from the point of view of outmoded military theory, such an operation in Spain was not called forth by conditions of necessity. One must assume that had it actually been employed, then it would undoubtedly have led to the war's different development.

The method of consecutive operations was, of course, not illustrative of a large modern war and for breaking through a powerful defense with large reserves. This is clear from the very conditions of the war in Spain, where the defense had no reserves at the final stage.

In these conditions the efforts of quantitatively limited means of struggle could be concentrated along a single line of front for immediate cooperation. This was quite sufficient for resolving the task of the breakthrough in Spain. Naturally, when means are limited one cannot detail them for independent employment.

All of the new means of struggle were employed in Spain on the basis of close and direct cooperation.

Aviation operated effectively on the battlefield, directly supporting the ground forces. There were instances of independent actions against important targets in the rear, although they yielded few positive results.

At the end of 1936 Franco dispatched 20–50 bombers against Madrid 30 times over the course of 52 days, in order to crush the Republicans' resistance. Each time the planes dropped about 50 tons of explosives. The goal, however, was not achieved. Madrid and Barcelona, despite systematic air attacks, held on for more than two years. At the end of the war the two railroad lines connecting Catalonia with France were subjected to daily raids and continued, however, to function to the end. The single-track line from Barcelona to Valencia worked over the course of two years until it was cut on the ground by the enemy, who had broken through to the sea.

The conclusion is often drawn from these examples that the air force cannot be a decisive factor in modern war. At the same time, however, they hardly take into account the fact that in the war in Spain the air force never reached such numbers that would have allowed it to undertake independent air operations. After all, along a front that was four times as large as the front in France during the First World War, aviation in Spain comprised only 12–15% of the number of planes that participated along the Western Front in 1918.

As regards the prospects of a large modern war, this aviation barely comprised 10% of the air power that the major states of Europe muster.

In this regard, the modest experience of Spain is very unconvincing.

Tanks were also employed in Spain only in direct cooperation with the infantry, without in any way breaking free of it. In the first period of the war they were usually employed in small groups, or altogether singly. As a result, they encountered concentrated fire and were often put out of action and burned.

A norm of 80–100 tanks per kilometer of front is possible given the mass employment of tanks in a modern war. In Spain up to 30 tanks per kilometer of front were employed only in rare instances; for the most part, there were considerably less (15 tanks per kilometer of front). Naturally, tanks could nowhere be employed for independently resolving operational tasks, because independently organized tank formations were lacking for this.

As a result, tanks in Spain only reinforced the infantry and were not able introduce anything qualitatively new into the nature of the engagement.

A truly modern tank attack, supported by all means of struggle, was not realized in Spain and could not be realized because of a shortage of the necessity quantity and quality of vehicles.

In this regard, the experience of Spain more likely showed very little, rather than much.

Fuller expressed himself on the employment of tanks in Spain: "On the whole, one can say that in this war tank tactics were absent."

As to the quality of the Italian vehicles that were employed, Fuller wrote: "Speaking without any exaggerations, the light tank that was employed is not capable of overcoming obstacles that can easily be overcome by a Scottish pony." He called this tank "an effectively moving coffin."

The conclusion is often drawn from the experience in Spain that the new means of struggle only supported the possibility of conducting a modern attack, but did not change anything in its character and forms.

From the point of view of the prospect of a large modern war and the massed employment of the new means of struggle, this conclusion was extremely short-sighted.

In the history of wars, in general, few are the cases when a new combat means immediately exerts a decisive influence on the nature of the struggle, for the art and ability of employing it is not usually born simultaneously with its appearance.

For the most part, when a new means of struggle is employed in a limited quantity it generally creates an incorrect impression as to its possibilities and then the prospect of its employment is often lost altogether.

The limited scope of a small war and the "penny's worth" experience of small combat events narrow to an extreme degree the perception of its possibilities and may direct thinking along a very narrow path, which closes its eyes to the prospects of a large war.

Engels has already pointed this out, speaking of the influence of a small colonial war on the minds of French commanders.

In his article, "The Possibilities and Prerequisites for a War by the Holy Alliance Against France in 1852," Engels wrote:

> As regards the French, they have even for a time lost the thread of the Napoleonic tradition of a major war, thanks to the small war they waged in Algeria. ... are not the generals commanding there losing the eye necessary in the conditions of a major war? There is no doubt of the fact that the French cavalry is being ruined in Algeria. It is losing its strength, is unlearning close-formation attack and is learning to operate in all directions, in which, however, the Cossacks, Hungarians and Poles will always be supe-

rior. Among the generals, Oudinot[11] has compromised himself at the walls of Rome and only Cavaignac[12] alone distinguished himself in the June fighting[13]; but all of this is far from *grandes epreuves* (major trials).[14]

We may also say that the war in Spain "is still not a major trial" and be correct in asking did it not yield a false impression of a large modern war, and did not the tanks "ruin themselves" in Spain; that is, did they not learn there to operate in individual scattered groups, instead of launching operationally-organized close attacks in the depth of the enemy's position.

The war in Spain undoubtedly yielded the first experience of the tactical employment of the new means of struggle on the fields of Europe and raised the first curtain on the modern battlefield.

This war was, however, more a workshop for the technical testing of individual types of modern weaponry, but was in no way a general rehearsal for a large war and of the new forms of struggle.

Thus one should treat the experience of the war in Spain very carefully.

In general, experience is often not important in and of itself. Far more important are the conclusions that are drawn from it.

Often the conclusions drawn from the experience of the war in Spain drew not at all cheerfully the prospects of a modern armed struggle. A positional front is inevitable; war is once again taking on the creeping character of a consecutive overcoming of frontal resistance; the system of attrition operations, and thus the strategy of attrition, is once again putting its inevitable imprint on the nature of waging war; the new means of struggle cannot change the nature of the modern engagement and operation and destructive attacks throughout the entire depth have no hope of being realized; one should not speak about any kind of new forms of the deep destructive operation—such are the sad refrains that inevitably flowed from many statements about the experience of the war in Spain. The return to the tried and true but hopeless methods of breaking through in 1918 has found a much greater recognition following the war in Spain.

And they say that nothing has changed.

The war in Spain was a complete repetition of the world war of 1914–1918. According to the overall course of the development of events, this is certainly true, and could not be otherwise. If one fights with old methods, then the old stories repeat themselves. The end of the war in Spain was, to be sure, different than during the 1914–1918 war on the French front. The attacker achieved his goal and the breakthrough led to a final result. But the reason for this in no way the result of the effectiveness of the old methods of struggle, but rather in the attacker's enormous and overwhelming superiority in men and materiel and in the Republicans' materiel weakness and absence of reserves. In such conditions the old methods were able to justify themselves, but they had very little to teach us as to the prospects of military art's development and a large modern war.

It was premature and shortsighted to say that the new forms of struggle, which require the deep striking of the entire depth of the enemy's resistance, have not justified themselves.

No one tried to, nor could they employ them, in Spain. There were no conditions for this, nor was there any real necessity. For many who have failed to understand these reasons and who are devoid of an understanding of the historical course of things, this has remained unclear. The experience of the experience in Spain became for them something all-embracing and exhaustive. They have truly turned into "a crutch for a lame mind."

The historical prospect of the immediate development of the new military art, which was already knocking at the door of history, remains unopened.

In the meantime, the entire negative aspect of the war in Spain indicated these prospects, if one examines events from the point of view of what has already arisen and was being developed.

There is much from the experience of the war in Spain that could still seem stable and retaining its significance. However, "For the dialectical method, what is most important is not what seems stable at a given moment but which is already beginning to die off, but that which is arising and developing, even if it seems unstable at a given moment, for it only that which is arising and developing is insuperable."[15]

The war in Spain was not yet a war of the **new forms of struggle in action**. Everything that is new is not everywhere immediately revealed in its entire specific content, and by no means in all conditions. Nothing new comes in and of itself. One must struggle for everything new in history. For the revelation of everything that is new, one requires the appropriate conditions, a progressive theory and a purposeful will. These conditions were not extant in Spain.

However, a half year following the end of the war in Spain, events took place in the eastern part of Europe that revealed different opportunities for waging the armed struggle.

II—The German-Polish War

1. Introduction

The German-Polish war was not in the true sense a full-bodied war of two politically equal forces, equally capable of resolving their tasks by forces of arms.

"A multi-national state not linked by bonds of friendship and equality of the peoples inhabiting it, but rather the opposite, founded on the oppression and lack of equality of national minorities, cannot represent a powerful military force."[16]

Therefore "…the Polish state proved to be so powerless and incapable of functioning that it began to fall apart during the first military setbacks."[17]

One should not try to explain the rapid defeat of Poland solely by the superiority of Germany's military organization and military equipment. We see how the same superiority from the very beginning did not yield such results in China, where the broad popular masses united for the defense of their country and organized effective resistance.

However, the military defeat of Poland will evidently still be the subject of a detailed historical study. According to its catastrophic outcome, it finds its equal example only in the defeat of Prussia by Napoleon I in the Battle of Jena[18] in 1806. Then Napoleon finished off his opponent in 19 days, counting from his entry into Prussia to the capture of Berlin. In September 1939 the Polish army was completely defeated in 16 days.

There is a lot in common in the foolhardiness and arrogance of Polish policy in the pre-September days of 1939[19] with the insanity of the bellicose ardor of the Prussian court circles on the eve of Jena.

As is known, Napoleon broke out in laughter at the Prussian ultimatum, which demanded the withdrawal of French forces behind the Rhine. He called the Prussian king's letter "one of those disgusting pamphlets that the English ministry makes them produce each year for 500 pounds sterling," and replied to it with the immediate assumption of the offensive. As a result, as Mehring[20] says, "...the Junker rabble stumbled into war more than it entered it; it was swept by the growing weight of its crimes to that downward slope along which it irresistibly rolled down into the depths of unprecedented shame."[21]

All of this is applicable in an amazing way to the riff-raff of the Polish military clique.

As to the Jena catastrophe, an official historian has written: "One can hardly find similar events in the entire course of military history."[22] History has found yet one more example of a similar event in the military defeat of Poland.

Of course, when an army suffers such a catastrophic defeat the reasons are always hidden in factors of political significance. In this regard, the battle of Jena was predetermined and, from a military point of view, only a formality.

In 1806 the Prussian army scattered, in Napoleon's expression, "like an autumn fog." The same thing happened to the Polish army in 1939. Clausewitz writes that Prussia would not have risked war in 1806 "if she had suspected that the first pistol shot would be the spark thrown in the magazine, from the explosion of which it would be blasted into the sky."[23]

The Polish military clique decided on such a move, because its unrestrained adventurist policy was incapable of foreseeing such a prospect. If the outcome of the German-Polish war was predetermined in such a way by the very correlation of political values, then this, however, cannot weaken interest in the military side of events.

Of course, one should not judge the character of a large modern war, its true intensity, duration and prospects of development according to the German-Polish war. In terms of the short duration and vigorous prosecution of its outcome in 16 days, this war more resembled a separate march or campaign, the content of which was a single overall strategic operation, carried out from beginning to end without a break and in a single maneuver development.

This war did not require the commitment of all the numerous and varied mainsprings of modern war and, in this regard, did not reveal all of its many aspects.

Thus it would be extremely frivolous to draw any kind of conclusions as to the all-embracing character of modern war according to the experience of the German-Polish war.

However, this war holds undoubted interest and has great significance from the point of view of such problems as:

    a) the nature of entering into a war;

    b) the conditions that give rise to a war of maneuver;

    c) the operational employment and possibilities of the modern means of struggle, particularly of aviation and motor-mechanized forces;

    d) the prospects for the maneuver development of the struggle up to the achievement of a decisive outcome, and;

    e) the methods of conducting operations.

If the German-Polish war cast a certain light on the possible resolution of these problems, then it acquires significance for the study of the nature of modern war.

From this point of view, the experience of the German-Polish war is all the more important in that it was waged between two organized regular armies, disposing of to one degree or another all the modern means of struggle, particularly on the German side.

2. Entering the War

The nature of entering a war usually determines the basic lines along which the war develops, at least during the initial period. And because any subsequent development flows from the proceeding one, in this way the nature of entering a war often determines its line of development as a whole. In order to get a correct understanding of war, it is necessary to comprehend how its opening came about.

In this regard, the German-Polish war represents a new phenomenon in history.

The political conflict between Germany and Poland, which flowed from the conditions of the Treaty of Versailles, according to which East Prussia was separated from central Germany by the so-called Polish Corridor, arose as early as the end of 1938. Its intensity increased over the long months. From the summer of 1939 an armed conflict was already coming to a head. And from the end of summer both sides were openly threatening one another, spoke of the inevitability of armed action, and were preparing for it.

However, when on September 1 the German army opened military operations with fully-deployed forces, crossing the borders of the former Poland along its entire length bordering on Germany, this occurred as a strategic surprise on a heretofore unknown scale.

No one can now say when mobilization, concentration and deployment occurred—acts that, according to the example of past wars and, in particular, the first imperialist war, were distinguished by quite definite boundaries in time.

The German-Polish war began with the very fact of Germany's armed invasion on the ground and in the air; it began immediately, without the usual preliminary stages in the practice of past wars.

History had run up against a new phenomenon. Following the first imperialist war, military literature put forward a theory, according to which war opens with an "invasion army," specially designated for this purpose; the country's main forces must then deploy and enter the fighting under its cover. According to this outline, the mobilization and concentration of the main mass of forces is carried out only after the beginning of the war; that is, just as it occurred in 1914. Entry into a war thus takes on an echeloned nature: first the invasion army moves out, followed by the mass of the main forces.

"The theory of the invasion army" was immediately subjected to serious criticism. It was essentially not accepted on faith by anyone.

As a counterweight to the invasion army, as the armed forces' first echelon, the German military press wrote:

"The strategy of tomorrow must strive to concentrate all available forces in the first days of the beginning of military activities. It is necessary that the effect of surprise be so shocking that the enemy is deprived of the materiel opportunity of organizing his defense." In other words, the entry into a war must take on the character of a stunning and overwhelming blow, which takes advantage of, as Seeckt wrote, "every ounce of strength."

For such a blow, even the tenet that it is unleashed during the war's first hours is inapplicable; quite the opposite, the first hours of the war ensue because this blow has been unleashed.

At the same time, the old tradition, according to which it is necessary to warn of this before striking, is cast aside. War is no longer declared. It simply begins with previously deployed armed forces. Mobilization and concentration are not related to the period following the onset of a state of war, as was the case in 1914, but are conducted unnoticed long before this. Of course, it is impossible to hide this completely. The concentration becomes known to one degree or another. However, there always still remains a step between the threat of war and entry into the war. It gives rise to doubts as to whether a military action is really being prepared or whether this is only a threat. And while one side remains in doubt, the other, which has firmly decided upon an attack, continues concentrating until finally, an enormous deployed armed force appears along the border. After this it remains only to give the signal and full scale war breaks out.

This is how the German-Polish war began. It revealed the completely new nature of the entry into modern war, and this was essentially the main strategic surprise for the Poles. Only the fact of the opening of military operations finally resolved the doubts of the Polish politicians, who most of all provoked the war through their arrogance, but who at the same time found themselves taken by surprise.

3. The Polish Command's Mistakes

The Polish command also made strategic mistakes and miscalculations, which cannot be placed exclusively in direct dependence on the internal political rottenness of the former Polish state. They lie in the amazing lack of understanding of the new conditions in which the start of a modern war may occur.

First of all, in this regard, the general staff lost the war and showed an example of a monstrous lack of understanding of the strategic situation and its deeply incorrect evaluation. The French general staff made an enormous mistake in its evaluation of the strategic situation upon entering the war of 1870. However, the Polish strategists far exceeded the woeful historical lessons of their teachers.

The Polish command's mistakes may be reduced to three main ones.

1. On the Polish side they believed that Germany's main forces would be tied down in the west by the entry of France and England and would not be able to concentrate in the east. They proceeded from the idea that about 20 divisions would be left against Poland and that the remaining forces would be thrown against the west and an Anglo-French invasion. Such was the great faith in the power and rapidity of the Allies' offensive. Thus Germany's strategic deployment plan in the event of a two-front war seemed completely wrong. Germany's capabilities in the air were evaluated the same way. Finally, they firmly counted on England's immediate and effective help with air and naval forces. The historical lessons of the past disappeared without a trace, which more than once

showed the true value of the promised aid from England, which has always known how to fight only with others' soldiers.

All of these false calculations make for even more false conclusions. They just about believed it possible to get by with just the single peacetime army. Thus they did not hurry to mobilize second-line divisions. However, they made this widely known, announcing the mobilization of a two-million-man army. They thought to scare off the enemy with such disinformation. However, this had the opposite effect, because the German command replied by concentrating even larger forces against Poland.

2. On the Polish side, they believed that regarding active operations by Germany, one could only speak of Danzig, and even then not about the entire Danzig corridor, and the Poznan area, which had been take from Germany by the Treaty of Versailles. Thus they utterly failed to comprehend the enemy's true goals and intentions, reducing the entire question of the long-maturing conflict to Danzig alone.

Thus they hardly concerned themselves with the Silesian direction, from which the German army's main attack actually came.

3. On the Polish side, they believed that Germany could not immediately attack with all of its forces designated against Poland, because this would require their mobilization and concentration. Thus they would be faced with an opening period which would give the Poles the opportunity during this time to seize Danzig and even East Prussia.

Thus Germany's mobilization readiness and its immediate entry into the war with all of its designated forces never occurred to the Polish general staff.

The Poles did not understand the strategic situation and this already meant the loss, at the very least, of the first stage of the war, and then of the entire war.

In this regard, the war was lost for Poland before it even began.

4. The Polish Strategic Deployment Plan

This profound misunderstanding of the entire strategic situation led the Polish command to a completely ephemeral strategic deployment plan. The Polish deployment against Germany was undoubtedly placed in very difficult conditions.

These conditions are more difficult than those in which the Russian army's strategic deployment unfolded in Poland in 1914. The Poles had to secure an 800-kilometer front from the Baltic to the Beskids (the western spurs of the Carpathians) against Germany. Moreover, to the north there still remained East Prussia, the border with which comprised 300 kilometers.

The winding and flanking outline of the border, which inevitably brought about deployment along various directions, and the unsecured eastern border with the Soviet Union, actually led to the creation of a deployment front of 2,500 kilometers. At least 200 divisions would have been required to completely secure such an enormous front. Poland, of course, disposed of no such forces.

The complexity of the Polish deployment against Germany was also determined by the circumstance that the former Poland, throughout the entire time of its woeful existence, prepared for a war not in the west, but in the east, against the Soviet Union. Its western border zone was not conceived as an operational base. It was more likely a rear base and completely unintended for the role of a theater of military activities. It had no kind of fortifications, although it was richly outfitted with rear bases and depots. Moreover, all of former Poland's military-economic targets and the center of Polish industry were located in the west. 95% of Polish coal extraction, ten zinc and lead factories, which delivered 100% of the zinc and lead (108,000 tons per year), and nitric factories, which yielded 50% of the entire Polish production of nitrates, were in Upper Silesia. On the whole, former Poland's entire economic base was located in the west. All of the communications and trade routes with Western Europe ran through the western areas.

Thus while deployed to the west against Germany, the Poles met the war with their rear, and not with their front.

It thus seemed that this circumstance alone would have forced them to approach a deployment in the west with particular care. However, if the matter concerned Danzig alone, then all of these conditions naturally remained outside the equation.

No less complex were the deployment's operational conditions. The operational directions in the west, particularly in the Danzig corridor, looked into each other's rear and were outflanked. The axis directly on Danzig was squeezed in a vise from two sides. The axis from the corridor into East Prussia was subject to a threat from the rear from Pomerania, and *vice versa*. In this regard, the employment of the Danzig corridor as an operational base was an extremely difficult strategic task. Weygand once wrote about it: "The corridor will place insoluble tasks before the Polish command, because its defense is a completely impossible affair."[24]

Finally, deployment in the Poznan area was outflanked: from the right from Pomerania, and from the left from Silesia. One the whole, a deployment on the Vistula's left bank was also outflanked: from the north from East Prussia, and from the south from Slovakia. Such were the general operational conditions of the Polish deployment in the west. This gave the general staff something to think about, which required from it a particular sagacity in the strategic art. However, in this situation one should speak least of all about art. They approached the strategic deployment plan on the Polish side with a poverty of thought that finds its equal example in history only in the Austrians' deployment in Bohemia against Prussia in 1866, when Benedek's army was also outflanked from different sides and defeated.

The basis of the Polish strategic deployment in September 1939 was an offensive plan, which set itself the task of seizing Danzig and East Prussia. Strategic arrogance, deprived of a real grounding, was reduced by this plan to the apogee of caricature.

Poland put up about 45 infantry divisions against Germany. Moreover, it had one cavalry division, 12 independent cavalry brigades, 600 tanks and about 1,000 operational aircraft overall. All of this brought the army's numbers to approximately 1,000,000 men.

Poland disposed of about three million trained soldiers, more than half of which had undergone training after 1920. However, an enormous part of this trained reserve was not employed at all. As a result, up to 50% of those suitable for military service remained outside the army in September 1939.

The Polish forces were deployed in approximately the following grouping:

Six infantry divisions in the Grodno—Bialystok area against East Prussia's southeastern boundary.

7–8 infantry divisions to the north of Warsaw, with their left flank resting on Modlin, against East Prussia's southern boundary.

2–3 infantry divisions in the northern part of the corridor, against Danzig.

4–5 infantry divisions in the southern part of the corridor, in the Graudenz area, against East Prussia's southwestern boundary.

7–8 divisions in the Poznan area, with orders to operate to the north into the flank of German forces from Pomerania, or to turn south, in the event of a German threat from Upper Silesia.

Eight divisions covered Lodz from the southwest, with their left flank reaching as far as Czestochowa;

Five divisions covered Cracow.

About 4–5 divisions were left in reserve in the Warsaw and Przemysl—L'vov areas.

The cavalry was mainly distributed among the northern groups of forces concentrated in the Poznan area.

Thus the entire Polish army, not counting the covering forces along the eastern frontier and the reserve inside the country, consisted of 6–7 separate groups, with its main part facing north, against Danzig and East Prussia. The powerful Poznan group of forces constituted a sort of strategic reserve and in the dreams of some of these strategic fantasists, was to evidently triumphantly enter Berlin, from which it was separated by only 150 kilometers. In reality, this group was doomed to waiting by its very designation. In this manner the Poles immediately renounced the independent offensive initiative in employing a significant part of their forces, inevitably predetermining that they were going to operate just as the enemy dictated. Such is the usual fate of those forces that are reserved without a definite assignment and active goal.

The entire mass of deployed troops was very poorly commanded and the operational groups' headquarters represented barely organized organisms. Finally, all of the troops remained in the open. There was no kind of fortified terrain, strong points and defensive lines, with the exception of the Kulm strong point along the Vistula River in the Danzig corridor and the fortress of Modlin near the confluence of the Vistula and Western Bug rivers. Nor was a serious attempt made to erect field fortifications during the days remaining before the outbreak of war. The Polish general staff carelessly declared that there was no need for this, as the war would be waged as a maneuver one.

Thus the Polish army marched straight toward the hurricane that was preparing to sweep it away.

In the view of some students of the German-Polish war, the Polish deployment is sometimes portrayed as something not lacking a definite strategic idea. It is even viewed as being founded on a definite strategic prospect of the war's development.

For example, the American Eliot[25] believes that the Poles organized their deployment in three echelons:

The first echelon comprised the covering forces immediately at the frontier and was primarily represented by the mass of cavalry.

The second echelon comprised three armies—in the corridor, in the Lodz area, and in the Cracow area, with an overall strength of 30 infantry divisions and 14 cavalry brigades.

The third echelon comprised the main mass of 50 divisions being mobilized east of the Vistula.

The second echelon was supposed to take the blow and win time through delaying actions until the third echelon could get itself ready, concentrate along the left bank of the Vistula and enter the fighting.

Neither facts nor the Polish command's subsequent actions indicate that they had such a deployment plan in mind or that it was even attempted.

In any case, the plan required an extremely developed mobilization system, which is just what the Poles lacked. It was precisely the fact that the mobilization of all forces in Poland was not organized and was clearly running late.

General mobilization was declared only on August 30; that is, on the eve of the German invasion. The mobilization was not fated to take place. It only introduced terrible chaos under the blows of the war's beginning. The railroads and dirt roads began to clog up with the mobilized reservists moving toward the troops that were already falling back. This entire sad picture showed that if the onset of a state of war catches a modern army in a non-mobilized state, then it is already completely impossible to count on the opportunity to mobilize and concentrate it and to enter the war in an organized fashion.

It was in this situation that on the morning of September 1 there followed the simultaneous air and ground invasion by the deployed German army along the entire front, particularly by the main forces from Silesia, from where they expected the enemy least of all.

There was no beginning period of war. There were no strategic prefaces and preliminary operations. The war began immediately on a broad scale and at top speed. It was precisely that moment of the surprise opening of military operations along a broad front and with all deployed forces that the Polish side miscalculated.

Given the Polish general staff's enumerated mistakes, this created a situation of complete strategic confusion, which quickly turned into general disarray. The Polish army was caught unawares by the very form of the surprise invasion by the German armed forces, and it was this that inflicted upon it an irreparable and most decisive blow.

5. The German Deployment

In studying the events of the German-Polish war, the question naturally arises as to how it was possible to secretly and unnoticed concentrate a nearly 1,500,000-man army on the Polish border and to deploy it for an invasion along the entire front.

In essence, there was nothing particularly secret in this. The concentration of German forces increased from month to month and from week to week. In order to determine the deadline for its beginning, one should look back to 1938 and the period following the annexation of Bohemia and Moravia to Germany. When troops are accumulated so gradually, first in one, then in another, and then in a third area—the process of concentration is not so independently expressed in time and is swallowed up by a number of other events accompanying it.

Upon putting one's finger at first into a vessel of cold water and then of hot water, one can immediately discern the difference in temperature. However, when putting one's finger into a vessel of water, which is gradually heated over a slow fire, it is very difficult to distinguish the gradual change in temperature.

Thus a concentration, compressed into a short period of time and causing high degree of intensity in the work of auto transport, becomes the dominant phenomenon in a given period and can be easily espied.

However, a concentration carried out gradually and consistently and stretched out in time is very difficult to calculate and is more likely to dissipate and dull one's attention. This is exactly the way the German armies were concentrated.

This concentration was no longer a single and unified act, limited in time, which begins and ends in a definite and previously calculated time and of a duration that may be approximately calculated by the enemy.

Concentration has taken on a deep nature. No one can fix its beginning at all. Its continuation always leaves doubt as to whether or not a real armed invasion is being prepared or whether this is only reinforcing a diplomatic threat. Only the very fact of an armed attack reveals its end.

Thus a modern war begins before the armed struggle.

Of course, as early as the beginning of 1939 the task of the Polish general staff was to tirelessly follow the accumulation of German forces in East Prussia, Danzig, Pomerania, Silesia, and Slovakia, to take note of every new fact of concentration, periodically sum up all the established facts and draw from them the necessary conclusions. If this was not done, then it is not surprising that one fine day Poland saw the German army's enormous deployed forces along its borders. However, one thing is sure: given the actual readiness of the entire military system still in peacetime, the secretly deployed headquarters, the short concentration routes, and the broad employment of auto transport, in modern conditions one can do a lot in secret and easily achieve a major surprise. As regards highly-mobile motor-mechanized troops, given their stationing in the forward theater of military activities, one ought to generally view the threat of their sudden concentration in the very fact of their existence. These troops on a motor, having completed a march of up to 100 kilometers the day before on during the previous night, find themselves on the border itself only at that moment when it has been decided to cross it and invade the enemy's country.

One should certainly admit that the German command managed to concentrate and deploy with great rapidity a powerful army during the final period before September 1.

The German command's strategic goal, of course, went considerably further than Danzig and included the complete defeat of the Polish army, the return of the provinces lost as a result of the Treaty of Versailles and the destruction of any threat to Germany from the east.

For this, the following was concentrated:

About 55 infantry divisions,[26] five tank divisions, four motorized divisions, and four light divisions; that is, a total of 13 mechanized and motorized divisions.

In all, this comprised about 1,500,000 men and 3,500 tanks.

The air force included two air armies, consisting of about 2,500 aircraft. On the whole, the deployed German forces had the following superiority over the Poles: 4:3 in infantry, 6:1 in tanks and 2.5:1 in aviation.

If one takes into account this correlation of forces, the favorable flanking situation and the Poles' unfinished mobilization, then the German side naturally enjoyed great advantages. However, these proceeded from the superiority of the Germans' military organization itself.

The German armed forces deployed against Poland consisted of two army groups—North and South, and five armies in the following approximate grouping:

Army Group South, under the command of Gen. Rundstedt[27]: Gen. List's[28] Fourteenth Army in Upper Silesia, along the Cracow axis; Gen. Reichenau's[29] Tenth Army, in the Kreuzberg area, along the Czestochowa—Radom axis (this army included Gen. Hoth's[30] motor-mechanized group), and; Gen. Blaskowitz's[31] Eighth Army in the Breslau area, along the Lodz axis.

In all, Army Group South included about 35 infantry divisions, three armored divisions, three motorized divisions, and three light divisions.

Army Group North, under the command of Gen. Bock[32]: Gen. Kuchler's[33] Third Army in East Prussia and Gen. Kluge's[34] Fourth Army in Pomerania against the Danzig corridor (this army initially included Gen. Guderian's motor-mechanized group, which later operated along the Third Army's left flank.

In all, Army Group North included about 20 infantry divisions, two armored divisions, one motorized division, one light division, and one cavalry brigade.

One of two air armies operated along the front of each army group, although they were united by a headquarters of an independent air force command.

Between army groups North and South there was a 200-kilometer free space against Poznan, which jutted as a salient into German territory and which was only observed by *Landwehr*[35] troops, which were to subsequently occupy the Poznan area.

Overall, the entire deployment front for the five German armies occupied up to 800 kilometers and formed a half-circle which deeply outflanked the western and central parts of Poland from the north and southwest. This was conditioned by the geographical shape of the borders and offered the German armies great advantages, because it enabled them to conduct the offensive along concentric axes, while throwing back units of the Polish army against each other in order to outflank and encircle them. The German armies began their general offensive from just such a flanking situation.

A similar deployment along a broad flanking front has already been observed once in history in the Austro-Prussian War of 1866. Then three Prussian armies, with an overall strength of about 300,000 men, deployed along a 400-kilometer front and along different axes and invaded Bohemia from Saxony and Silesia, achieving the defeat of an Austrian army of equal strength. Within a little more than 70 years and with completely new combat means, five armies, numbering 1,500,000 men in all, deployed along a front twice as large—800 kilometers, while disposing of entirely new opportunities for carrying out a concentric offensive from different directions.

At the heart of the German deployment lay a unified operational plan that relied on a rapid war of maneuver. There were definite prerequisites for this, which consisted of the complete absence of peacetime fortifications in the western part of Poland, in the accessible and generally open terrain, in significant air superiority, and in a large core of highly-mobile troops. Of course, the enemy still disposed of a number of powerful natural lines, among which was such a serious barrier as the line of the Narew, Vistula and San rivers. The Narew River, with a broad, wooded and swampy valley and individual fortifications still remaining from the time of the first imperialist war, represented a particularly large barrier against operations from East Prussia. This line really did delay the Germans longer than all the rest. However, the extent to which the enemy could manage to take advantage of all the opportunities presented to him by terrain conditions depended upon the rapidity and very method of conducting the operation.

In assigning several consecutive tasks, the German command's operational plan purposefully led to the encirclement and destruction of the entire Polish army.

It was first planned to cut off the Polish forces in the Danzig corridor and occupy it through the joint operations of Army Group North's Fourth Army and part of the

Third Army. The Fourth Army was to force the Vistula at Bromberg and, upon linking up with the Third Army attacking from East Prussia, continue the offensive together with it along the right bank of the Vistula toward Warsaw from the north and to outflank it from the east.

At the same time, the offensive was to develop from the south along the Czestochowa—Radom axis to outflank Warsaw from the south. The main role was entrusted to Reichenau's most powerful central army.

The offensive by Reichenau's army was to be supported: from the south by List's army along the Cracow axis, and from the north by Blaskowitz's army, which had the task of attacking in the general direction of Lodz.

Thus the entire Polish army, which was deployed to the west of the Warsaw meridian, was to be outflanked from the north and southwest and encircled on the left bank of the Vistula. In particular, through the actions of the Fourth Army from Pomerania and Army Group South's armies, the Polish Poznan Army, which was not even being attacked from the front, was to be seized by pincers from the north and south. It was also planned that Slovak troops would lengthen the Fourteenth Army's right flank for a deeper attack against the Poles' left flank along the shortest route to Sambor and L'vov.

Throughout the course of the operation, whenever a few units of the Polish army never the less managed to slip out of an encirclement along the western bank of the Vistula, the Third Army's outflanking attack from East Prussia and that of Reichenau's army from the southwest, along with Guderian's and Hoth's motor-mechanized groups along the enveloping flanks, spread further to the east over the Vistula to the Western Bug, where the encirclement ring closed. This was but the natural development of the initial plan, on the basis of which a single strategic operation unfolded and which was conducted from beginning to end according to a single operational idea.

The entire German plan was thus a broadly envisaged operation along external lines, which pursued the goals of completely encircling and destroying the enemy.

The entire vigorous development of the German offensive, which led to the complete elimination of the Polish army within 16 days, may generally be broken up into three stages.

The first stage occupied four days, from September 1–4 and was marked along the entire front of the German invasion by a decisive frontier battle, which led to the defeat of and the start of a retreat by individual groups of the Polish army along the entire front.

The second stage occupied the period September 5–10 and was marked by a decisive pursuit of the already scattered groups of the Polish army, leading to their encirclement in different areas.

The third stage continued during September 11–16 and was marked by fighting to encircle and destroy, leading to the elimination of the main mass of the Polish army.

Let us briefly examine the course of events in all three stages.

6. *The First Stage*

At 0545 on September 1 the German armed forces along their entire deployment front invaded the confines of former Poland on the ground and in the air. If one counts all the separate air and ground operations that carried out a series of individual but

Part II—6. The First Stage   261

unified tasks, then the German invasion opened with 17 different operations, unified, on the whole, by a common plan of attack.

Two air armies immediately hurled themselves on the Polish air force's airfields and the most important railroad lines, extending their attack on the first day as far as the line Bialystok—Warsaw—the Vistula—the San. The struggle for air superiority was waged with all decisiveness.

As early as the first 48 hours following the opening of military operations no less than a third of the Polish air force had been destroyed, caught off guard on its airfields. Within a few days a large part of it lay in ruins by its hangars.

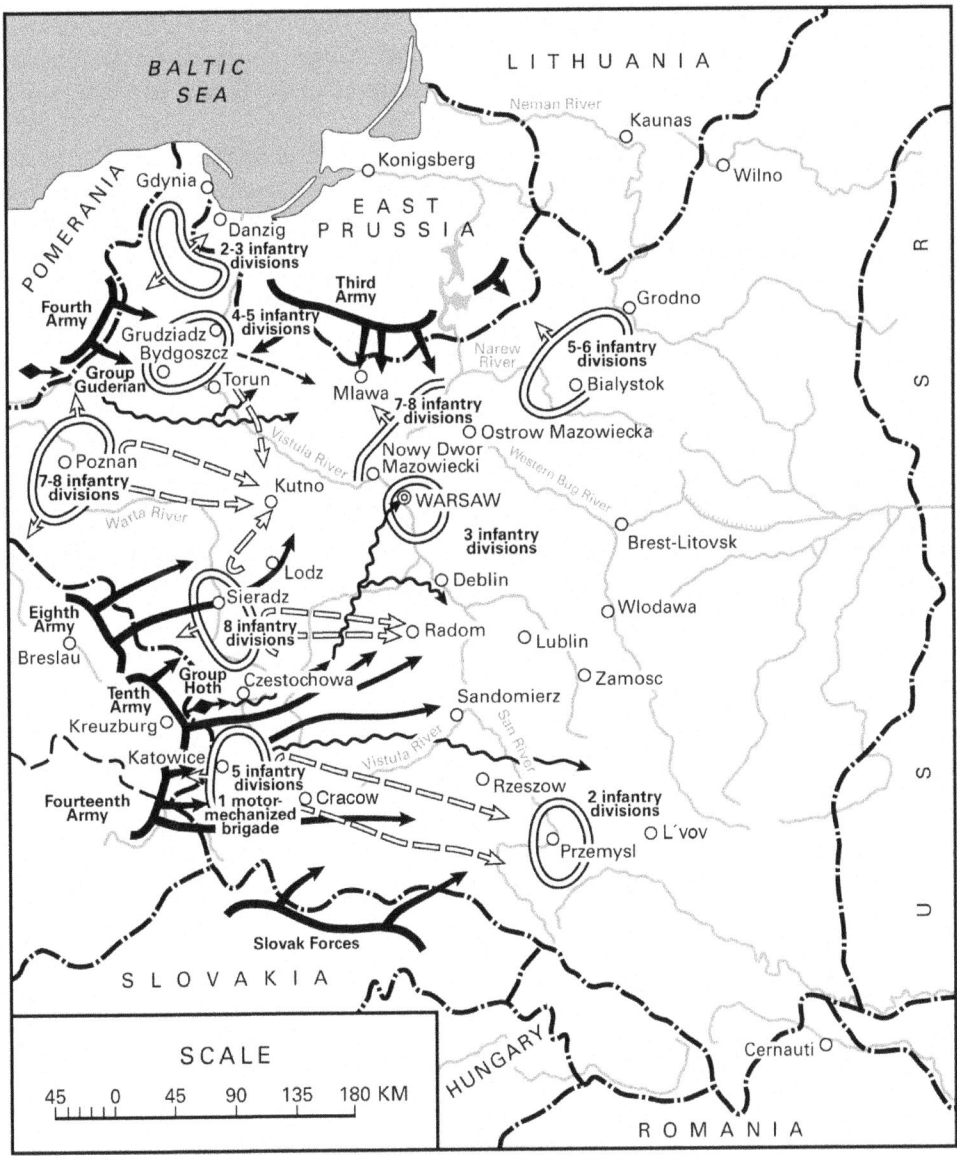

Map 2. The Sides' Deployment on 1 September 1939 and the Development of the Initial Operational Plan.

This immediately gave the German air force complete air superiority, which it then undisputedly held right up until the end of the campaign.

Railroad transport was simultaneously paralyzed and before long not a single train could arrive from the eastern bank of the Vistula River to the west.

The Polish army, deployed along the left bank of the Vistula, was isolated from the center of the country and its sources of supply, if such remained to the east of the Vistula.

This crushing blow from the air played a decisive role for the development of the operation and should scatter any doubts as to the significance of aviation's deep independent operations in modern war, if only these operations are purposefully directed against those targets upon which the viability and stability of the enemy army depends. It was namely the viability and stability in which, according to political conditions, the Polish army was generally deficient, that were paralyzed in the first days of the war by the German air force's actions.

At the same time, on the ground five German armies, supported by another part of the air force, attacked the Polish forces along the entire front. A frontier battle of enormous breadth took place, which according to its scale, decisiveness and significance and the forces completely committed to it, took on the true character of a new and modern general battle.[36]

The Poles made a desperate attempt to meet the blow and put up resistance. Along some axes, particularly near Mlawa and near the Warta River, the fighting became fierce. However, stunned by the attack from all directions, poorly led and not having any kind of prepared positions (not only in the engineer sense, but also simply in the sense of their timely selection), the Polish forces were unable to resist the German armies. It was only along the approaches to the Narew, to where units of the German Third Army were attacking (its right-flank units were attacking into the Danzig corridor to link up with the Fourth Army, which was advancing from Pomerania), and where the terrain conditions made the rapid development of the German offensive difficult did the Polish resistance assume at first a more organized and stable character. However, on the whole, the general frontier battle was lost by the Polish army and following the first 3–4 days it was forced to fall back along the entire front.

On September 4 the Germans took Mlawa. The Danzig corridor and the important railroad junction of Dirschau were seized. The Fourth Army, preceded along the right flank by Guderian's motor-mechanized group, forced the Vistula near Kulm. The Polish forces in the Danzig corridor were thus cut off. Along Army Group South's front, Reichenau's army took Czestochowa and broke through the Polish lines along the Warta River after a two-day fight. Simultaneously, as early as the third day, Hoth's motor-mechanized group had penetrated to a depth of 100 kilometers, outflanking the Poles' position along the Warta River. This maneuver to the southwest of Czestochowa destroyed the Polish 7th Division and its commander, Gen. Gasiorowski,[37] was captured.

As one should have expected, the Poles' Poznan group had nothing to do during these days. It was not attacked from the front; the entire German attack from Pomerania and Silesia in no way disturbed it, while flowing around it from the north and south. The Poznan group could have attacked to the north into the German Fourth Army's flank, but there it would have had to encounter Guderian's motor-mechanized group. It could have turned to the south into the flank of Blaskowitz's army, but this army was already outflanking it. Of course, there was as yet nothing terrible in either circumstance and it required only operational decisiveness and a definite plan of operations. But there was neither one nor the other. What the Polish command's plan of operations once the Germans' invasion along the entire front had been uncovered remains unknown. Most likely, there was no such plan. Thus the will of chance decreed that the Polish Poznan group remain a passive observer of the general battle along the border and took absolutely no part in it. It didn't do anything and, after standing around for a couple of days, began to slowly fall back on September 4 to the west, when it was basically already too late, because by this time it had already been outflanked from the north by Guderian's group and from the south by Blaskowitz's army. Such is the usual lot of a strategic reserve army, which has been moved up to the front with a very indefinite mission. The lateness of the Poznan army's withdrawal played a decisive role for the entire subsequent development of events and essentially marked the beginning of the ripening catastrophe.

The loss of the frontier battle meant an enormous defeat for the Poles and, in their circumstances, the beginning of the collapse.

First of all, their strategic deployment plan, which was based upon an offensive into East Prussia and Danzig, collapsed. A completely different strategic situation arose. A maneuver wave crashed in from all directions and, as a result of the lost frontier battle it threatened to flood into the depth. It was necessary to adopt an entirely new strategic decision which, in the new conditions, required, first of all, the creation of a front, the organization of resistance and the halting of the approaching wave. The occupation of an organized defensive front along a favorable natural line was the sole method of struggle that the Poles could choose.

It's quite clear that an army, poorly motorized and mechanized and weak in the air, must not allow itself to be drawn into a war of maneuver against an enemy with powerful motor-mechanized forces and air superiority. The sole possibility for fighting such an army is the employment of natural obstacles and organizing a defensive line. Up until now this was usually possible. This was possible in 1914 along the Marne and was possible in Spain in 1936 on the Manzanares River. The Poles disposed of no less favorable lines. The Narew—Vistula—San line was a sort of natural barrier. Upon occupying it, they could still have halted the revolution of the German maneuver wave. However, it was first of all necessary to want this and, secondly, it was necessary to fight for this and, thirdly, it was necessary to have the opportunity to do this, and, fourthly, the enemy had to allow it.

If the first two conditions were of a subjective character, then the latter two depended more on factors of objective significance.

The German-Polish war showed how fundamentally these factors had changed in modern conditions under the influence of the new means of struggle and their response time.

The Polish army did not manage to create a front and halt the revolution of the maneuver wave. The reason for this lay, first of all, mainly in its lack of combat capability, which proceeded from the bankruptcy of the Polish state. However, on this overall background, against which the German-Polish war unfolded, some new regularities in the development of modern operations, waged by the new means of struggle, were revealed.

7. The Second Stage

On September 5 the entire Polish front, which had been torn and disrupted, faltered and began to fall back. The retreat began without any kind of plan, without any kind of established intentions, and without any kind of prospect. It thus took on an unorganized character and unfolded of its own accord.

The German armies took up a vigorous pursuit, which developed without pause to an increasingly greater depth. Numerous motorized columns of engineer, pontoon and communications troops were thrown forward, who repaired bridges and roads, laid down crossings and communications lines behind the attacking units. Nowhere was the enemy accorded the opportunity to gather his forces, or to organize and create a resistance. In the end, he was forced to flee and, commanded by no one, began to break up under cruel blows from the air and ground.

The German air armies continued to suppress the Polish air force, rear, transport, and command centers with part of their forces, pushing their attacks further and further to the east, as the front line on the ground advanced. However, for the purpose of directly supporting the ground forces, a significant part of the air force sharply limited the radius of its operations and now shifted its attacks to the retreating Polish forces, inflicting enormous losses on them. Squadrons of assault aircraft and fighters scattered the Polish forces' retreating columns, cutting off their paths of withdrawal.

All this led to events that determined the beginning of the defeat of the entire Polish army on the ground.

To be sure, along the left flank of the German front, where only the left-flank units of the German Third Army were still fighting, the offensive was developing slowly. On September 5 Ciechanow was taken; however, the enemy put up particularly stubborn resistance while falling back to the Narew. The Poles even undertook a cavalry raid into East Prussia along their extreme right flank. This attempt was quickly eliminated. Nonetheless, all of this had great significance for the Poles, as it secured the most important right flank, where the Germans could very quickly create a direct threat to Warsaw and the entire rear of the Polish army along the most direct axis from East Prussia. It seemed as though this circumstance could be profitably exploited in order to also create a front south of Warsaw.

However, events along the remaining directions developed with such speed that this opportunity was increasingly lost in the conditions of the Polish command's general incompetence.

The situation was indeed taking on a catastrophic character.

In the Danzig corridor the fortifications near Kulm were taken on September 6 and the Polish 9th and 27th divisions, which had been cut off from the south, were destroyed, along with a tank battalion, two Jaeger battalions and the Pomorze Cavalry Brigade. At the same time, about 15,000 prisoners and 90 guns were taken. The Fourth Army's forces, which had crossed the Vistula, linked up with the Third Army and now attacked along a common front to the southeast along the right bank of the Vistula. They were preceded by Guderian's motor-mechanized group which, having passed through the Third Army's rear along the East Prussian border, had passed to the latter's left flank. It covered 200 kilometers in 1 1/2 days and on September 7 had concentrated for a deep attack over the Narew toward Brest-Litovsk. Thus having carried out its first task of occupying the Danzig corridor and eliminating the Polish forces concentrated there, Army Group North deployed for a deep maneuver to bypass the Warsaw area and the Vistula River from the east. Thus this maneuver developed somewhat later.

Along Army Group South's front, the pursuit immediately achieved a broad scope and a great depth. Particularly great results were achieved by Reichenau's army, which was preceded by Hoth's motor-mechanized group. To its right, List's army took Cracow on September 6 and, pushing forward its motorized units, continued to advance to the San in large leaps. The eastern Silesian industrial area was completely occupied by German forces. To the left, Blaskowitz's army, having seen the Poles' Poznan army, which it had already preempted from the south, retreating to its north, carried out a sharp movement to the north with its right shoulder and completely outflanked the enemy. Thus about eight Polish divisions and three cavalry brigades, to which individual Polish units had attached themselves, which had managed to fall back from the corridor, ended up being pressed to Vistula in the Kutno area and outflanked from the north along the river's right bank by the German Fourth Army, and from the south by Blaskowitz's army. In the meantime, Reichenau's army was developing a deep pursuit directly east and northeast, in the general direction of Radom, flowing around the Poles' Silesian group from the north. Hoth's group preceded it. It could have been dispatched for a less deep envelopment directly into the rear of the Poles' Poznan group. However, such an employment of highly-mobile forces was evidently considered to be too limited. Given the presence of the enemy's capital in the depth, the seizure of which would undoubtedly have had a decisive significance, Hoth's group was given another assignment. It was thrown directly at Warsaw as early as from the Warta River. It overtook the retreating Polish forces and got as much as 150 kilometers ahead of Reichenau's army.

On September 8 the armored units of Hoth's group reached Warsaw. Along the way Gen. Reinhardt's[38] tank division pushed aside the Polish 21st Division and a cavalry brigade. This tank division was the first to break into Warsaw, having penetrated into a western suburb of the city. However, it encountered barricades erected on the streets and was bombarded and had to halt. An attack was set for September 9. It was conducted along two of the city's streets by two tank regiments, supported by infantry from the division's rifle brigade. The tanks broke through four lines of barricades and reached Warsaw's main western train station. But here they were halted again by barriers and fire. The supporting infantry was too weak to support a further advance by the tanks within the city. Thus the order went out to fall back and the tank division abandoned the city. The new highly-mobile means had not yet been adapted for independent combat within its confines. However, the breakthrough to Warsaw had enormous strategic significance. This was the first example of the independent employment of armored forces, thrown far ahead of the front.

A spearhead of armored vehicles plunged deeply into the enemy's country and into the body of his army. This spread horror and confusion. At the first reports that that a tank column was racing toward Warsaw, the Polish government fled to Lublin as early as September 5. The general staff hurried behind from Warsaw. All of the communications lines were broken. Polish radio communications showed their complete inability to work on its own. After this no one received any orders, nor did anyone know where to go or what to do and people were left to their own devices. From this point the Polish command lost complete control and remained a headquarters without troops. Such were the strategic consequences of the breakthrough by Hoth's group to Warsaw. These consequences proved to be even more significant, when part of the armored troops from Hoth's group moved down the left bank of the Vistula to the south, outflanking the Radom area from the north and east, and when another group of motorized forces, which had been thrown far forward, simultaneously reached Sandomierz, on the Vistula, and to the south even broke through to the eastern bank of the river as far as Rzeszow. This broad and deep maneuver had enormous operational results.

First of all, all of the Poznan group's routes of retreat to Warsaw had now been cut and it was completely surrounded. Second, two of the Polish army's major groups (Poznan and Silesia) had been cut up and completely isolated. Third, the Polish forces retreating from the Warta River to the east, while being attacked from the rear, had ended up in the Radom area. Finally, fourth, a threat had appeared to the L'vov area in the deep rear.

However, much more important was the fact that thanks to the breakthrough by the mass of German mobile forces to a great depth, the Polish army, which had been torn to pieces and taken apart, no longer represented an organized and controlled mass. Three of its groups were already encircled: one in the Danzig corridor, another, the largest one, in the Kutno area, and the third in the Radom area.

Only two groups along the flanks—to the south of the Narew and along the road to L'vov—still maintained a certain freedom of action, but they were also being thrown back under the blows of the German troops' superior forces. On September 10 the German forces managed to break through the line of Polish fortifications along the Narew and occupy Ostrow Mazowiecka. Following the breakthrough by Reichenau's army to the Warta River, this was the second of two frontal breakthroughs that took place in the German-Polish War.

Now the very real threat of Warsaw's encirclement from the north and east arose along the right flank. Along the left flank, the Silesian group of the Polish forces, which were being vigorously pursued from the front and threatened from the right by the Germans' motorized units that had broken through, were hurriedly falling back to the San and further to the east.

To top it off, Slovak forces appeared at this time from the south, through the Carpathians. Such was the situation ten days following the start of the war. In these conditions any possibility for creating a front had been lost. To this threatening situation one must add that, of course, there was no aid of any kind from England. The war that had begun in Western Europe could in no way alter the situation; all communications with the West through Pomorze had been cut and the Polish navy's insignificant forces had been destroyed.

Such was the situation in the Polish theater of war by September 10, when the German general staff announced that "the fighting in Poland is approaching its culmination point."

8. Why the Poles Could Not Create a Front

The entire course of events showed that the Poles had no opportunity to create a front of organized resistance and to halt the revolution of the German offensive's maneuver wave.

But just how did such a situation come about?

Why was it impossible to carry out that which had usually been accomplished in all past and recent wars, in which in the final analysis the attacker's maneuver wave had always smashed into an organized front, before which it halted. After all, it was possible to achieve this in the war in Spain, even with inconsequential forces, before the walls of Madrid. And any front, if only it is organized, begins to manifest the power of modern fire and creates those conditions that are necessary for even a temporary stabilization of the situation. Even the attacker's temporary halt offers such opportunities to the defense and so alters the situation that the end of the war of maneuver often ensues and it is replaced by a difficult and exhausting war of position. After all, such was the strongly established regularity in the development of the course of military operations, which, it is possible, may triumph more than once, if the war is waged with the old methods.

Of course, the weakness of the entire Polish state and military system and the inability of its army were that decisive reason that determined the entire outcome of the German-Polish war.

However, this outcome, with the same final result, could have taken different forms. After all, there have been cruel defeats in past wars as well, when an army that had, however, maintained a certain combat formation, was simply defeated by its enemy on the battlefield.

The Polish army was defeated in another way. In essence, it was not even able to assume a formation in which it could have been beaten as an organized force. It was simply torn to pieces after ten days and taken in detail from all sides.

Such a catastrophic situation and the loss of any kind of opportunities for organizing resistance could, of course, ensue only under the weight of such overwhelming attacks that bring about general confusion and chaos and strike the very brain of the army, which ceases to be an organized force.

The question then naturally arises for military research, how was this possible and what is the reason behind it?

Let us concern ourselves for a minute with those forms that the war in Poland took in September 1939.

Let us suppose that the Polish army's rear had not been suppressed from the air, that transportation had not been paralyzed and that the command centers had not been put out of action.

Let us then suppose that the Germans' highly-mobile forces had not broken through to Warsaw and Sandomierz, that they were not threatening Warsaw with a deep envelopment from the north, that they had not overtaken the columns of the retreating Polish forces, and that they had not appeared in the deep rear along their path of retreat.

We will leave everything else in the resulting situation just as it happened following the lost frontier battle, when the Polish army was forced to fall back along the entire front.[39] In other words, let's suppose a retreat situation for the Poles, as was the case for the Anglo-French following the Battle of the Frontiers in 1914 and the Spanish Republican army's weak forces when they were falling back on Madrid in 1936.

In such a situation, the Polish forces, while they had been defeated and suppressed, but pressed only from the front, were able to carry out, relatively unhindered, a withdrawal and occupy, finally, an organized defensive front along the Narew—Vistula—

San line and, possibly, even further to the west, along the Rawa—Bzura—Pilica line. In this case, evidently, only the group of forces remaining in the Danzig corridor could not have avoided encirclement. However, the group to the north of Warsaw could have strongly occupied the powerful line of the Narew River. The Poznan group could have fallen back to the middle Vistula, and the Silesian group to the San River. Then the front would have been organized and all the reserve divisions could have been mobilized and the rear strengthened under its cover.

However,

a) command and control were paralyzed and put out of action; there were no longer any communications with the troops;

b) transportation was paralyzed; all the major railroad junctions are under systematic bombardment from the air, there was no supply and complete chaos in the rear, and;

c) chiefly, the spearheads of the tank formations had deeply penetrated into the body of the entire army; they had broken through into the deep rear between the groups of retreating troops, all the way to the capital, and had long since bypassed the retreating columns, gotten into their rear everywhere and preempted them along all the most important lines all the way to the Vistula and the San.

In these conditions, all the opportunities for organizing resistance fall away. A front cannot be created, because it has already been blown up from the rear. After all, one cannot erect a fence if all its foundations have been undermined from within.

The deep operation, as the simultaneous strike against the entire depth of the enemy's operational base and as the rapid spread of the attack into the deep rear, has realistically shown its enormous and efficacious significance. It has created the opportunity for the unbroken development of the maneuver wave and deprived the retreating party of any conditions for gathering his forces and organizing a defensive front of struggle.

The decisive role in the achievement of these results belonged to the new method of employing the modern means of struggle, chiefly aviation and independent motor-mechanized formations.

The German air force was employed in two ways: for independent operations of strategic importance—against the enemy's airfields, railroad junctions, supply routes and important military targets in the deep rear, and against the enemy's forces for the direct tactical support of its own troops.

Depending upon the course of the fighting on the ground, the air force's operating radius would increase and then shrink. At times, when the locus of combat operations would break out along some sector of the front, the entire mass of aviation would appear on the battlefield and, with overwhelming attacks from the air against the enemy's combat formations, would help to crush his resistance.

Not a single concentration of Polish forces could be carried out but that it would be immediately discovered by the German air force and suppressed. Thus all of the Poles' attempts to organize a counterattack were foiled each time. Bombers and assault aircraft scattered the Polish forces before they were in a condition to begin operations or meet an attack. The Polish cavalry suffered particularly from this.

As a rule, the activities of the armored formations were supported all the time by aviation. At the same time, the closest interaction of ground and air was achieved. When, for example, the armored units of Guderian's group were met with the enemy's heavy artillery fire upon reaching the Narew, bombers summoned from East Prussia appeared over the battlefield within 20 minutes. At the same time, a group of "Fieseler Storch"[40] aircraft landed in the immediate rear of the battlefield and operated from field landing strips.

Cooperation with the air force was undoubtedly one of the chief reasons for the German motor-mechanized forces' success. In two cases, when the tank units attacked without air support, they were unsuccessful.

Thus **the air force's direct assistance to the success of the ground forces showed its enormous significance in the modern engagement**.

However, this should by no means be opposed to the independent employment of aviation, which played a decisive role in the German-Polish war.

The first strikes by the German air force were directed against the enemy's aviation and rear. At the same time, the first successes were achieved not in air battles, but in operations against ground targets and in bombing airfields and military installations.

Distant targets of strategic importance were always being attacked from the air, sometimes with greater intensity, and sometimes with lesser. Transportation, communications and command and control were paralyzed by bombing to such a degree that they were unable to function in any kind of normal fashion. This is what created chaos and confusion in the rear. Important economic installations were subject to air attack, including oil deposits in Galicia. The German air force bombed them for ten straight days, until the Poles were finally left without oil.

Thus the entire rear was suppressed and paralyzed, and this deprived the front of any kind of stability and capability of fighting. This decisive result was achieved through independent air force operations.

The German-Polish war showed that if a modern war is not won with the aid of aviation alone, then, in any event, there is no way it can be won without it.

Without it the deep operations of highly-mobile forces on the ground would also not be able to fully develop, because they would inevitably encounter a prepared and unmolested resistance in the depth.

At the same time, the German armored and motorized divisions encountered each time in the depth the enemy's units already suppressed and scattered. Only this offered them the freedom of maneuver for the independent resolution of their combat tasks in the depth and at a great remove from the remaining mass of troops.

Thus the employment of aviation showed its decisive significance for carrying out the new deep forms of struggle, which deprived the Poles of any opportunities for creating an organized front of resistance.

However, of course, this result could not have been achieved without the simultaneous employment of highly-mobile formations for deep independent operations.

The independent employment of armored and motorized divisions for resolving operational tasks in the depth, far ahead of the combined-arms infantry formations' front, found its practical employment in the German-Polish war and immediately imparted to the fighting a character deeply different from the combat activities of past wars. The theory of this question, as the essence of the new forms of the deep operation, had already been elaborated in the preceding years. This concept flowed from the very nature of the highly-mobile means of struggle and required only their corresponding organization and employment.[41]

The German armies' operational formation along the main directions of their offensive consisted of two echelons:

The first echelon, which one may call the vanguard, consisted of armored and motorized formations, which independently broke the enemy's first line of resistance, flowed around his flanks, broke through between the gaps, and broke into the deep rear, and

the second echelon, which one may call the main one, consisted of the main body of combined-arms infantry formations, which followed rapidly behind the first echelon, took upon itself the fight against the main mass of the enemy and completed his defeat at the same time that he was already being attacked by the armored units that had broken through.

This was the case with the Fourth Army's offensive, which was preceded in the corridor by Guderian's motor-mechanized group, as well as of the Tenth Army, preceded by Hoth's motor-mechanized group.

The Polish front was not continuous and highly-mobile formations had plenty of opportunities to break through into the depth in the free spaces. At the same time, they did not concern themselves with clearing territory of the enemy and destroying the remaining pockets of resistance. All this was left to the infantry following behind.

The highly-mobile formations were immediately pushed forward up to a distance of 100 kilometers and poured into the enemy's depth. They were ruled by one desire—to move further forward, which in the final analysis would decide the outcome.

Delivery and supply did not present insuperable difficulties, about which people have spoken so much. The delivery of fuel and ammunition was organized by air, and this played an enormous role. "Junkers"[42] transport aircraft were dispatched daily to the front, loaded down with fuel cans, which were dropped by parachute.

The armored units overtook the enemy's retreating columns, while shooting them up from the march. They did not get involved in extended fighting with them, did not take prisoners and did not leave behind them mounds of killed and wounded. They preempted the retreating enemy along important lines, got across his path of retreat and nowhere allowed him the opportunity to organize a front of struggle, because they were everywhere in the enemy's rear. Thus they created an incomparably greater threat to the enemy, because they deprived him of the opportunity of accepting battle with the main body of combined-arms formations attacking from the front.

They preempted the battle and rendered it impossible or useless.

For example, they showed that not only the engagement, but also movement, may be a decisive factor in war, and with their speed they replaced the attack in that they preempted the formation of a front, which requires force to break through it.

The motor-mechanized formations "**replaced action with threat, the attack with maneuver, and the pursuit with preemption**," as one foreign observer said.

But what is even more the case; the motor-mechanized formations set the entire tone of the operation: they traced its pattern on the ground, imparted new forms to it and directed the entire course of its development.

And this is despite the fact that they comprised the army's smaller core. But just the same way that a small helm on a large ocean-going craft directs and turns its enormous body, thus the relatively small core of highly-mobile formations directed and turned the entire course of the operation.

To continue the comparison, then the difference consists only in that the helm directs the ship from the rear, while the highly-mobile formations directed the course of the operation from the front.

Operating ahead, the motor-mechanized units enabled the German command to lord it over the situation from beginning to end, to dictate its will, to hold the initiative in its hands the whole time, and to each time tear it from the enemy.

Does this mean that the role of the combined-arms infantry formations is now playing second fiddle? Of course not. Non-motorized formations, which still comprise the main body of forces, retain their enormous significance. The German-Polish war showed this as well. The frontier battle was basically won by them. Without them, the highly-mobile formations would not have had a basis for their activities. The mass of infantry, following behind the motor-mechanized formations, was everywhere ready to break the front, if it had not been possible to prevent its creation. Finally, it also completed the enemy's defeat tactically; that is, through the engagement it resolved that situation which operationally; that is, through maneuver, had been prepared for it by the motor-mechanized forces.

Thus the operational interaction of two combat arms found its resolution. This imparted to the fighting completely new and unusual forms.

First of all, the offensive, which in the past usually acquired the character of an even advance of the entire front line along a given direction, acquired the form of a deep penetration into the enemy's territory along various directions.

Secondly, this offensive immediately acquired the character of a pursuit, which overtook the retreating party, preempted him along important lines and got into his rear.

Thirdly, the fighting did not unfold along some kind of general front, as had been the case in past wars, but immediately spread to a great depth; it thus did not take linear forms and acquired a deep character.

As a result, the Polish forces were divided and torn apart in detail; preempted everywhere by German motor-mechanized units in their rear; they were unable to accept battle with the German infantry front facing them.

Thus the defensive front could nowhere be organized and the German offensive's maneuver wave could continue its uninterrupted revolution.

Such were the results of the deep independent employment of aviation and highly-mobile formations together with it—on the basis of the new methods and forms of conducting operations.

9. The Third Stage (The End of the War)

A description of the events of the German-Polish war could essentially be concluded on September 10, when the German highly-mobile formations broke more than 200 kilometers into the depth of Polish territory and everywhere began to dominate in the rear of the Polish forces, which had been encircled into individual and scattered groups controlled by no one.

From the time when that happened; that is, within ten days after the start of the campaign, the catastrophe of the Polish army no longer afforded any doubts and shone forth in its entirety.

Everything that took place in subsequent days, resulting in the inglorious end of the Polish state, was an inevitable and logical consequence of an already-ripened situation.

However, in order to comprehend the new nature of armed struggle, the third and final stage of the German-Polish war, which occupied the September 10–16 period, acquires enormous interest. As regards its link to the preceding events, it was the exploitation of an already-accomplished operational maneuver.

The content of this stage was the grandiose battle to destroy the enemy's already-encircled groups.

This battle unfolded in completely new forms and revealed the true character of the modern deep operation.

The fact that in many places the enemy had already ceased resistance and was surrendering does not alter the significance of the final events of the German-Polish war and only shows that, in the end, the new methods of conducting operations place the enemy in such a situation in which he no longer presents an organized force and is incapable of resisting.

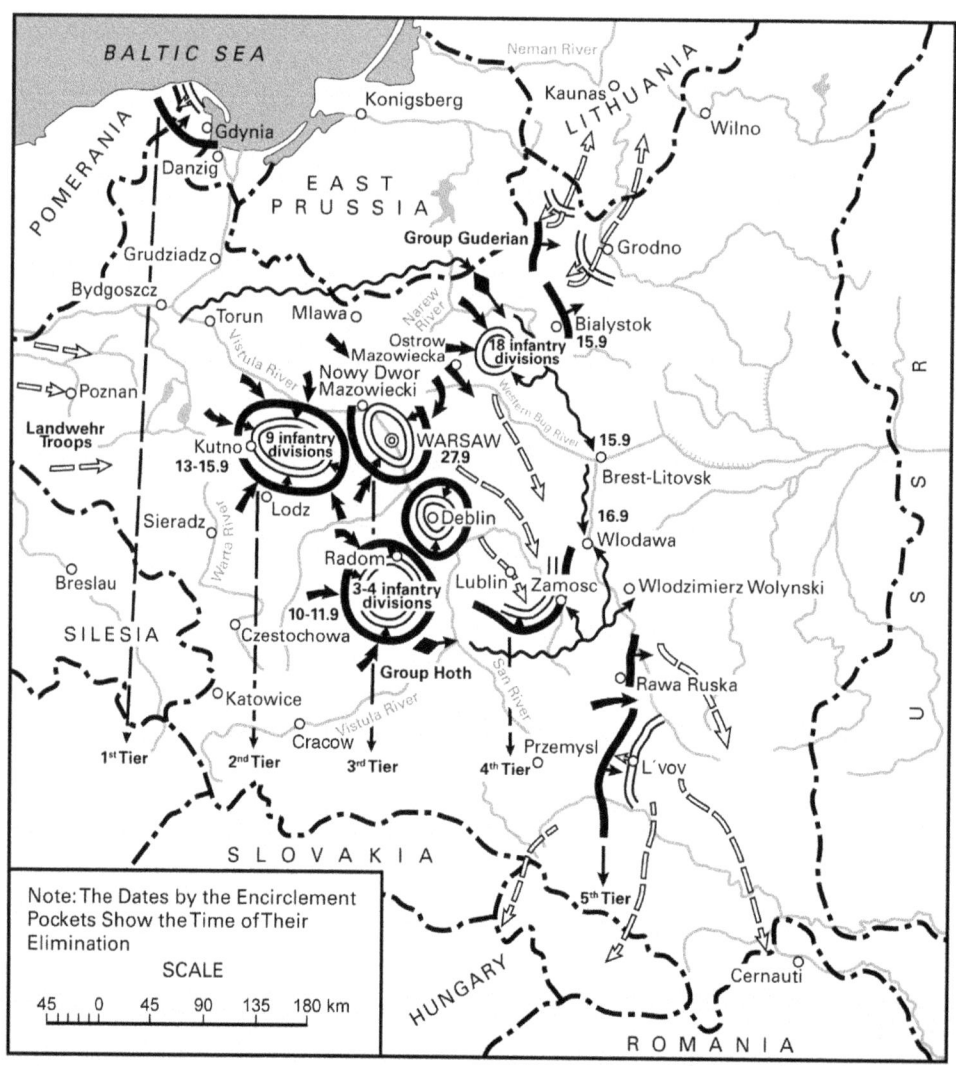

Map 3. The Development of the German Offensive Plan from 10 September to the End of the War.

By September 10 the Polish army had been encircled into individual groups and isolated in different areas in a large territory.

The final destruction battles did not present a picture of a single battle within the confines of a single overall territory. There was no longer an overall front of struggle. The struggle had broken up into individual pockets, operationally unconnected and completely independent according to their tactical significance.

One may name at least five such pockets as of September 10:

The first pocket was in the Danzig corridor, where a small group of Polish forces had been pressed to the sea and was still defending the Gdynia area and the Hel peninsula.

The second pocket was in the Kutno—Lowicz area, where pick Polish forces, which had fallen back from Poznan, had been encircled. The remnants of forces, which had managed to break out of the corridor, had joined them. Finally, another part of the forces, which the German Eighth and Tenth armies were driving ahead from the Sieradz area and the Warta River, had fallen back on this group. All of this amounted to about ten infantry divisions and three cavalry brigades under Gen. Bortnowski.[43]

A third pocket in the Radom area, where about three divisions had been encircled.

A fourth pocket in the Warsaw area, where a large garrison was putting up stubborn resistance.

A fifth pocket in the L'vov area, to where all of the forces along the Cracow direction had fallen back.

The new battle took on such "pocket" forms at its concluding stage.

The Poles still made desperate attempts at resisting and breaking through in a number of pockets.

However, nowhere were they successful in doing this. Moreover, there was nowhere to break through to. The German forces' actions deeply flooded the entire territory of Poland; thus, having broken out of one encirclement ring, the Poles must immediately fall into another.

The Radom group was the first to surrender as early as September 10–11. Here about 60,000 Poles were taken prisoner, including the commanders of the 3rd, 7th and 19th divisions.

After this, following a number of unsuccessful attempts to break through to Warsaw, the largest major group surrendered near Kutno. It fought for five days within the encirclement ring and had been completely suppressed by artillery fire and numerous air raids. Here 300,000 men were taken prisoner and 12,000 guns captured.

The groups in Pomorze and Warsaw held out longer. But these were already individual episodes in the entire epic, leading to its inevitable end.

However, the war's final stage played out not only around the encircled groups of Polish forces. It was marked by the subsequent broadening and development of the German maneuver, which crowned the entire campaign by achieving a complete strategic encirclement.

Although a significant part of the Polish army had already been encircled in detail in various areas to the west of the Vistula, never the less its path to the east had not yet been entirely barred. The German Third Army's offensive from East Prussia to the south, which would soon enough lead to the complete encirclement of Warsaw, was, in general, developing more slowly as a result of the stubborn resistance being put up by the Poles along the Narew River.

Therefore, the way from Warsaw to the east still remained open. Warsaw itself continued to stubbornly hold out, evidently counting on playing the role of a gate for breaking through to the east for that part of the Polish army that had been encircled in the Kutno area.

Finally, the routes to Lublin and further east and southeast which were, to be sure, clogged by a string of refugees, could also still be used for withdrawing the Polish army's units that had broken through over the Vistula River to the east.

Although the German motorized units, upon arriving at L'vov, had already neutralized this direction to a certain degree, never the less the routes leading to the east and Romania remained the sole salvation for the remnants of the Polish army.

In this situation the encirclement of the entire Polish army to the west of the Vistula River had already become impossible.

The operation's initial plan required its subsequent development in accordance with the new conditions of the situation. Thus it was decided to shift operations to the eastern bank of the Vistula, in order to completely close the encirclement ring there.

Such was the natural development of the German offensive plan once again showing that "strategy is a system of props"; that is a system of continuous "propping up" initial decisions in accordance with new information about the developing situation.

In carrying out a new maneuver over the Vistula River to the east, the highly-mobile formations were to once again play the decisive role. To be exact, without them a new maneuver could not likely have counted on its timely realization and could not likely have rapidly played its role.

The German tank and motorized formations were now free to carry out a new maneuver of strategic significance, because the arriving infantry had taken upon itself the elimination of the Polish forces' internal encirclement pockets.

Two major motor-mechanized groups—Guderian's from the north and Hoth's from the southwest—immediately spread their turning maneuver to the east of the Vistula and launched an offensive toward each other, in order to close the encirclement in an overall strategic ring.

On September 13 the German motor-mechanized units reached the Western Bug River: Guderian's units 40 kilometers to the north of Brest-Litovsk, and Hoth's units to the east of Zamosc. At the same time, units of Guderian's group cut the Polish 18th Division's path of retreat to the north of the Western Bug River and captured it along with its headquarters.

Thus in 13 days the new motor-mechanized formations, operating without pause from the start of the campaign, had penetrated 400 kilometers into the depth of the enemy's territory and, together with all their flanking movements and turns, had covered about 600 kilometers. This amounts, on the average, to 46 kilometers per day. Actually, on some days more than 100 kilometers were covered.

The armored motor on tracks brilliantly passed its test.

On September 15 Guderian's units broke into Brest-Litovsk, while Hoth's units captured Vladimir-Volynskii. Simultaneously, Bialystok was occupied in the northeast.

If in their jumping-off position the flanks of the German deployment were 800 kilometers from each other, now the two motor-mechanized groups, which were sweeping along the external flanks, were separated only by a distance of 100 kilometers. This was still at least a 2–3 days' march, but 100 kilometers no longer represents its former distance for the motor.

On the following day, September 16, the forward units of both German motor-mechanized groups, while continuing to move toward each other—one from the north and another from the south—linked up near Wlodawa along the Western Bug River.

Thus the maneuver wave had achieved its final goal in a single and unbroken revolution.

Now the encirclement ring had closed fully at the strategic level, and this took place not along the Vistula, as had initially been outlined by the German plan for the campaign, but 150 kilometers to the east, along the Western Bug River.

In this fashion was concluded the entire offensive plan, which led to its logical conclusion in a single unbroken development of the maneuver. This end presented a picture of a large strategic encirclement, carried out for the first time in the history of military art in such a way and on such a scale.

This encirclement was essentially carried out more for the completion of a strategic maneuver. Even without this, the Polish army represented only bits and pieces. However, from the point of view of achieving a strategic result, this maneuver had enormous significance.

Warsaw was now completely encircled from the east. All of the Polish forces' attempts to break out to Siedlce had been beaten off. Yet another encirclement pocket had formed in the Lublin area. Finally, the Polish government and the commander-in-chief, along with the general staff, which were wandering along the southeastern extremities of Poland, were now completely isolated from their army and territory.

Such were the results of the final maneuver by the German motor-mechanized formations, which concluded the entire campaign.

Only two groups of forces remained outside the encirclement ring, comprising no more than the ruins of the former Polish army. One of these groups in the northeast, following the occupation of Bialystok, ended up being isolated in the Grodno area and began to cross over into Lithuania. The other group to the southeast of L'vov had already been drawn into the general panic, which had broken out in this final piece of territory belonging to the collapsed Polish state.

Everything that ended up outside the overall encirclement ring was now streaming into Hungary and Romania.

The Polish government was the first to run away. Following prolonged wanderings and pursued everywhere by German aircraft, it crossed the Romanian border on September 17. The Polish commander-in-chief Rydz-Smigly,[44] along with the entire general staff, arrived behind it in Romania. Behind this ran the bureaucrats, officers, gendarmes, and the bourgeoisie. The remnants of the Polish air force scattered: about 500 aircraft flew to Romania, while the remainder landed in Latvia and Lithuania. Only the soldiers, who were left to their own devices, were not inclined to abandon their homeland. However, part of them was never the less tricked and forcibly removed to Romania and, in the north—to Latvia and Lithuania. About 20,000 Polish troops even ended up in Hungary.

And, when in a situation of the complete collapse of the Polish state and its army, in the east on the morning of September 17, along the 800-kilometer long Soviet border there appeared numerous columns of the Red Army—this was a mighty act of liberation from the Polish landowners' oppression of the peoples of Western Belorussia and Western Ukraine, who had been thrown to the arbitrariness of fate and who were now united to their native family of the fraternal peoples of the Soviet Union.

The remnants of the Polish army, which had ended up in the east among the few divisions outside the German encirclement ring, surrendered to the Red Army after a series of combat collisions.

The war had ended.

And although Warsaw and its garrison of about 100,000 men capitulated only on September 27 and individual Polish garrisons continued to offer resistance in Pomorze and along the Hel peninsula, in Deblin and a few other places, one may consider September 16 the day the campaign ended, when the forward units of the two motor-mechanized groups, who with their flanking movement toward each other closed the strategic encirclement ring around the main body of the entire Polish army along an enormous 185,000 square-mile part of the former Polish state, linked up near Wlodawa along the Western Bug River.

Thus this bourgeois country, which was riven by internal contradictions, based upon the oppression of its constituent nationalities and lacking neither the vitality nor the unity for the struggle, fell apart.

The military results are known and are only a formal result—the rout of the entire Polish army, 694,000 prisoners, 1,900 captured guns and 800 destroyed and captured aircraft.

According to official German data, the German army lost a total of 10,500 men killed, 30,300 wounded and 3,400 missing in action.

If this data is correct, then they speak to the fact that a victory can be achieved with significantly fewer losses than before through the new means and methods of struggle.

10. The New Forms of Struggle in Action

In September 1939 events played out on the Polish plains that were completely unusual for the history of the military art of the past.

Even if these events had taken place during peacetime maneuvers, then even in this case they should have attracted the special attention of military research.

And the German-Polish war was still a war, even though it was waged against a state that lacked internal strength for resistance.

To pass indifferently past the events of this war, only in order to not upset one's established conception of the old "classical" forms of struggle; to reduce everything to the fact that this was only an isolated case and that nothing new has happened; to dispassionately describe events, while only formally fixing the facts—means to not understand anything in the new manifestations of historical development and to imitate an ostrich, having chosen his tactics so convenient for military conservatism.

To sum things up, one may entirely lose the sense of everything that is new and maintain that, in general, nothing new happens in history.

The theory of the deep forms of struggle was at first met with condemnation. It was considered a romantic invention of military theorists.

When these forms were first employed in action they began to maintain that there was nothing new in this.

It often happens that a new concept is at first condemned as a fantasy and poetry; and then when it is realized in this or that form, they begin to indifferently maintain that nothing new has transpired.

The German-Polish war was, of course, a new type of war. It is, of course, an individual case, for any war is a specific situation through and through, which always takes place in special conditions peculiar to it alone.

However, in each war phenomena are reveled which are characteristic and logical for wars of a given era.

In this regard, the German-Polish war cannot comprise any kind of exception and, certainly is not one.

Quite the opposite, to judge by the vividness and integrity of the events that played out during the war, it revealed too much that was new in the forms and methods of waging modern war.

These forms and methods proved to be completely unexpected for the backward Polish command. Lurking in them was the entire strategic surprise for the Poles, in whose power they were held from the beginning to the end of the stormy drama that played out in 16 days.

During this time forms and methods of struggle were employed that had never once been tested in action.

The German-Polish war was the first war of the new forms of struggle in action.

In this, despite all of its special conditions, lie its historical significance and its role in the history of the development of military art.

First of all, the German-Polish war is of particular historical and theoretical interest as regards its maneuver nature in which it unfolded from beginning to end. The experience of this war is important, in the sense that it showed the possibility of a modern war of maneuver, in general, and revealed the conditions necessary for this.

It was not the correlation of space and the numbers of the armed forces that proved to be the decisive factor that determined the nature of modern war in this sense.

From the point of view of the front's width and the density of deployment, the conditions in Poland were to a significantly lesser degree conducive to the maneuver nature of the war than in Spain. The front in Spain stabilized and became one of position, with a length of 1,500 kilometers, which was occupied by an army of 500–600,000 men. Nowhere in Poland did the front stabilize, although it was narrowing all the time and by the end was 400 kilometers in length (L'vov–Brest-Litovsk–Bialystok), along which armies of 1,000,000 men were deployed.

Thus given an almost four-fold smaller front by the end of the war and an army twice as large, the war in Poland never the less took on the nature of an expressed and unbroken maneuver.

This shows that the roots of a maneuver and positional war are in the modern age hidden in other conditions—in the means of struggle and the forms and methods of their employment.

The German-Polish war revealed not only the conditions in which a modern war of maneuver is possible, but also showed:
- the opportunities available for the waging of a war of maneuver;
- the methods that are necessary to employ for this, and
- the forms that the struggle must adopt for this.

Of course, one needs definite objective conditions for the possibility of a war of maneuver, consisting of the geographical outline of the front, the nature of the terrain and the broken nature of the deployment.

However, all of these conditions were present in Spain to an even greater extent than in Poland. None the less, a war of maneuver did not develop there.

Besides objective conditions for a war of maneuver, one also needs the opportunities for exploiting these conditions. These lie in the modern means of struggle. If there are generally objective conditions for the maneuver development of the struggle, then a war may become one of maneuver only given the presence of powerful aviation, dominant in the air, given the presence of powerful highly-mobile formations (tank and motorized), breaking through into the depth, and given the presence of a boldly attacking infantry, echeloned in depth and supported by large amounts of artillery of various calibers, and tanks and planes for supporting the infantry, thus guaranteeing the attack's striking power.

However, these means alone are still insufficient, given the corresponding conditions for this, for a war to become one of maneuver. It is necessary that all of these means find a new and deep employment and that the forms and methods of struggle cross over from the age of the outmoded linear strategy to the new era of the deep forms of struggle.

Only given the presence of all these conditions, particularly the latter one, which has decisive significance, may a war take on a maneuver nature and be transformed into an unbroken revolution of the maneuver wave right up to a decisive outcome.

The experience of the German-Polish war shows that it is necessary to strictly distinguish the conditions for maneuver, in general, and the possibilities for exploiting these conditions.

The conditions existed in Spain for a war of maneuver, but there were no opportunities.

The conditions existed in Poland for a war of maneuver, although to a lesser degree than in Spain, although there were more opportunities for it there.

Both one and the other are necessary for a war to become one of maneuver.

Of course, the situation in which the conditions and opportunities for a war of maneuver may find their true significance are quite diverse and have no limit as to their variety.

In this regard, the experience of the German-Polish war cannot be mechanically transferred to cover all events that remain to be played out upon the battlefields.

However, this war revealed the new opportunities for waging a maneuver struggle, and in this lies its great significance as a war of the new forms of struggle in action.

If one examines the German-Polish war from the point of view of the general nature of an accomplished strategic maneuver, on the whole, then at first glance there is nothing new in such a form of a strategic offensive.

The German offensive, which led to the rout of the Polish army, was an already historically-known example of an enveloping concentric offensive along external operational lines. Such operations took place in a number of wars in the XIX and XX centuries and, each time, were conducted when separate armies could, from various sides, outflank and attack an enemy occupying, according to the geographical conditions of his deployment, an internal position. The most typical example of such an offensive in the last century was the Austro-Prussian War of 1866, when three Prussian armies from different sides enveloped and attacked Benedek's Austrian army in Bohemia.

In the 1914–1918 war the concentric offensive from different sides was evident only in a few separate operations, although with a completely different result.

To these belong the offensive of the Russian 1st and 2nd armies into East Prussia in 1914, which ended in the defeat of each of this armies in detail; the successful offensive by Mackensen's[45] German Eleventh Army and the Bulgarian First Army against Serbia in 1915; the offensive by Falkenhayn's German Ninth Army against Romania through the Carpathians and Mackensen's from the south, over the Danube, during 1916–1917, which ended in the rout of the Romanian army.

A number of offensive operations were like this in our civil war of 1918–1921, particularly the offensive by the 12th and 1st Cavalry armies against Rydz-Smigly's[46] Polish Third Army in June 1920 and comrade Frunze's operation in the Tavria[47] in the fall of that same year, which led to the defeat of Wrangel.[48]

The Japanese carried out a series of operations against the Chinese army in 1938, although nowhere did they achieve its encirclement.

The Red Army's operation in Mongolia in the summer of 1939 had the same character of an offensive by two separate enveloping groups.[49]

However, it is necessary to take into account that in the XX century, particularly in the 1914–1918 war, the offensive along external operational lines was limited, as a rule, to the scope of local, individual operations.

On a large strategic scale, this offensive form could no longer find employment, because the formation of a continuous front, in the conditions of which the linear strategy had reached its own contradiction and, as a rule, put an end to operations by troop masses in different groups and along different axes.

The front became continuous and the deployment of forces took on the character of a single overall unbroken line that cut through the entire theater of military activities. The very expediency of an offensive by separate groups along different axes was placed in grave doubt in the new conditions of the XX century.

Schlieffen wrote that "it is no longer profitable to attack a powerful opponent from different directions, which are separated by great distances and exclude the interaction of the separated groups of forces."

Actually, many operations conducted in the war of 1914–1918 along external lines failed to yield a result, because the cooperation of the separated groups of forces, which were attacking along different axes, could not be achieved, and the enemy occupying an internal position retained the advantage of attacking his enemy's separate groups in detail.

The more the operation's spatial dimensions grew; that is, the further separately attacking groups of forces stood from each other in their jumping-off positions, the less one could count on achieving their cooperation.

Thus, it seemed that the concentric offensive along external operational lines should have died out and, together with it, the possibility of encirclement (Cannae), to which such an offensive led in its most complete form.

But here the German-Polish war showed that this situation has changed substantially.

In the same theater, where in 1915 a joint concentric offensive by the German and Austrian armies—one from East Prussia, and the other from Galicia, failed to achieve a result and did not lead to the encirclement of the Russian army—in September 1939, in approximately the same operational situation, but in new conditions and with new means of struggle, this was achieved against the Polish army.

The motorization and mechanization of the army and aviation and new com-

munications equipment (radio) have once again made the concentric offensive by separate groups of forces along external lines possible and, what is more, as the events of the German-Polish war showed, with a significantly more rapid and decisive result than in the past.

Such an offensive has now acquired a more vigorous development and has achieved the encirclement in its most complete and decisive form.

The chief reason behind this is that the speed of the modern means of struggle has changed the former significance of space in the operation.

There existed earlier definite boundaries of cooperation for separately attacking groups of forces. Motorization has greatly expanded these boundaries. Thus that which was impossible earlier has become achievable in modern conditions.

In 1866 the Prussian deployment's flank corps were located at a distance of about 400 kilometers from each other. In a concentric and enveloping maneuver they approached to a distance of 4–5 kilometers on the battlefield at Königgrätz, having covered 125 kilometers in 12 days; that is, at an average pace of 10.5 kilometers per day.

In 1939 the German deployment's flanks against Poland were in the jumping-off position at twice the distance, or 800 kilometers from each other.

In a concentric enveloping maneuver, the units of two flank motor-mechanized groups—one attacking from Pomerania and the other from Silesia—came together along the Western Bug River at a distance of about 100 kilometers in 13 days, having covered during this time up to 600 kilometers. This amounts to an average of about 50 kilometers per day.

Thus in approximately the same time of 12–13 days, 125 kilometers were covered in 1866 and about 600 kilometers in 1939.

Thus the pace of the operation's development has changed in the new conditions and this has become possible only through the introduction of the motor into the army.

The risk that was earlier associated with dividing the army into separate groups in space, remains in force, in certain conditions even now, but it has been significantly reduced.

The events of the German-Polish war showed that the great mobility of the modern motor-mechanized army enables us to significantly more quickly unify the separated groups of forces at the decisive spots and thus achieve their interaction.

The army's deployment in different groups has thus become less dependent upon space and its capacity.

Air superiority enables us to operate by surprise. The new means of communications support firm control and the constant interaction of the separated groups of forces. In these conditions, operations along external lines gain new opportunities, allowing for the conduct of rapid and decisive maneuvers against the enemy's flank and rear.

The enormous significance of airborne infantry, which enables us to carry out surprise and vertical turning movements by landing it from the air in the enemy's depth, fills out the enormous possibilities for operations along external lines.

A landing by airborne infantry at the decisive point in the enemy's rear may be a double envelopment, carried out by two separate groups of forces, and may rapidly and unexpectedly lead to a complete conclusion and, in this regard, plays the role of a lock, which firmly closes the encirclement ring.

At the same time, the side occupying an internal position is now losing those advantages that he had earlier. He is always more limited in space; he always occupies a closer and more concentrated position; his entire operational base is narrower and his communications more vulnerable. This, to a significantly greater degree than before, increases the threat to him particularly from the air; it restricts his mobility and subjects him to heavy losses.

It is becoming all the more difficult for the side in the internal position to win time and hold off the enemy's separate groups, attacking from various sides, until a decisive victory is achieved over one of them. After all, in this lie the only possible tactical actions for the side in an internal position. For him there is nothing more hopeless than to get involved in or be drawn into prolonged and hopeless fighting along one axis until the enemy resolves his task along another. It is precisely this that enables the side attacking along external lines to narrow the range of fighting, to bring together his enveloping flanks and, finally, to completely trap the enemy in an encirclement ring.

Of course, one needs to have a superiority in men and materiel and a higher level of command skill for operations along external lines. This was a decisive condition in the success of the German offensive against Poland.

However, if these factors are at hand, then a concentric offensive from different directions along external lines gains all the advantages in modern conditions.

However, the very possibility of such an offensive depends to a significant degree on the geographical conditions of the deployment.

However, the conclusion already suggests itself that given a continuous, occupied front, which requires a breakthrough, attacks, concentrically directed from various directions may, in modern conditions, acquire all the advantages over the ramming breakthrough of the front along a single main axis.

The events of the German-Polish war have, for the time being, left that question unanswered and it remains for history to prove this in practice.

However, one thing has become obvious: the conduct of decisive encirclement and destruction operations has acquired new opportunities. The strategy of destruction has acquired new prerequisites for its most decisive realization.

The significance of the German-Polish war lies in the fact that it has shown this.

The strategy of destruction was always the highest manifestation of military art in all ages. However, the forms in which it was realized were quite varied and changed along with an alteration in the overall conditions of struggle.

In the German-Polish war the strategy of destruction found its realization in forms, in which according to their qualitative content were quite distinct from all that the history of wars had known up to that time. This requires special examination.

The German armies' campaign began along a salient-shaped front with an overall length of 800 kilometers. There were significant gaps between the separate armies: 200 kilometers between the Third and Fourth armies and 300 kilometers between the Fourth and Eighth armies. As the operation developed concentrically, the front constantly narrowed. The entire campaign concluded along the line L'vov—Brest-Litovsk—Bialystok, with a length of 400 kilometers, or twice as small as the front from which it began.

By the sixteenth day of the operation the front had been fully extended to this line (some places along this line had been reached even earlier).

If one draws a straight line from Poznan through Warsaw to Brest-Litovsk as the strategic axis which determined the overall depth of the entire campaign, then this will amount to 500 kilometers.

On the whole, as of September 1 the Poles stood 150 kilometers from the German capital (the Poznan–Berlin line). After half a month (16 days) they were 150 kilometers behind their capital (the Warsaw–Brest-Litovsk line).

This comprises 500 kilometers of the strategic front moved back in space and means that the average daily pace of the operation's development was approximately 30 kilometers (500 divided by 16).

Along the main directions the forward front line (the fighting front in some areas remained in the depth) advanced 25–30 kilometers per day. This was the infantry's usual daily march which, while trying to catch up to the highly-mobile formations operating ahead, carried out marches of up to 50 kilometers on certain days.

However, when on September 16 the operation's final front halted at a length of 400 kilometers it still had a depth almost to the operation's initial front; that is, to a depth of 500 kilometers. At this depth and in many and very diverse places, the fighting was still going on.

Thus by the close of the campaign the depth of the front was greater than its width.

This was a completely new phenomenon in the forms of armed struggle, which most vividly expressed its new deep nature.

Before then the front of struggle was always even, with an overall line of infantry moving forward, cleaning out entirely the enemy left in the territory behind. Only besieged fortresses, remaining in the front's rear, constituted an exception in this regard.

Now the front has been moved ahead by the separate deep attacks by highly-mobile formations along different axes, while still leaving behind themselves a series of pockets of struggle.

In this regard, the conclusion of the grandiose battle in Poland revealed the highly unusual picture of a deep, multi-tiered battle.

On September 16 the fighting was still being waged along an enormous space of 185,000 square kilometers. Its pockets were echeloned to a depth of 500 kilometers and formed at least five lines or five tiers.

Beginning from the rear, the first tier forms a pocket of fighting near Pomorze and in the Poznan area, where the German *Landwehr* troops only occupied this province.

The second tier formed an encirclement pocket near Kutno; the third tier was the junction near Warsaw; the fourth tier was the pocket in the Lublin area, and; finally, the fifth tier was the operation's final front along the line L'vov–Brest-Litovsk–Bialystok.

Along these five lines, between them and alongside the main pockets of fighting, where the battle would flare up in a bright flame, numerous individual fires from less important combat events continued to burn in various places.

It seemed that an enormous space had flamed up in different places. And a few days would pass, while the fire would die down everywhere, leaving in its wake the ashes and ruins of the destroyed Polish army.

This was the picture of the new, deep battle, if one takes in at a single strategic glance its enormous depth of 500 kilometers, from west to east.

This was a picture of the deep operation of complete ("total") destruction. It resulted in an entire system of "Cannaes": some of them had already concluded, while others were maturing, and others were only just beginning.

This system of battles for encirclement and destruction that have arisen and played out in the most varied places in space to a great depth is characteristic of the new forms of a large deep operation. These forms, which mean the complete crushing of the enemy's army in detail, were the result of the fact that the forces and means of struggle were no longer employed along a single line of applying combat efforts and, in accordance with their possibilities and speeds, were being employed immediately throughout the entire depth of the arena of struggle; that is, they acquired deep operational employment.

This was the first realization of the new, deep forms of struggle in action and for the first time showed their new qualitative significance.

In this lies the significance of the German-Polish war for the history of military art.

May one immediately take on faith all the conclusions of the German-Polish war?

The victory was none the less achieved over an enemy unequal as to forces, quality and equipment. The war was waged none the less in an arena where there were no previously fortified lines and which least of all was favorable to positional warfare.

The front was not continuous by the beginning of the war and the foolish enemy subsequently did everything not to create it.

Could the new deep forms of struggle have justified themselves in such conditions when faced with an enemy equal in strength and equipment, given borders girded with permanent fortifications, and given the presence of large reserves in the depth?

History has left all of this unrevealed in action. All of this yet remained in question. With all the more impatience we now had to await the unfolding of the events of the armed struggle in Western Europe, where the war was only beginning in September 1939 and where for a lengthy period it was not prepared to reveal its true character.

Only more than half a year later events unfolded in the West that showed the subsequent paths of development of the new military art at a higher level of a major modern European war.

June–July, 1940

End of Part Two

Six

"The Development of the Theory of Soviet Operational Art in the 1930s" (1965)

G. Isserson[1]

I

Soviet military theory, based on Marxist-Leninist teaching and employing the rich military experience of the past, came into being as early as the civil war years and covered a great path of development in the two decades before 1941.

In this development, of particular interest are the military-theoretical views of the 1930s, with which we entered the Great Patriotic War. If in the 1920s our military-theoretical thought rested primarily on the experience of the First World War and was, to a significant extent, focused on the past, then from the 1930s it faced forward, toward the study of the problems of a future war and the means of waging it.

This period has particular significance for the development of our military theory. It yields a colorful picture of a great deal of research, broad-ranging and creative thought and important and fundamental decisions. It was precisely during these years that the fundamentals of the deep engagement and the deep operation were elaborated, which opened a new page in the theory of operational art.

The deep forms of struggle were conditioned by the entire socio-economic development of the Soviet Union and the Red Army's reconstruction. They were necessary for the resolution of the problems of conducting destructive operations, overcoming the continuous front and its breakthrough throughout the entire operational depth; that is, the achievement of that goal that was not and could not be achieved during the First World War.

The history of the problem of the deep engagement. For the sake of historical truth, it is necessary to mention that the problem of the deep engagement was first raised by the English military theoretician Fuller at the end of 1918. In anticipation of a decisive offensive in 1919 (The Entente was not counting on the triumphant conclusion of the war in 1918), Fuller then suggested organizing an attack by rapidly moving tanks throughout the depth of the enemy's tactical position, alongside the tank attack of the forward edge of the defense. To be sure, the concept of a long-range tank group had not yet been formulated by him, although all the tactical prerequisites were already present in this proposal.

However, Fuller's theoretical views on the problem of the deep engagement ended with this. The conditions of the capitalist development of bourgeois armies forced him to adopt the theory of small, professional armies, for which the problem of the offensive would be resolved in another way altogether. This theory, which reflects the class character of the capitalist military system, stood in obvious contradiction to the actual nature of modern war. For Fuller, the deep engagement was not a combined-arms engagement. He wrote that "uniting tanks with the infantry is the same as harnessing a tractor to a draft horse."[2] Such a point of view was, of course, completely unacceptable for us.

Foreign manuals from the 1930s utterly failed to contain instructions on the deep engagement in the sense of the simultaneous suppression of the enemy's entire tactical depth. This idea belonged to our military-theoretical thought.

If one turns to the origins of our first specific conceptions about the deep forms of struggle, it's impossible not to mention two documents from 1928–1929, which are of great significance.

The first document was M.N. Tukhachevskii's[3] memorandum on the reconstruction of the Red Army and its outfitting with new and modern weapons, chiefly tanks and aviation.⁴ Having laid out in it a major program rearming the army, Tukhachevskii concluded that the new materiel-technical base will enable us to renounce the former exhausting forms of struggle for each part of the enemy's combat formation in detail and to go over to new and more effective forms and methods of waging the engagement, while simultaneously suppressing the entire depth of the enemy's position.

The second document is V.K. Triandafillov's memorandum on employing tanks in the offensive engagement in three echelons, according to their operating range—NPP, DPP and DD,⁵ which break through to various depths all the way to the enemy's artillery positions and headquarters and thus suppressing, in conjunction with long-range artillery and aviation, the entire tactical depth of his position.⁶ This method of employing tanks was already the practical and concrete expression of Tukhachevskii's idea that the new, modern means of struggle—tanks, long-range artillery, aviation, and airborne landings enable us to renounce the old and drawn-out methods of consecutively striking the enemy in detail and to go over to the new forms of the simultaneous deep strike. Triandafillov, in his memorandum, laid down the specific fundamentals of the new forms of the engagement and presented a basic outline for its organization and conduct.

Thus in these two documents, Tukhachevskii and Triandafillov laid down for the first time the idea of the deep engagement and thus exerted an enormous influence on the development of our army and the formation of our military-theoretical thought's basic views.

This idea was reflected in the 1929 Field Manual, which looked far ahead and which was the most advanced of the European manuals of the time. The manual's article 191 spoke of detaching special battalions to be thrown directly against the second defensive zone. Article 207 very precisely established the concept of the long-range tank echelon, which was designated to enter the defensive depth simultaneously with the attack against the forward edge. Thus PU-29⁷ already contained the first prerequisites for adopting the deep tactics, based on the combined-arms actions.

Tukhachevskii's and Triandafillov's contribution to creating the theory of the deep forms of struggle lies in the fact that they did not dally in the rear of altered historical conditions, but foresaw the possibilities of the new technical means of struggle, when our army did not yet have them and had not been reconstructed.

K.B. Kalinovskii⁸ (the first chief of the motor-mechanized forces) thoroughly worked out the tactics of the tank groups'—NPP, DPP and DD—actions and thus laid down a practical basis for the entire concept of the deep engagement. Thus one may consider the foundation to have been established in 1930.

The concept of the deep engagement was first recognized in academic circles. As early as the beginning of 1930 the Frunze Military Academy⁹ was resolving tactical problems on maps and in the field on the new basis of the deep engagement and played a major role in spreading them throughout the army. R.P. Eideman[10] (the chief of the academy), N.Ya. Kotov,[11] K.A. Chaikovskii,[12] P.I. Vakulich,[13] S.N. Krasil'nikov,[14] P.G. Ponedelin,[15] I.P. Kit-Viitenko,[16] R.S. Tsifer,[17] and others carried out significant work in this area in the academy.

Soviet military theoreticians were pioneers in this field, when they had not even begun to talk about the tactics of the deep engagement in the West.

At the beginning of the 1930s, proceeding from the experience of conducted exercises and maneuvers, Tukhachevskii wrote in one of his reports: "Modern means of suppression, employed on a massive scale, enable us to achieve the simultaneous attack and destruction of the entire depth of the enemy's tactical position.

"These means, particularly tanks, enable us:

"a) suppress the enemy's defensive fire system in such a way that a large part of the artillery and machine guns cannot take part in repelling the attack and the penetration by the attacking infantry and NPP tanks into the depth of the defensive zone;

"b) to disrupt the command and control system and tie down and isolate the enemy's reserves in order to defeat in detail the various echelons of the enemy's combat formation during the fighting in the depth."

Thus Tukhachevskii quite clearly defined the tasks of the deep engagement. However, they were not understood by everyone. K.Ye. Voroshilov spoke out against Tukhachevskii during a plenum of the RVS[18] of the USSR, His criticism revealed a clear failure to understand the essence of the problem, which Voroshilov reduced to a single type of engagement—the attack against an enemy who has halted.

Of course, deep tactics were primarily elaborated for the most complex type of engagement—the attack against the enemy's defense. In essence, however, deep tactics were not a type of engagement, but a new form and new method of conducting the engagement and were to be employed in any kind of attack.

Tukhachevskii patiently explained this to Voroshilov in a special memorandum, in order to eliminate the disorder that had arisen in the minds of the command element.[19] Such representatives of the high commands as I.E. Yakir,[20] I.P. Uborevich[21] and S.S. Kamenev[22] supported him and the correct understanding of the essence of deep tactics as a new form and method of the modern engagement was affirmed.

The first initial tenets of the deep operation. The establishment of the bases of the deep engagement was only a half measure. Tactical breakthroughs had succeeded during the First World War with the aid of the old methods of the engagement. The chief essence of the entire problem was how to complete the tactical success with an operational development of the breakthrough and, having penetrated through the broken front and gained freedom of maneuver, to destroy the enemy's forces on the operational scale.

Thus the idea of the deep engagement immediately touched upon the most cardinal problem of operational art and outstripped it with a new solution.

The tacticians were triumphant, while the operational artists went around thoughtful and anxious. And then there occurred a great misfortune. V.K. Triandafillov and B.K. Kalinovskii perished in the summer of 1931 during a plane crash. The family of operational artists was orphaned and at first operational thought was unable to find a new path.

Having expressed his anxiety that "our military theory is lagging far behind the country's successful fulfillment of the party's general line," Tukhachevskii said that "due to the growth of our socialist economy, we cannot remain at the previous level of military-theoretical thinking and must achieve the decisive development of our military-theoretical thought on the basis of Marxism."

Following the death of Triandafillov, Tukhachevskii continued to earnestly work on the deep forms of struggle. In 1932 he completed the first part of his broadly planned work, *The New Problems of War*, in which he studied the influence of modern technical means of struggle on the changes in the forms and methods of conducting the engagement and operation.²³ The first part of this work, however, chiefly contained technical and tactical questions. Tukhachevskii was evidently preparing to expound on operational and strategic problems in the second and third parts of the work, in which he planned to study the bases of modern war and the struggle against imperialist coalitions. He was not fated to complete this work.

At the same time, it was quite clear that changes in tactics must be reflected in operational art. Everyone realized the necessity of a decisive step along the path of creating a new theory of conducting operations. In pointing out the importance of this task, Tukhachevskii wrote that "the reconstructed army will call forth new forms of operational art."²⁴ The first grain of truth had ripened for this in the concept of the deep engagement; behind it, the new operational thinking was already breaking into consciousness. Our army had achieved such a degree of development and such opportunities for its employment, which were authoritatively putting forward demands for the new employment of men and materiel in major, decisive operations on the ground and in the air.

First of all, we had to review all of the basic problems of operational art, as the study of the conduct of operations, in light of view of completely altered conditions. Such a formulation of the question put forth a number of new problems, opening an enormous field for Marxist-Leninist study. We had every reason to maintain that our operational art's main task consisted of creating new forms and methods of decisive and destructive operations in the new historical conditions, with a new army and on a new materiel-technical base.

Of course, there could be no complete analogy with the tactical resolution of the problem, because the engagement (tactics) and the operation (operational art) had qualitative differences, which are determined by the scale of space and time and the difference in the operational organization of troops from their tactical combat formation, which represents a unified and connected system of direct cooperation. In this regard, the organization of the deep operational strike must be significantly different from the organization of the deep engagement and put forth a number of new problems. Nor could the simultaneity of the deep strike find such an immediate expression at the operational level.

However, one thing was for sure: aviation, airborne landings and mechanized and motorized formations, organized in the appropriate manner for independent operational employment, could also spread their long-range striking power into the enemy's operational depth, which is measured approximately at 50–60 kilometers; that is, to the line of his operational reserves, forward airfields and army headquarters. At the same time, the question was not only that the modern long-range and highly mobile means of struggle enable us to organize a deep strike against the enemy, but also that this is necessary in order to radically resolve the problem of the operational breakthrough, which cannot be resolved without the striking of the entire operational depth.

It was necessary to transfer the basic outline of the deep engagement to the operational scale. For this it we needed, first of all, motor-mechanized formations, which were capable, by virtue of their organization and armament, of resolving independent operational tasks. Secondly, the problem was how to carry these formations' efforts into the enemy's operational depth. Thus the main problem of organizing the deep operation essentially came down to the resolution of this problem: how to transform the tactical breakthrough into an operational one; that is, how to push independent motor-mechanized formations through the completed breach of the tactical defense.

Such were the first initial tenets of the theory of the deep operation. However, for the time being these were only general discussions, which required theoretical grounding and specific formulation. The great deal of work that had been started along these lines in 1931–1932 was linked to the creation of the Frunze Military Academy's operational department, which played a definite role in the development of our operational art.

The Frunze Military Academy's operational department. The very fact of the department's establishment signified a new step in the elaboration of the theory of operational art, which had grown out of its narrow bounds. The profound study of the problems of conducting modern operations was required. Aside from this, an insistent demand had arisen for well trained and broadly educated operational workers for the higher staffs. The Frunze Military Academy's operational department, which was called upon to resolve these tasks, began its work in the autumn of 1931. It began the review of operational art's fundamentals and put into play a great deal of scientific-research work, putting forth and resolving a number of new problems The deep operational forms of struggle became the guiding idea. They now began to acquire their theoretical grounding and specific formulation.

A collective of smart and capable instructors worked in the department, including old military specialists[25] who deeply understood the necessity for reviewing their views on the character of modern operations and who were imbued with the new ideas on the new forms of struggle. The department's collective included the instructors A.V. Fedotov[26] (my assistant in the department), S.N. Krasil'nikov, Ye.N. Sergeev,[27] A.M. Peremytov,[28] and specialist technical leaders—A.N. Lapchinskii,[29] D.M. Karbyshev,[30] I.I. Trutko,[31] and B.K. Leonardov[32] (from the Military-Medical Academy). Lapchinskii worked on questions of employing aviation in the operation. Karbyshev made it possible to thoroughly study the conditions of the deep breakthrough through his profoundly brilliant and detailed elaborations on the organization of modern defense on the part of the enemy. Trukto was in charge of elaborating rear-area questions and Leonardov on working out the organization of medical evacuation. Having scientifically grounded his calculation of possible losses in the modern operation and the demand for hospital and evacuation equipment, he drew up an entirely new plan for medical evacuation in the deep operation. Unfortunately, many of the enumerated men from the department's leadership are no longer alive. A.N. Lapchinskii died as early as 1938. D.M. Karbyshev died a hero's death in a fascist camp, and Fedotov, Sergeev, Peremytov, and Trutko became victims of lawlessness during the years of Stalin's cult of personality.

It is necessary to note that extremely favorable conditions for the operational department's work were created by the chief of the academy, R.P. Eideman, who knew how to value and respect the young and creatively working cadres, and to protect and help them.

M.N. Tukhachevskii's and A.I. Sedyakin's[33] (then the chief of the military training directorate) instructions had a great deal of invaluable significance for the directing the operational department's work. The broad scope of Tukhachevskii's operational thinking and Sedyakin's inquisitive mind guided us to many problems and pointed out ways to resolve them.

A.I. Yegorov[34] (then the chief of the RKKA Staff) also adhered to progressive views on the new character of modern operations. He loved and supported any kind of new thinking. As early as 1931 he had delivered a major report at the academy on the "spatial operation" (as he called the deep operation). The report was illustrated by a mobile diagram, which had been drawn up by academy instructor V.I. Mikulin.[35] The term "spatial operation" was, of course, imprecise, because any space has two dimensions in a single plane—along the front and in depth. Operations had already achieved their maximum spatial expansion along the front as early as the First World War. Their spread into the depth was characteristic for the development of operational forms in a future war. Thus the difference between the operations of the past was better and more accurately expressed by the concept of the deep operation. This definition became established in Soviet military theory and was then picked up by all of the bourgeois military literature.

The operational department received the broadest support from Tukhachevskii, Sedyakin and Yegorov, who occupied leading posts in the army, and this was particularly valuable, for it is known how difficult it is to forge a new path and to make the first breach in already established ideas, to which many old military specialists in the academy clung. I recall with what disbelief and ironic comments they first met the operational department's elaborations. Of course, this made our work more difficult, but could not delay the development of our military theory. Some of the old specialists simply remained on the sidelines of this process. However, the majority of them soon understood the progressiveness of the ideas of the deep operation and firmly set out on the new path, bringing a great deal of benefit to the development of our operational art. Among these were N.Ye. Varfolomeev,[36] Ye.A. Shilovskii,[37] N.N. Shvarts,[38] P.F. Shafalovich,[39] A.I. Gotovtsev,[40] and a number of others. Even A.A. Svechin[41] finally agreed with the inevitability of shifting to the new forms of struggle and supported the concept of the deep operation, while viewing it, however, within the confines of the strategy of attrition.

Not everything went smoothly in the department's work and by no means did the recognition of the new ideas happen immediately.

At first there was no complete unity on questions of the new character of modern operation even in the RKKA Staff itself. Some of the operational directorate's workers (S.A. Mezheninov,[42] S.P. Obysov[43]) did not support the basic fundamentals of the deep operation. They objected, in particular, against the independent employment of motor-mechanized troops ahead of the front and in the depth of the breakthrough, operating apart from the combined-arms formations. However, as far as these questions were concerned, the operational department enjoyed the full support of the chief of the RKKA Staff, Marshal Yegorov.

The elaboration of the theory of the deep operation. The entire elaboration of the theory of the deep operation was conducted at the operational level; we were unable at the time to work on questions of military strategy as the waging of armed struggle at the level of war as a whole.

The suppression of the enemy's operational depth undoubtedly touched upon the strategic sphere of armed struggle, but this required first of all the resolution of such practical problems of employing motor-mechanized formations and their cooperation with aviation and airborne landings, which were limited by operational scope.

At the beginning of the 1930s we had three mechanized corps, which were insufficient for uniting into larger groups (or armies) for front purposes. Therefore their employment was at first conceived in the form of individual corps, along with motorized divisions and cavalry at the army level. Thus the beginning stage in the elaboration of the theory of the deep operation was the army operation as the operation of a shock army.

While elaborating this theory, we had in mind two possible situations: first, when the enemy is on the march and is free to maneuver, and; second, when, having adopted an organized operational disposition, he has constructed a closed front of opposition.

In the first case, given the presence of a spatial gap between the sides, it was considered possible to organize a deep strike along a selected axis by pushing ahead a group of rapid and mobile troops (motor-mechanized formations and cavalry), supported by aviation. This group, in conjunction with aviation and an airborne landing in the rear, was to attack and tear from the approaching front a certain part of the operational formation and form a breach in it with exposed flanks, thus causing it to waver. The chief task was to prevent it from forming a closed front and digging into the ground. Thus it must knock out those supports upon which a continuous front is built and is maintained. The group operating ahead of the front was called the **vanguard echelon**.

The arriving combined-arms formations, which comprise the **main echelon**, could be directed into the newly-formed flank and attack with a decisive goal. At the same time, the operational depth must not remain undefended, because it can be subjected to the same degree to a deep breakthrough on the part of the enemy. Thus it was considered necessary to move 2–3 marches behind the main echelon an army reserve group, which was given the name of the **reserve echelon**.

Thus the army's entire operational formation during an offensive against an advancing enemy consisted of three echelons: vanguard, main and reserve and could occupy up to 200 and more kilometers in depth. The deep forms of the operation acquired quite a complete expression in this formation.

The resolution of the problem of the deep operation against the consolidated front of an enemy, who has taken up the defensive, was more complex.

Four problems required practical resolution for achieving the goals of the deep breakthrough operation: a) what was to be the operational organization and operational employment of the different combat arms (chiefly motor-mechanized, combined-arms, aviation, and airborne landings); b) to what operational depth should and can one push forward operational efforts, bearing in mind their support (this problem primarily concerned the allowable depth of the motor-mechanized group from the combined-arms formations' front); c) how to organize the operational development of the breakthrough, so that the tactical breaking of the front grows immediately into an operational breaking throughout the entire operational depth and complete destruction, and; d) how to isolate the enemy's broken front in his operational depth, in order to prevent the concentration of new reserves capable of hindering the operational development of the breakthrough and to prevent the restoration of the broken front.

The theoretical and practical elaboration of these questions in a number of examples on maps led to the following decisions, which formed the basis of the initial concept of the deep operation:

a) the army's operational organization for the breakthrough should consist of two echelons: the attack echelon (EA), consisting of combined-arms formations, reinforced with artillery and tanks, for breaking through the tactical defense, and a breakthrough development echelon (ERP), consisting of rapid and mobile mechanized, motorized and cavalry formations, for developing the breakthrough through the broken tactical breach in the defense into its operational depth;

b) the breakthrough development echelon should be committed into action immediately following the breakthrough of the first defensive zone, if it has been penetrated along a sector 6–8 kilometers in width and, in a favorable situation, even earlier. In this case it itself suppresses the last resistance in the tactical depth of the enemy's defense. In any event, it should seize the enemy's second defensive zone before he can fall back on it or occupy it with his reserves;

c) the entire operational development of the breakthrough at the army level is conducted to a depth of 60–100 kilometers, as far as the line of the enemy's forward depots and army headquarters;

d) army aviation (light bomber and assault) is employed in preparing the breakthrough and subsequently for operational cooperation with the breakthrough development echelon, in order to deprive the enemy's reserves of the opportunity of operating and putting up resistance in the depth;

e) front aviation (long-range bombers) is employed to completely isolate the enemy's broken front from his strategic rear and to prevent the arrival of his strategic reserves;

f) an airborne landing is made to the depth of the enemy's forward depots and army headquarters to cooperate in the operational depth with the breakthrough development echelon.

Such, in general, was the initial basic outline of the deep operation, which had been adopted in the academy as early as 1932. The first operational map exercise was developed on its basis on the theme of "the shock army's deep offensive operation," which was issued and sent out to other academies and military district staffs.

In 1932 lectures on the tactics of the deep engagement were read in the department. In them the concept, the fundamentals of which Triandafillov had established, were deepened. These lectures were published by the academy. In the same year lectures were read on the new problems of the modern deep operation. At first these were of a general theoretical nature, but in 1933 they acquired a definite and calculated formulation. In the work *The Fundamentals of the Deep Operation*, which was issued by the military academy, the already applied theory of the forms and methods of conducting the deep operation and its development into the depth to a decisive and final outcome were expounded. The chapters on the work of the army staff in the control of the deep operation at each stage of its development acquired great significance in this work.

On orders from M.N. Tukhachevskii (then the deputy defense commissar), a commission from the RKKA Staff, under the chairmanship of A.I. Yegorov, reviewed this work. The commission recognized the necessity of distributing this work, as an unofficial textbook, to all the academies and military district staffs. *The Fundamentals of the Deep Operation* was published by the Frunze Military Academy in a run of 100 copies and for the next few years became a study guide on operational art and played a certain role in the formation of our military theory's views.[44] In this work the theory of the deep operation was for the first time clothed in specific forms and acquired an applied exposition. It was still being used as late as 1936 as a textbook in the newly created General Staff Academy.

Naturally, *The Fundamentals of the Deep Operation*, as the first work in this field and written during the first Five-Year-Plan, when our army's technical reconstruction was still in its beginning stage, was far from foreseeing and, even more so, resolving all of the complex problems of organizing and conducting the deep operation. However, the initial basic fundamentals had been established. In subsequent years the theory of the deep operation acquired its further theoretical development and a number of substantial corrections were made to it in the brilliant operations of the Great Patriotic War.

In 1933 a large two-sided war game was conducted in the operational department. A difference of opinions was reflected as to individual and principled problems of the deep operation. The dispute was chiefly about the possibility of the motor-mechanized group's independent actions ahead of the front and in the enemy's operational depth, apart from the combined-arms formations.

Under the influence of the commanders from the RKKA Staff's operational directorate, who were present, the student who was playing the role of the commander in the game refused to push his motor-mechanized group ahead of the front in order to attack the approaching enemy with a decisive aim. It required the insistent interference of the leadership, in the role of the front command, for the course of events to take on the desired direction for the game's goal.

Marshal Yegorov was always present at the game, which lasted three days. He attentively followed the course of the game and with his leading questions supported the bold and enterprising employment of the motor-mechanized group for resolving independent operational tasks. In the marshal's closing remarks during an analysis of the game, he said that this was the first time the problems of the deep operation had been so fully and completely elaborated in a war game. This was a recognition of the definite results that had been achieved in the development of the new fundamentals of our operational art.

Practical work in the army. However, military theory is never created by theoretical research alone. It is born in the practice of training the army in peacetime and during its actions in wartime. Thus it would be quite incorrect to assume that the theory of the deep operation arose and was created in a single, closed collective in the military academy's operational department.

The deep forms of struggle had so matured with the appearance of the new weapons that this theory arose simultaneously in the army on the initiative of a number of military men. They worked, independently of the operational department, on the theory of the deep operation, particularly in the armored academy, the Zhukovskii Air Academy[45] and the chemical defense academy, as well as in the military districts, particularly in the Belorussian and Ukrainian districts and the OKDVA.[46] I.P. Uborevich (the commander of the Belorussian Military District) and I.E. Yakir (the commander of the Ukrainian Military District) and their chiefs of staff—Bobrov[47] and D.A. Kuchinskii[48]; then the deputy chief of armored forces, I.K. Gryaznov,[49] the chief of the armored academy, M.Ya. Germanovich,[50] the chief of the chemical troops, Ya.M. Fishman,[51] and other comrades from the armored and artillery academies introduced a number of new tenets into the theory of the deep operation and thus added to and developed it.

Experimental exercises with tanks and airborne landings, conducted by Uborevich, Yakir and Gryaznov, added a lot of value. In this regard, troop maneuvers in the '30s and military district war games were a major school and yielded a number of valuable theoretical and practical conclusions.

It should be noted in particular that the combat formation of the motor-mechanized group on entering the breakthrough and its activities in the operational depth was elaborated by Uborevich and his staff. Uborevich also resolved in a new way the problem of the engagement of the vanguard, reinforced by tanks, until the arrival of the main forces. Yakir and his staff specially worked out the problem of the motor-mechanized groups' cooperation with the airborne landing in the operational depth. Problems of the deep engagement, applicable to the conditions of the Far East, were worked out in practice by V.K. Blyukher,[52] I.F. Fed'ko[53] and M.V. Sangurskii.[54] Under Gryaznov's leadership, a number of exercises with tanks were carried out in the Trans-Baikal area. A great deal of work, which enriched and deepened our military theory, was carried out among the troops.

In general, if the elaboration of questions of the deep forms of struggle inevitably took on a more theoretical cast in the military academy's operational department, then in the military districts this theory's problems took on more concrete expression and acquired a more practical elaboration.

Thus our military theory developed and became enriched from the beginning of the 1930s, which during subsequent years, with the completion of the army's reconstruction and the delivery of more improved weapons, took on the character of a complete concept of the deep forms of struggle in the fields of tactics and operational art.

Many comrades, aside from those listed, took part in this work, and among them one should single out S.N. Bogomyagkov,[55] V.D. Grendal',[56] A.V. Kirpichnikov,[57] V.K. Mordvinov,[58] P.D. Korkodinov,[59] B.L. Teplinskii,[60] and a number of others. These were mainly young cadres, full of great enthusiasm and faith in the success of their work in developing Soviet military theory.

During these years the troops began to retrain on the basis of the new principles of conducting the engagement. As early as 1931 Uborevich issued the first unofficial instruction on the deep engagement. In 1933 the RKKA Staff's official instructions, drawn up by A.I. Sedyakin, appeared.

The 1929 Field Manual was already out of date and under the leadership of M.N. Tukhachevskii a new 1936 Field Manual was drawn up, which for the first time fully reflected the fundamentals of the deep tactics.

One of the manual's leading articles declared that "...the enemy must be pinned down throughout the entire depth of his position, encircled and destroyed" (PU-36, article 164).

During these years the first significant scientific-research works on strategy and operational art and the independent employment of motor-mechanized formations, of aviation and airborne landings, appeared.[61]

Turning now to the brilliant deep operations conducted by the Soviet Army in the recent war, we must recall that the period of the '30s under consideration. It was precisely then that the fundamentals of the deep operation were first time elaborated and formulated. Of course, this was only the beginning and, of course, the initial concept of the deep operation was still far from perfect and required still greater elaboration. However, a beginning had been made and it established a secure foundation for the subsequent development of our operational art. This took place as early as the second half of the 1930s, in the intense situation of the events of 1937 and the eve of the Great Patriotic War.

II

During the second half of the 1930s the development of Soviet military theory unfolded in a situation of the increasing threat of war and a number of military events in Europe and other parts of the world. This stage, according to its content, was complex, contradictory and changeable. The negative influence of Stalin's personality cult, which brought the Red Army difficult trials, told in this. However, during these years our military theory continued to improve and deepen.

In 1936 the problem of the Red Army's reconstruction and rearmament had been basically resolved, although this process, due to the constant development of equipment and the appearance of new and improved means of struggle, could never be considered finished.

The new deep forms of the engagement and operation continued to be improved and developed. These forms acquired all the more recognition and affirmation by the entire course of the events of the 1930s, which were characterized by the enormous growth of the armed forces on the European continent. Now western, bourgeois military-theoretical thought, obviously having copied this concept from us, began to speak about it in definite terms and develop it on the pages of its official press. Behind these general discussions the already quite specific views of the German-Fascist command on the employment of tank formations and the deep echeloning of the combat formation were coming to light.

In 1936 the fascist theoretician of tank warfare, Gen. Guderian, established the following order for employing tank formations in the offensive: the first echelon passes directly through the tactical depth of the defense and attacks its reserves (this was our breakthrough development echelon); the second echelon attacks the enemy artillery (this was our long-range group); the third echelon attacks the infantry within the confines of the tactical depth of the defense (these were our groups of direct and long-range infantry support). At the same time, in Guderian's opinion, the employment of armored divisions must manifest itself with particular effect, when the defense will already be opened along a particular sector and the sudden appearance of tanks at this moment will enable them to immediately penetrate beyond the defensive zone and into the open maneuver space. All of this system of attack was, in a somewhat altered sequence, a copy of our outline of the deep engagement, which had been adopted as early as 1932–1933.

Thus the deep organization of the operation and the operational depth of combat activities were more and more recognized as characteristic for modern conditions. However, we had already passed from these basic principles to a higher class of mastering the art of conducting deep operations. This task was brought about the necessity of creating a special institute for the deepened elaboration of operational art and the training of educated commanders for the higher staffs. The confines of the M.V. Frunze Military Academy's operational department were too narrow for this and as early as the beginning of 1936 the question of creating a special military academy, as a higher operational school, was raised. It was created in the autumn of 1936, under the name of the General Staff Academy. This had great significance for the subsequent development of the theory of operational art. The training of our command cadres was elevated to a higher level and entered a new phase.

The General Staff Academy. The elaboration of the theory of the deep operation in the new academy was further developed, although it retained its operational scale, while the academy's study plan pursued the goal of training experts in the organization and conduct of modern operations. This would have essentially transformed the academy into a technical school for training cadres for the higher staffs. From the point of view of the army's needs during the formation of the new forms of the operation, this was correct. But a negative side was hidden in such a narrow establishment of the task. In 1936 the theory of the deep operation had reached such a level of development that it was already impossible to exclude the strategic sphere of its employment and when only strategic scales and the situation in the entire theater of military activities could impart to it an intelligible and justified, in the given conditions, purposeful significance.

In the operational department, which was the first stage in the elaboration of the new operational forms of struggle, the deep operation could still be studied as such, without regard to the overall strategic situation. However, after the basic outline of the deep operation had been drawn up, another approach was required. It was necessary to view the deep operation in the General Staff Academy as a means for carrying out a definite strategic task and to endow it with a specific purpose, depending upon that situation that might arise and develop in the given theater of military activities. In other words, in order to transform the developed outline of the deep operation into a real phenomenon, it was necessary to place a definite strategic foundation under it and imbue it with strategic content.

All of this was quite clear when the General Staff Academy began its work. However, the slightest hint at the necessity of introducing into the academy a course on strategy, in one form or another, into the academy, as a basis for operational art, ran into objections from above. When this question was raised during one of the conferences before the academy's opening, Marshal Yegorov, the chief of the General Staff, asked the academy's representatives, with a hint of irritation: "Well, what are you going to study in strategy? The war plan? Strategic deployment? Or the conduct of the war? No one is going to allow you to do this, because this is a matter for the General Staff!"

Of course, when the question was put this way, no one could object, and the academy chief, D.A. Kuchinskii, a man with a very lively and practical mind and a great organizer, agreed with the marshal and renounced the introduction of a course on strategy in the academy. However, of course, it was not a question of elaborating in the academy practical problems of a strategic nature, which was the competence of the General Staff. Rather it was to move the course on operational art closer to the real military-political situation that had arisen as a result of the deployment in the center of Europe of fascist Germany's large and aggressive army. For this, it was necessary to evaluate the new correlation and disposition of forces along our western frontier; to analyze and study the possible situation for the outbreak of war and the nature of its beginning period. All of this would have brought the course on operational art closer to the scale and problems of strategy and demanded a great deal of research work in this area.

M.N. Tukhachevskii pointed out the importance of this task. He believed that it was impossible to answer the question about the character of an entire future war, because as it develops the war changes its forms and nature and it is impossible to guess them beforehand. However, he pointed out that "The first period of the war must be correctly foreseen in peacetime and correctly evaluated in peacetime, and one should prepare oneself for this."[62] Unfortunately, such work did not find a place in the academy. This was one of the reasons that our military-theoretical thought was not able to acquire a correct and flexible strategic orientation before the war as regards those possibilities and conditions in which military activities could begin along our borders. The representatives of the high command also declined to read lectures on strategic problems in the academy, and only Tukhachevskii spoke once on general problems of modern war at the beginning of 1937. One should note, however, that the military district commanders I.E. Yakir and I.P. Uborevich spoke in the academy with reports on problems of the deep engagement and the employment of motor-mechanized forces in the operation and conducted war games with the instructors. Corps Commander G.P. Sofronov,[63] who at the time was the chairman of a commission on elaborating problems of employing airborne forces, carried out exercises with the students and acquainted the instructors with this new problem.

Thus the creation of the General Staff Academy in 1936 changed nothing in the system of our higher military education as regards strategy. The true roots of this situation lay in Stalin's cult of personality, in which questions of policy and strategy were considered the exclusive competence of the higher political and military leadership. The negative consequences of this were evident in the beginning of the war in 1941, when many high command levels (fronts and armies) were faced with the necessity of independently trying to understand the situation on a large scale and adopt important decisions of strategic significance. A certain confusion and inability to comprehend the difficult situation as a whole, to adopt an expedient decision on a large scale and to subordinate the entire course of events to it were, to a significant degree, the result of the lack of strategic orientation and readiness to think in large categories of strategic significance. We played a heavy price in 1941 for our narrow view on the tasks of training cadres and for the insufficient development our military-theoretical thinking in the field of strategy.

The academy's leading departments (operational art and the tactics of higher formations) understood the importance of strategic questions for the correct orientation of their work. In the winter of 1936–1937 their chiefs approached the first deputy defense commissar, M.N. Tukhachevskii, with a request to elucidate a number of questions of a strategic nature. The conversation with Tukhachevskii touched upon important problems of modern war, its beginning period and the methods of conducting modern operations.[64] It had great significance for organizing the course on operational art in the academy and introduced clarity in the understanding of many important problems and showed in which direction our military-theoretical thought should develop. Of course, the plan for the academic course was laid out in the General Staff's directives, but Tukhachevskii played a large role in setting forth a number of operational problems.

The department of operational art had already drawn up operational assignments based on the realistic prospect of the beginning period of war, as it could have arisen at the time. Of course, this representation was far removed from the situation that arose in 1941 and which was impossible to foresee then. To be sure, the forces of fascist Germany and its possible allies were viewed as the main enemy, although the strategic conditions and operational forms of deployment along our western frontier during the beginning period of war were far from having been sufficiently studied.

It was assumed that a continuous front would form during the initial strategic deployment, which would require a breakthrough and make a frontal attack inevitable. From the point of view of calculating forces and the theater's capacity, this was generally true. However, at the same time the new capabilities of motor-mechanized forces to break through the front before it could manage to organize and establish itself and thus cause it to waver to a great depth on either side were not taken into account.

The development of a maneuver course of operations was, of course, foreseen, but generally following the breakthrough of the front. Maneuver activities in the operational depth were supposed to achieve their greatest development and a decisive outcome. However, in order to gain this opportunity, it was considered necessary to first break through the front. Attention was chiefly concentrated on the resolution of this more difficult task.

According to the accepted views of the time, which Marshal Yegorov adhered to, it was planned during the beginning period of the war to invade the enemy's border territory through active operations in the air and on the ground, to disrupt his mobilization and concentration and thus secure the deployment of one's main forces. These tasks were to be carried out along the most important operational directions by invasion groups, made up of motor-mechanized and cavalry formations and border troops, supported by powerful aviation. The actions of these invasion groups were to thus unfold in separate operations, carried out before the deployment of the main forces. As to their nature, they reminded one of the old methods of activity similar to the invasion of the German group of Belgian territory for the purpose of capturing Liege at the beginning of the First World War.

Such was the initial point of view. M.N. Tukhachevskii spoke against it with great authority. According to his weighty considerations, individual actions by invasion groups, given the presence of fortified borders and the powerful composition and high readiness of the border troops, could not count on success and must result in great losses. In 1934, in one of his reports, Tukhachevskii wrote that "the conduct of war according to old methods; that is, in previous forms of strategic deployment, will prove to be impossible" and that "old and familiar views on the concentration of mass armies by railroad to the frontiers and the mass character of frontier battles no longer correspond to actual conditions."[65] Foreseeing the great vulnerability of frontier theaters of war to the enemy's aviation, he considered the entire accepted system of mobilizing and concentrating mass armies outdated and requiring radical changes. Tukhachevskii proposed maintaining powerful forward armies along the frontier zone as the main forces' first operational echelon. In his opinion, these armies should as far as possible secretly concentrate during the threat of war in areas occupying a flanking position in regard to those directions along which the opening of military operations is most likely by the enemy.

Tukhachevskii attached great significance to fortified areas built along the border. According to his thinking, the fortified areas were supposed to be a shield and take upon themselves the enemy's attack and, for the secretly concentrated forward armies, a hammer for launching a flank attack against him. However, by no means should the fortified areas be assigned a passive-defensive significance. They were, in Tukhachevskii's opinion, an operational factor, organically linked with the field armies' active operations and a support for their maneuver in an overall strategic operation.

Such were the basic theoretical views as to the nature of operations in the beginning period of war. Unfortunately, we did not have the opportunity of employing them, due to the completely different political-strategic situation which found us in June 1941.

Proceeding from these views, which at the time carried great weight, the General Staff Academy in 1936 began work on developing our military theory and training our higher command echelon. The academy's basic operational task embraced the consecutive development of the army offensive operation in the Belorussian theater of military activities. It was later studied in the academy for two-three years in a row. The entire collective of the department of operational art took part in drawing up this assignment, as well as a student with the academy, M.V. Zakharov[66] (now Marshal of the Soviet Union), who had been seconded to the department.

Questions of the deep operation acquired their most thorough and versatile elaboration. Three variants on the commitment of the breakthrough development echelon (ERP) were foreseen:

In **the first variant**, involving a weakly occupied defense and the absence of the enemy's major reserves, the ERP is committed at the very beginning of the attack, or before the complete breakthrough of the tactical depth of the enemy's defense. In this case, the ERP must itself make a breach in the defense and break through into its depth. Of course, such a variant promised the most rapid course of the offensive, but could be employed only against a weak enemy.

The second variant was considered the most common case; the ERP is committed after the tactical depth of the defense is penetrated and a breach opens in it. It was assumed that given a defense of average strength and the presence of sufficient offensive means that one could achieve this as early as the close of the first day's fighting.

The third variant was the most difficult, when one must break through a heavily fortified zone and the very breakthrough of the tactical depth of the defense may lead to several days of heavy fighting. In this case, the commitment of the ERP was not excluded for strengthening the tactical attack in the depth and the complete smashing of the defense together with the attacking forces. Such a variant for employing the ERP was considered the least desirable, because it would lead to the expenditure of its forces even before the start of carrying out its main task in the operational depth. However, this variant could not be excluded while breaking through a permanent fortified area.

Several variants of the ERP's activities in the operational depth were also elaborated.

The first, so-called **short** variant: given the absence of any kind of significant enemy reserves, a comparatively weak ERP, having seized the second defensive zone, is immediately directed to the rear of the defense, in order to encircle and destroy the defending garrisons in conjunction with the forces attacking from the front. In this case, only the forward motorized detachments and reconnaissance are pushed forward up to 50 kilometers into the operational depth.

The second, so-called **deep** variant: a powerful ERP will immediately fall upon the enemy's operational reserves for the purpose of attacking and destroying them in conjunction with aviation and an airborne landing, made in the deep rear. In this case, the entire attack may spread to a depth of up to 100 kilometers, while individual blocking detachments of motorized infantry are left in the rear of the enemy garrisons still defending from the front.

Finally, the third, **combined** variant: the ERP cooperates with another development echelon, committed by a neighboring army. In this case, the two breakthrough development echelons, operating toward each other along different axes, must close the encirclement ring around a large enemy group and destroy it.

In one or another form, all these variants were employed in the Great Patriotic War.

Then the entire theme of the offensive operation was broadened by the commitment into the fighting of a cavalry-mechanized army, which was controlled by the front command.[67] Thus the elaboration of the deep operation had already acquired a strategic character. However, since only the goal of studying the activities of a cavalry-mechanized army, consisting of several mechanized and cavalry corps and motorized divisions, supported by aviation and an airborne landing, was being pursued, then the sphere of the strategic operation was by no means fully comprehended.

In any event, the deep operation in the General Staff Academy acquired its subsequent elaboration, which placed a number of new problems before our military-theoretical thought. In this regard, 1936–1937 were years of its new upward flight and animation. Unfortunately, this ascent did not continue for long.

A difficult period. In the spring of 1937 there began events that shook the Red Army to its foundations. The arbitrariness and illegality fostered by Stalin's cult of personality spread to a large part of the higher and senior command element. Honored and experienced cadres became their victims and the army was essentially beheaded. These cadres had for many years guided the army's training and the operational instruction of its command element and had moved Soviet military theory, indicating the path of its development. Now they were declared "enemies of the people" and the military-theoretical teachings on the new forms of the engagement and operation were put in doubt and were practically declared to be sabotage. All of the textbooks, official and unofficial military literature, the authors of which were repressed, were removed, and no one knew what one could or could not be guided by in military theory. Even in the General Staff Academy they began to sound the alarm against the main questions of the deep operation, objecting to the activities of motor-mechanized formations ahead of the front and to their employment for developing the breakthrough into the depth. And all of this took place a year before the maneuver operation fully revealed its new nature in the German-Polish campaign in the autumn of 1939.

The incorrectly understood and generalized experience of the war in Spain also had a negative influence on the recognition of the new ideas. From it the deeply mistaken and historically shortsighted conclusion was drawn that the new means of struggle only support the possibility of conducting the modern attack, but had not changed anything in its nature and forms.

Following the war in Spain and our liberation campaigns[68] into western Belorussia and western Ukraine, the mechanized corps, those main shock forces of the deep operation on land, were disbanded, and the development of bomber aviation—the main strike force in the air—was cut back. Such an unjustified measure deprived the theory of the deep operation of the chief materiel basis upon which it had been developing. The disbanding of the mechanized corps caused the army enormous harm.

All of these considerations could not but tell on the development of our theoretical views. Creative initiative was severely restricted for a time. The seed of doubt was planted in military thinking and instead of deepening and developing the theory of the deep operation, which was already knocking at the door of history, they began to disavow it and quietly put on the brakes.

Of course, this could not but bring about a certain confusion in the minds of the young command element, which after 1937 was moved into high command posts and which in 1941 was to take upon itself the first blows launched by the German-Fascist command precisely in the style of the deep operation. However, these young, honest and brave commanders were not able to correctly act in the whirlpool of events in the beginning of the war in which they had been suddenly been thrust, and this can be explained to a certain degree by the fact that they had been insufficiently oriented in the new character of deep operations, with which they had to contend.

Summing up, in 1937–1938 there occurred a certain deviation from the correction line of development of our military theory, which brought about a certain stagnation and lack of clarity in this field. And while this relapse left harsh consequences, it proved to be, however, only temporary.

A new upswing. The cult of personality was unable to delay the overall forward development of Soviet military theory. As early as 1939 military-theoretical thought was taking new steps in its further development, taking into account the experience of ongoing military events. To be sure, at times the "phony war"[69] in the West and the Soviet-Finnish War in the winter of 1939–1940 disguised the true forms of a major modern war and could even lead one into error. The Maginot Line[70] still seemed impregnable and inevitably defined the positional nature of war. The war in Finland seemed to confirm this.

Thus the forms of the deep operation remained undiscovered. Only the German-Polish war in September 1939 was the first realization of the new forms of struggle in action. Of course, this was only a single campaign and the conclusions drawn from it could not have final significance.[71] However, within half a year events unfolded in the West that fully revealed the nature of deep operations at the high level of a major European war.

The first events of the Second World War in Poland and France already showed that Soviet military-theoretical thought had been on the right track and had correctly foreseen the deep forms of modern operations. However, their sharply expressed maneuver nature and the unprecedented scope of depth exceeded all of the most optimistic assumptions. The 1939 and 1940 campaigns in Poland and France revealed the new nature of the beginning period of modern war. They showed that military activities begin with an invasion by the main mass of the armed forces, concentrated ahead of time. This imparted to the beginning period of war a picture of unexpectedly unfolding operations on a strategic scale and required their examination from the strategic point of view. In these conditions operational art was not only coming up fully against strategy, but was merging with it in an organic interrelationship.

However, our operational art was to a certain degree confined in its own framework, while the strategic sphere of war remained, unfortunately, basically beyond the reach of military theory. The necessary attention was not fixed on unveiling the beginning period of war and all the necessary theoretical conclusions applicable to our western theater of military activities were not made. This was doubtlessly a gap in our military theory and, of course, told in the beginning of the war in 1941.

In the last years before the war the forms and methods of conducting operations continued to be studied in the General Staff Academy, chiefly at the operational level and without reference to the strategic situation that might arise at the beginning of a war. However, under the influence of ongoing events, definite shifts occurred in military-theoretical thought. First of all, the study of maneuver operations came to occupy a significantly greater place. Secondly, the problem of defense at the operational level attracted general attention. Young commanders, who remained at the academy following their completion, introduced a fresh current in creative work. They comprised the basic cadre of the teaching corps. Among them were: I.Kh. Bagramyan[72] (now Marshal of the Soviet Union), F.P. Isaev,[73] V.Ye. Klimovskikh,[74] N.V. Korneev,[75] A.V. Sukhomlin,[76] N.I. Trubetskoi,[77] A.I. Shimonaev,[78] P.G. Yarchevskii,[79] and a number of others. The older generation of specialists—A.I. Gotovtsev, A.V. Kirpichnikov, S.N. Krasil'nikov, F.P. Shafalovich, N.N. Shvarts, Ye.A. Shilovskii, and others—also contributed their great experience and knowledge to the resolution of new problems in these years.[80]

The theory of operational defense. In 1938, for the first time during the existence of the General Staff Academy, the problem of the defensive operation was raised. The reasons for this were not openly discussed in academic circles. However, each of the operational workers understood that in a collision with German fascism's powerful and aggressive army along certain sectors of the front and during certain periods of time, defense will be a logical and, in certain conditions, inevitable method of action, in order to restrain pressure from a powerful enemy and to wear him out. At the same time, defense on the operational scale represented the least researched problem. Throughout the entire history of the M.V. Frunze Military Academy and the General Staff Academy, the theme of "the army in defense" had never once been studied. Tactically, we had elaborated the defense well and it occupied in all our field manuals the place and significance appropriate to it. However, at the operational level, to speak of the army's defense along a significant sector of the theater of military activities was somehow considered indecent and nearly in contradiction to our offensive doctrine. At the same time, it was not taken into account that the latter does not exclude defensive operations as a type and method of military activities. One may adhere to an offensive doctrine and possess a theoretically well developed defense. In the opposite case, one may actually adhere to a defensive doctrine and ignore the thorough elaboration of defensive questions at the operational level, as did the French. Such is the dialectic of this question, which, unfortunately, was not properly elucidated.

In the First World War, the defense, despite its extremely powerful engineer-tactical development, did not acquire operational organization. Everything came down to the fight to retain the tactical defense zone. The reserves were designated only for resolving this task through counterattacks and counterblows. When the tactical zone was broken through, the defense shifted back and resistance was organized along a new line. All of the defense's operational possibilities were exhausted with the fall of the tactical zone and it required the concentration of new reserves, which either restored the previous situation, if they could manage it, or created a new defensive front.

Now it was necessary to resolve in a new way the problem of defense and its forms. Depending upon the methods of attack, the defense at the operational level had to adopt a deep nature and be capable of holding out in case of a breakthrough by the enemy's tank formations into its rear. For this purpose, it was proposed to organize a tactical defense zone within the confines of the army, consisting of two defensive positions, linked by a number of switch anti-tank lines, and in the army rear, bound by the army defensive zone, and to transform each inhabited locale and each convenient terrain sector into an anti-tank "fortress." The depth of the army defensive area could occupy up to 75–100 kilometers. The idea was that the enemy's tank group, which had broken through the tactical defensive zone, should fall into a labyrinth in the operational depth, which had been built up with anti-tank areas ("fortresses"), and smash itself against them. The very organization of the defense was supposed to force the enemy to develop the offensive in depth not in the way he had planned, but in a way that had been predetermined by the entire system of built-up lines and anti-tank "fortresses." It was precisely in this regard that the defense was to impose its will on the attacker and be waged to a great depth and represent a unified operational system.

The theory of the operational defense was enunciated in the academic work, *The Fundamentals of the Defensive Operation*. A major operational map exercise was also elaborated in the academy on the theme "The Army's Defense and the Launching of a Counterblow." The elaboration of the defensive operation undoubtedly enriched, in a new way, Soviet military theory and had the same significance for the development of defensive forms that the deep operation had for the development of offensive forms.

On the eve of the war. Thus in the last years before the war the circle of operational questions expanded considerably. This told notably on the revival of our military-theoretical thought.

The session of the Higher Military Council,[81] which took place in December 1940 and during which the results of the events of 1939 and the summer of 1940 were discussed and important reports were delivered on the nature of modern operations and the decision made to create motor-mechanized corps, had great significance for the development of Soviet military theory on the eve of the war.

At the end of the 1930s a great deal of work was conducted to publish new, official instructions, manuals and directions. The draft of the new field manual, which had been drawn up in 1939, made corrections to PU-36 on the basis of recent experience and significantly broadened the concept of the deep engagement. For example, a new article (article 294) was added to the draft manual on the development of the breakthrough, with instructions as to the tasks entrusted to the combined-arms formations at the moment the ERP's units pass through the broken defense. Also new to the PU-39 draft was the chapter on "The Fundamentals of Troop Control in the Engagement," which laid out the principle questions of making an operational decision and carrying it out. While clearly expressing offensive ideas, the draft manual simultaneously set aside a significant place for the defense, particularly in pointing out the necessity to deeply echelon it.

The 1939 draft field manual, which was revised in the spring of 1941, was the last manual before the war. It completed the process of great work on manuals, which reflected the stormy development of our military-theoretical thought. Four field manuals appeared between 1925 and 1940 (1925, 1929, 1936, and the draft PU-39). In each of these, the forms of the deep engagement were more broadly developed and the results of a specific stage in the development of our military theory summed up, which clearly reflected the entire nature of its forward motion.

The elaboration of a textbook on the problem of conducting operations. It was significantly more complex to draw up a textbook on the conduct of operations. The necessity of such a textbook, which had not existed in the past, was brought about by the nature of the deep operation as a complex system of employing qualitatively heterogeneous combat efforts in a single, centralized and unified cooperation on the ground and in the air. In 1934 A.V. Fedotov drew up a draft textbook on the conduct of operations on Marshal Yegorov's orders, but the draft was not accepted by the General Staff. At the end of the summer of 1936, on orders from A.I. Yegorov, a new draft textbook was drawn up. The task was complex and its utility raised doubts. The theory of the deep operation was still in the development stage; it had not yet established itself to such a degree that it could be finally codified. Aside from this, it was necessary to erect a foundation of a specific strategic conception under a textbook on the conduct of operations, at least for the beginning period of the war, and to forecast the main and principle lines of its further development. However, this complex and higher field of strategy had not been well researched. Thus the draft textbook proved only to be an elucidation of the techniques of conducting operations. The completed draft remained with A.I. Yegorov throughout the winter of 1936–37 and then, in connection with the events of 1937, stayed in his safe.

The work on the textbook helped us to once again think through all of the deep operation's basic tenets, to more precisely formulate them and to draft them with greater accuracy. In the draft textbook they acquired a more mature, clear and grounded expression. One copy of the draft was sent to the General Staff Academy and became the foundation for teaching the academic course of operational art. Certain of its sections were published for use as an unofficial textbook under the title of *The Fundamentals of Conducting Operations*.

The attempt to publish a textbook on the conduct of operations was not renewed before the war. As a result, we did not have before the war any kind of operational instruction or official textbook on operational art. Nor was there such a textbook in a single one of the European armies, including Germany's. In general, it is doubtful that a stable official textbook on the conduct of operations could have played a positive role during that period of great changes in the forms and methods of waging armed struggle. It was far more important to continue the comprehensive study of the problems of operational art in their further development and to inculcate the new ideas of the deep forms of struggle to a broad contingent of operational workers. The General Staff Academy, the students of which directly studied and were educated in the spirit of the deep operation's concept, which lay at the heart of the heart of their military thinking, performed this task. From their milieu there came a series of outstanding military leaders and organizers of victorious deep operations during the Great Patriotic War.

The higher command element was also familiar with the fundamental tenets of our operational art through instructions and major military games and maneuvers carried out in the military districts. Thus despite the absence of an official guidebook on the conduct of operations, the fundamentals of the deep operation were well known to the higher command element, and this fully revealed itself when the Soviet army, following the difficult period at the beginning of the war, went over to the conduct of decisive offensive operations. It was precisely then that the enormous significance of the great creative work in the field of military theory, which had been conducted before the war, revealed itself. However, at the start of the war, due to the conditions of the time, this theory could not be efficacious and applicable.

Conclusion. The study of the history of the development of Soviet military-theoretical thought would be incomplete and would not have achieved its goal if the question of why our military theory, which foresaw so much and so correctly the nature of future operations, was unable to play a positive role in the difficult situation that arose during the beginning period of the war, was not revealed and explained. Serious conclusions should be drawn from this.

Despite the fact that a series of operational-strategic problems, including problems of the beginning period of war, remained unstudied, we possessed on the eve of the war a progressive military theory, according to its chief tenets. It proceeded, first of all, from the correct forecast of a future war as an attack on the Soviet Union by a coalition of capitalist countries and a decisive struggle with them to the death.

It foresaw the stubborn and prolonged nature of this struggle, requiring the enormous exertion of all the country's moral and materiel forces. "We have to take into account the fact that we will be faced with difficult and prolonged wars, and we must be able to distinguishe the war's periods and to be able to consecutively break up capital's coalition," wrote Tukhachevskii.[82] It was precisely this starting tenet that demanded of our military theory the clear resolution of problems of the consecutive conduct of offensive operations with the most decisive aims, all the way up to the rout of the enemy on his territory, which imparted to our military doctrine a clearly expressed offensive nature. However, bearing in mind the prolonged and intensive course of the struggle, with its inevitably changeable flow, our military theory foresaw a series of consecutive stages of the most varied operational-strategic nature and content in the war's campaigns.

We by no means thought to finish the war in a single, lightning attack, and it was precisely this realistic point of view, like many others, that distinguished our military theory from the fascist and adventurist strategy of lightning war. The entire course of the Great Patriotic War showed the correctness of our point of view and completely confirmed it by its real development from beginning to end.

In the field of operational art, our military theory constructed the conduct of the operation on the deep strike against the enemy, achieved through the joint employment of the combat arms and various weapons, to each of which, depending upon the given specific situation and available equipment, was imparted greater or lesser significance. The reliable defeat of the entire operational depth expressed the chief idea of our theory of operational art.

Our theory recognized the offensive operation as the main type of operation. However, the deeply echeloned defense found its place in theory and was fully elaborated in the draft of the 1939 field manual. Military-theoretical research generally yielded a satisfactory basis for conducting different types of operations: the breakthrough, maneuver activities, the envelopment, the envelopment, activities in the operational depth, and various types of defense, as well as the escape from encirclement. The command of the armies and fronts could find in our theory of operational art a sufficient basis for expedient operational-strategic decisions and the organization of activities in the most complex conditions of a situation. It was another question of which kind of operations and what forms and methods of activities should have been selected. This depended on the specific situation and required its correct comprehension and a great deal of mental agility, unbound by any dogmas. However, it was precisely here that our school of operational art was not fully up to the task.

We were bound by specific tenets of a declarative nature as to the offensive conduct of the war: that our army would be the most offensive army; that we would shift military activities to the enemy's territory, etc., etc. These tenets were delivered from above as immutable guiding directives for our military policy and strategy and formed the basis of the command element's entire military thinking. During the period of Stalin's cult of personality they acquired the significance of law and were not subject to discussion in theory. As a result, all of our military mentality imagined a future war as nothing less than the immediate assumption of the offensive. Any other strategic possibilities were excluded and were not examined by theory.

Even the events that played out in Poland in 1939 and in France in 1940 did not change these reigning official views and did not cause them to waver. However, in the depths of their conscience, the General Staff's higher officers understood that the situation during the beginning period of war might develop in another manner altogether.

In some circles in the General Staff and the General Staff Academy they even spoke about this quite specifically, with the corresponding calculations in their hands. However, these conversations were conducted only behind closed doors and did not go beyond their offices.

Thus the situation in which the Great Patriotic War began in June 1941 proved unexpected for our high command's entire subjective strategic and military-theoretical orientation, which gave rise to a definite sense of confusion and inability to understand events, to subordinate them to one's will and to seize the initiative. The orientation of military-theoretical thought, upon which our command had been reared for years, continued to influence military speculation by inertia, although it had long before come into contradiction with the real factors of the strategic reality that had arisen along our western frontiers, at least from the autumn of 1940, when Hitler began to concentrate his forces in western Poland and East Prussia.

The situation that arose in the beginning of the Great Patriotic War demanded a completely different strategic orientation. However, the rapid alteration of the mentality of the high command, which had already entered into a struggle to the death with the enemy, had not been secured by the inculcation of flexible thinking, unsubordinated to any kind of declarations and free to adopt those operational decisions it considered necessary in the developing situation. It was precisely here that the reasons lay for the fact that the command of the higher formations failed at the beginning of the war to draw from our progressive military theory that benefit that it could have brought.

Besides, the old and experienced commanders, who had created Soviet military theory and could have put it into practice with great skill, were no longer there and there was an obvious shortage of operationally trained commanders at the start of the war. Thus the difficult drama that burst upon us in the summer of 1941 had profound reasons of political and strategic significance connected with Stalin's personality cult. The consequences of this were incalculably grave. They required enormous sacrifices and caused enormous losses.

However, the heroic Soviet people, led by the great Communist Party, were able to overcome the difficult consequence of the war's first period. And when that occurred, the Soviet army opened a brilliant chain of deep operations of unprecedented strategic scope. These operations were so majestically realized, because alongside the other decisive factors, there basic foundations had been elaborated even before the war by progressive Soviet military theory. They enriched it, introduced much that was new to it, and created the rich stock of Soviet military art.

Chapter Notes

Foreword

1. United States and William J. Clinton. *A National Security Strategy for a Global Age* (Washington, DC: The White House, President of the United States, December 2001), Preface.
2. As examples see B.J.C. McKerrcher and Michael Hennessy, eds., *The Operational Art* (Westport, Connecticut: Praeger, 1996); Shimon Naveh. *In Pursuit of Operational Excellence: The Evolution of Operational Theory* (London: Frank Cass, 1997); Michael D. Krause and R. Cody Phillips, eds., *Historical Perspectives of the Operational Art* (Washington, D.C.: Center for Military History, 2005); and John Andreas Olsen and Martin van Creveld, *The Evolution of Operational Art: From Napoleon to the Present* (Oxford: Oxford University Press, 2011).

Chapter One

1. Editor's note. The RKKA (*Raboche-Krest'yanskaya Krasnaya Armiya*), or Worker's and Peasant's Red Army was the official name of the Soviet army until after World War II.
2. Editor's note. This refers to the defeat of Gen. Aleksandr Vasil'evich Samsonov's 2nd Army at Tannenberg in August, 1914. One of Isserson's earliest works dealt with this battle. See his *Kanny Mirovoi Voiny (Gibel' Armii Samsonova)* (Moscow, 1926).
3. Editor's note. Gen. Erich Friedrich Wilhelm Ludendorff (1865–1937) joined the German army in 1885. He rose to fame in World War I as the chief staff officer of Gen. Paul von Hindenburg and was virtual dictator of Germany during the last two years of the war. However, his political-military strategy ultimately failed and he was forced to resign. Following the war, Ludendorff briefly supported Hitler and wrote on military subjects.
4. Editor's note. This refers to the German army's great offensive (operation "Michael," March 21-April 5, 1918) against the British army in Flanders. This offensive was the first in a series of major German attacks against the Allies during the spring and summer of 1918. Isserson previously wrote about this operation in his *Martovskoe Nastuplenie Germantsev v Pikardii v 1918 Godu. Strategicheskii Etyud* (Moscow, 1926).
5. Editor's note. Carl von Clausewitz (1780–1831) served in the Prussian and Russian armies and took part in several of the major campaigns of the Napoleonic era. Following the Napoleonic Wars, he headed the War College in Berlin. His posthumously issued *On War* is considered one of the most influential military works ever published.
6. Editor's note. This refers to the articles of the 1919 Treaty of Versailles that severely restricted the size of the postwar German armed forces and the weapons they could field.
7. Editor's note. Field Marshal Alfred Graf von Schlieffffen (1833–1913) served as chief of the German General Staff from 1891 until his retirement in 1906. He is best known as the author of the so-called "Schlieffen Plan" for waging a two-front war against France and Russia.
8. Editor's note. Gen. Karl Eduard Wilhelm Groener (1867–1939) occupied a number of responsible staff and command positions during World War I. Following the war, he briefly served as chief of the General Staff. He later served in a civilian capacity in the Weimar Republic and wrote on political and military subjects.
9. Editor's note. Frederic Culmann was a French artillery officer who wrote in the prewar and postwar eras.
10. Editor's note. Friedrich Immanuel was a German author who wrote on military subjects before and after World War I.
11. Editor's note. Gen. Horst von Metzsch (1874–1946) joined the imperial German army in 1891. During World War I he served in various staff positions and commanded an artillery regiment. He served in various staff positions after the war and wrote on military subjects following his retirement.
12. Editor's note. Gen. Edouard-Jean Requin (1879–1953) served on the staffs of marshals Joffre and Foch during World War I and was also a member of the French military delegation in the U.S. During World War II he commanded an army. Requin was the author of several military works.
13. Editor's note. Maj. Gen. John Frederick Charles Fuller (1878–1966) became chief of staff of the British tank forces during World War I. Following the war, he wrote extensively on the subject of a future war, stressing the importance of armor and calling for the creation of small professional armies.
14. Editor's note. Vladimir Kiriakovich Triandafillow (1894–1931) fought in the Russian Civil War and later held a number of command and staff position before dying in an airplane crash. His most famous work is *Kharakter Operatsii Sovremennykh Armii* (Moscow, 1929).
15. Editor's note. Isserson is here referring to the results of the first Five-Year Plan (1928–33) for the

industrial and agricultural transformation of the country into a modern industrial power. The greatly increased levels of industrial production also helped transform the Red Army into a modern force.

16. Editor's note. Field Marshal Helmuth Carl Bernard Graf von Moltke ("the elder") (1800–91) served as chief of staff of the Prussian army from 1857 to 1871 and chief of the German army's General Staff from 1871 until his retirement in 1888. He is best known for his victories in the various wars of German unification against Denmark (1864), Austria (1866) and France (1870–71), where he perfected the art of the turning movement against the enemy's flanks.

17. Editor's note. The Comintern (*Kommunisticheskii Internatsional*), or III International, was established in 1919 to replace the discredited II International, which had collapsed over its member parties' support for their countries in World War I. The Comintern was from the beginning dominated by the Soviet Communist Party and throughout its existence faithfully carried out Moscow's directives, often to the detriment of the member parties. The Comintern was abolished in 1943, although its disciples continued with their work.

18. Editor's note. Friedrich Engels (1820–95) was an early German communist and chief collaborator of Karl Marx. Engels wrote extensively on military subjects in an attempt to integrate military affairs with his political ideology. Following the Russian Revolution, the young Red Army viewed him as one of its theoretical bulwarks.

19. Editor's note. This is the translation of Isserson's term *frontal'nost'*, which he often used to describe the phenomenon of two armed fronts facing each other.

20. Editor's note. Gen. Sigismund von Schlichting (1829–1909) spent most of his career in the Prussian and German armies command postings. He was one of the first theorists to write about the operational level of war.

21. Editor's note. This refers to Frederick II ("the Great") (1712–1786), king of Prussia from 1740 to 1786. In his various wars, he raised the linear formation to its apogee. However, his methods proved unsuitable during the wars of the French Revolution and against Napoleon.

22. Editor's note. The battle of Marengo was the culmination of Napoleon's second Italian campaign. He defeated the Austrian general Melas here on June 14, 1800, and signed a peace treaty the following year.

23. Editor's note. The battle of Borodino was fought on September 7, 1812, between Napoleon and Field Marshal Kutuzov, who commanded the Russian forces. The battle, one of the largest and bloodiest of the Napoleonic wars, was technically a Russian defeat, although Kutuzov managed to save his army by retreating and abandoning Moscow to the French.

24. Editor's note. The battle of Ulm was fought on October 16–19, 1805. Napoleon defeated the Austrians under Gen. Mack, taking several thousand prisoners. Ulm was the prelude to the larger battle of Austerlitz.

25. Editor's note. The battle of Regensberg (Ratisbon) was fought on April 23, 1809. Here Napoleon defeated the Austrians under Archduke Charles, south of Regensberg. The Austrians then fell back to Vienna.

26. Editor's note. The battle of Leipzig (October 16–18, 1813) saw Napoleon arrayed against the Russians, Prussians and Austrians. The French were defeated and had to fall back across the Rhine.

27. Editor's note. The campaign in France took place during the winter and spring of 1814, following the French defeat at Leipzig. Despite scoring a number of impressive victories over the scattered Allied forces, Napoleon was finally driven out of Paris, after which he abdicated.

28. Editor's note. The 17.5-mm Dreyse rifle, popularly known as the "needle gun," was used by the Prussian army from 1848 to 1871. It could fire 10 to 12 rounds per minute and had an effective range of 600 meters.

29. Editor's note. The 11-mm chassepot rifle was used by the French army from 1867 to 1874. It could fire 8 to 15 rounds per minute and had an effective range of 1,200 meters.

30. Editor's note. This refers to the breech-loading, cast steel guns manufactured by Alfred Krupp (1812–1887) for the Prussian and German armies during the Wars of German Unification and afterwards.

31. Editor's note. Field Marshal Ludwing August Ritter von Benedek (1804–1881) commanded Austrian forces at Königgrätz (1866), where he was defeated by the Prussian forces under von Moltke.

32. Editor's note. Marshal Francois Achille Bazaine (1811–1888) had a distinguished career in the French army until the Franco-Prussian War. There he surrendered the last French army to the Prussians at Metz on October 27, 1870.

33. Editor's note. Marshal Marie Esme Patrice Maurice de MacMahon (1808–1873) surrendered the French army as a result of the battle of Sedan (August 29–31, 1870).

34. Editor's note. Louis Napoleon Bonaparte (1808–1873) was a nephew of Napoleon I. He was elected president of the Second French Republic in 1848, but overthrew the government in 1851 and proclaimed himself Napoleon III and established the Second Empire. He was, in turn, overthrown in 1870, following France's defeat in the Franco-Prussian War.

35. Editor's note. The Battle of Königgrätz (Sadowa) was fought on July 3, 1866) between Prussian and Austrian forces. The Prussian victory forced the Austrians to sue for peace and made Prussia the dominant force in Germany.

36. Editor's note. During the Battle of Sedan (August 29–31, 1870) the Prussians pinned the French army against the Belgian frontier and forced it to capitulate.

37. Editor's note. The Battle of Cannae was fought on August 2, 216 BC, in southeastern Italy during the Second Punic War between Rome and Carthage. At Cannae the Carthaginian forces under Hannibal defeated the Romans with a double envelopment. The idea of "Cannae" was later elevated by Schlieffen and others to the ideal of military art.

38. Editor's note. Gen. Genrikh Antonovich Leer (1829–1904) served as chief of the Russian General Staff Academy from 1889 to 1898. His scholarly works sought to find permanent laws of military art.

39. Editor's note. Leonidas (c. 540–480 BC), as king of Sparta led Greek forces against the Persian king Xerxes during the latter's invasion of Greece. The Spartans and their allies held off the Persian forces at Thermopylae, but were defeated and wiped out as the result of an enveloping maneuver.

40. Editor's note. The Battle of Mukden (Shenyang) saw the Japanese forces under Marshal Iwao Oyama

defeat the Russians under Gen. Aleksei Nikolaevich Kuropatkin in the last major land battle (February 20-March 5, 1905) of the Russo-Japanese War.

41. Editor's note. The Battle of the Marne (September 5-12, 1914) pitted the German right-wing armies in France against an Anglo-French force along the Marne River, east of Paris. Here the Allies counterattacked and drove the German armies back to the Aisne River, thus setting the stage for four years of trench warfare.

42. Editor's note. Gen. Friedrich Adolf Julius von Bernhardi (1849–1930) served in the Prussian and German armies during the Franco-Prussian War and World War I. His book, *Germany and the Next War* was highly influential in the years before World War I.

43. Groener sheds a good deal of light on this question in his works *Das Testament des grafen Schlieffen* and *Der Feldherr wider Willen.*

44. Editor's note. The Battle of the Frontiers (August 22–24, 1914) was the first major clash of the belligerent armies during World War I on the Western Front. The French Fifth Army was roughly handled by the German right wing between the Sambre and Meuse rivers, but managed to extricate itself and fall back.

45. Editor's note. The campaign to the Vistula River began with the counteroffensive by the Western Front under Mikhail Nikolaevich Tukachevskii against Polish forces in Belorussia in July 1920. The Soviet forces pursued the Poles nearly 600 kilometers to the west, but were in turn defeated by a counteroffensive along the Vistula River in August.

46. Stalin, *Politicheskii Otchet TsK XVI S"ezdu VKP(b).*

47. Editor's note. Lt. Col. Richard Hentsch (1869–1918) was dispatched by chief of staff Helmuth von Moltke ("the younger") in September 1914 to the German armies along the right wing during the Battle of the Marne. His estimate of the situation led to the German armies' retreat back to the Aisne River and the onset of positional warfare.

48. Editor's note. This refers to the so-called "Race to the Sea," which began in September 1914, following the German retreat from the Marne River and which continued for two months afterward. This period was characterized by repeated attempts by both sides to outflank the other, until the open flank ended at the North Sea and the continuous front was formed.

49. Editor's note. Lt. Col. Gaston Duffour (1875–1953) joined the French army in 1897 and commanded a battalion during World War I and later wrote historical studies of the conflict. During World War II he commanded French forces in their unsuccessful defense of the lower Seine River.

50. From lectures read in the French academy.

51. Editor's note. Gen. Marie-Eugene Debeney (1864–1943) joined the French army in 1886 and taught at the Ecole de Guerre. During World War I he served as an army chief of staff and commanded a corps and an army. He also served as chief of staff to the commander-in-chief of the French army, Gen. Petain. Following the war, Debeney commanded the Ecole de Guerre and served as chief of the General Staff.

52. From lectures read in the French academy.

53. Editor's note. The Entente refers to the original major Allied powers (Britain, France and Russia) at the outbreak of World War I and later joined by the United States.

54. Editor's note. Marshal Ferdinand Foch (1851–1929) was an influential professor at the prewar French military academy. During World War I he commanded an army and an army group, served as chief of the General Staff, and, in 1918 was appointed commander-in-chief of the Allied armies.

55. Culmann, *Strategiya.* Library of Foreign Military Literature, p. 64.

56. An unusually vivid example of this is the outcome of the Austro-Prussian War of 1866.

57. Editor's note. George Soldan took part in the compilation of the official German military history of World War I. He also wrote articles on military topics.

58. Editor's note. Gen. Johannes Friedrich von Seeckt (1866–1936) served in the German army during World War I on both the Eastern and Western fronts, as well as in the Near East. He later became chief of the postwar *Reichswehr,* where he advocated combined-arms tactics for his small force.

59. Editor's note. Mikhail Vasil'evich Frunze (1885–1925) was an early Bolshevik organizer who rose to high rank in the Red Army during the civil war. Frunze commanded an army and *fronts* during the civil war and held a number of administrative posts afterwards. He was appointed war commissar in 1925, but died under suspicious circumstances several months later as the result of a medical operation.

60. Editor's note. The White Guards was the collective name for the anti–Bolshevik forces during the civil war. The name, preceded by the appropriate adjective, was, or a time, appended to almost any anti–Soviet force.

61. Editor's note. Kliment Yefremovich Voroshilov (1881–1969) joined the Bolshevik faction early on and served as a commander and commissar with various units during the civil war. Following Frunze's death, Voroshilov served as defense commissar until 1940 and is chiefly responsible for the Red Army's lack of preparedness for the German invasion in 1941. Following a brief and unsuccessful period of command during the war, he was relegated to ceremonial positions for the rest of his life.

62. Editor's note. VLKSM was the acronym for the All-Union Leninist Young Communist League (1918–1991), also known as the *Komsomol,* which served as the youth auxiliary of the Communist Party of the Soviet Union.

63. Editor's note. Narodnyi Komissariat po Voennym i Morskim Delam, *Polevoi Ustav RKKA [1929]* (Moscow-Leningrad: Gosudarstvennoe Izdatel'stvo, Otdel Voennoi Literatury, 1929), p. 9.

64. Editor's note. Here Isserson is undoubtedly referring to the Japanese invasion of Manchuria, beginning in 1931, which brought Japan into direct and antagonistic contact with the Soviet Union.

65. Editor's note. Here, Isserson evidently has in mind the Baltic states of Estonia, Latvia and Lithuania.

66. See part I, The Evolution of Operational Art During the World War.

67. In our literature the concept of a future operation is often defined by the term "spatial operation." This should be considered as inexact. Any space has two dimensions in a single plane—along the front and in depth. The maximum spatial envelopment along the front of the operation has already been achieved in the World War. Characteristic of the operation's evolution is its maximum spatial spreading in depth. Thus its characteristic distinguishing

feature may be better defined by the concept of the deep operation.

68. Editor's note. This is probably a reference to the Maginot Line, built at the instigation of French minster of war Andre Maginot (1877–1932). The line, which stretched from the Swiss border to the border with Luxemburg, proved a failure in World War II.

69. Editor's note. The five-year plans were a feature of Soviet economic policy from 1928 to the collapse of the USSR. Each plan established the production goals for the economy for the period in question and was the barometer by which the country measured its degree of economic development. The second Five-Year Plan referred to here was from 1933 through 1937.

70. Editor's note. Otto Eduard Leopold von Bismarck (1815–1898) became minister president of Prussia in 1862 and chancellor of the German Empire in 1871. Bismarck's skillful policy of "blood and iron" during the German Wars of Unification led to a united Germany.

71. Editor's note. Gen. Hippolyte Langlois (1839–1912) joined the French army in 1858 and fought in the Franco-Prussian War. He wrote extensively on the employment of artillery and was a professor of artillery at the Ecole de Guerre and later commanded the academy. Langlois retired from active duty in 1904.

72. Editor's note. Gen. Jules Louis Lewal (1823–1908) joined the French army in 1846. He served in Italy, Mexico and during the Franco-Prussian War. He was appointed head of the Ecole de Guerre in 1877 and later served as minister of war. Lewal was the author of several works on military affairs.

73. Editor's note. This is a reference to the German offensive along the Aisne River in May, 1918.

74. Duffour. Lectures read at the French higher military school.

75. Editor's note. The French attack from the Pratzen Heights at the battle of Austerlitz (December 2, 1805) won the battle against the Russian and Austrian forces.

Chapter Two

1. Editor's note. The RKKA Military Academy (previously the RKKA General Staff Academy) was renamed the Frunze Military Academy in 1925, upon the death of the war commissar of the same name. For some reason, Isserson refers here to the academy by its old name.

2. Editor's note. Gen. Maurice Gustave Gamelin (1872–1958) served with distinction during World War I and was chief of staff of the French army during the interwar period. During World War II he was appointed commander-in-chief of the French army, but was dismissed in May 1940, following the German breakthrough at Sedan.

3. In modern conditions, following the coming of fascism to power in Germany, this is changing significantly.

4. Editor's note. The Poles'ye, or Pripyat' Marshes, is a large expanse of swamp and forest stretching eastward from the Western Bug River to the Dnepr River.

5. Editor's note. Gen. Raymond Adolphe Sere der Revieres (1815–1895) was a French military engineer. Following the Franco-Prussian War, he supervised the construction of an extensive system of fortifications along France's eastern frontier in anticipation of a German invasion.

6. Editor's note. This refers to the fighting between Japanese and Chinese forces following the Japanese army's conquest and occupation of the Chinese province of Manchuria in 1931.

7. The reconnaissance and screening front here is according to articles 16 and 214 of the *Boevoi Ustav Konnitsy*, part III.

8. Editor's note. The T-26 was a Soviet light tank produced during 1931–41. The 1933 model weighed 9.6 tons and carried a crew of three; it was armed with a 45-mm gun and a 7.62-mm machine gun. The BT series of high-speed light tanks was produced during 1932–41. Several models appeared, of which the most numerous were the BT-5 (11.5 tons, crew of three, a 45-mm gun, and a 7.62-mm machine gun) and BT-7 (14 tons, crew of three, a 45-mm gun, and a 7.62-mm machine gun).

9. {This implies the existence of two refuelings in the mechanized corps' units.}

10. Editor's note. The T-27 tankette was produced during 1931–33 and used chiefly for reconnaissance. It weighed 2.7 tons, had a crew of two and was armed with a single 7.62-mm machine gun. It could also be transported by aircraft. The D-8 was a light armored car produced during 1932–34 and used chiefly for reconnaissance. It weighed 1.58 tons, had a crew of two and was armed with two 7.62-mm machine guns.

11. Editor's' note. The ANT-14, the USSR's first all-metal aircraft, appeared in 1931. It had four engines and could carry a crew of three, as well as 36 passengers. It had a maximum speed of 236 km/hr and a range of 900 kilometers.

12. Editor's note. The front is a Russian-Soviet major field force equivalent in its organization to an army group.

13. {It is presented here only in general numerical indices.}

14. Editor's note. This refers to the three wars (against Denmark in 1864, against Austria in 1866 and against France in 1870–71) waged by Prussia and its allies to unite Germany.

15. Editor's note. Isserson here is probably referring to the Japanese occupation of Manchuria during 1931–32.

16. Approximately two rail lines with 30 trains each, for concentration.

17. Editor's note. Gen. Giulio Douhet (1869–1930) joined the Italian army in 1892, later switching to the nascent air branch. During World War I he was court-martialed and imprisoned for criticizing his superiors' failure to employ air power effectively. He spent the rest of his life after the war theorizing on the use of air power, especially heavy bombers. His most influential work was *The Command of the Air* (1921).

18. Editor's note. The TB-3 was an early Soviet heavy bomber, which first appeared in 1930 and saw sporadic service in World War II. One model carried a crew of four and had a speed of 212 kilometers per hour and a range of 2,000 kilometers. The TB-3 was armed with 5–8 7.62-mm machine guns and could carry a bomb load of up to 2,000 kilograms.

19. {The technical development of aviation is leading to the further growth of this range}.

20. Editor's note. Gen. Alexander Heinrich Rudolph von Kluck (1846–1934) joined the Prussian army in 1866 and took part in the Austro-Prussian and Franco-Prussian wars. At the start of World War I he was ap-

pointed to command the First Army along the extreme right wing of the German advance into France, but was forced to retreat as a result of the Battle of the Marne.

21. Such an extension will not be achieved in space, because the entire operational {formation} of the approach will stretch out from a single line of deployment. However, it retains its complete significance in time.

22. The composition of these reserves will, of course, be different in each case. Usually, one should expect 2–3 divisions in the defense's army reserve. This question, in any event, is one of the most important for intelligence during the preparation for the breakthrough, because it determines the possible correlation of forces in the depth and the prospects of the ERP's activities.

23. This means that the defense has not been moved right up to the frontier.

24. Only the equipping of mechanized formations with tanks with significantly increased armor, firepower, speed, and cross-country qualities could raise the question of the ERP's breaking through the defensive front at the same time as the beginning of the EA's attack. Such a resolution of the problem is not realistic at the contemporary level of the quality of mechanized equipment.

25. As was already shown, such a development of the initial offensive operation is by no means mandatory for all cases. The initial offensive operation may begin immediately with a breakthrough, if the enemy has immediately gone over to the defensive.

26. See *The Evolution of Operational Art*, part II.

27. This length is equal to 400 kilometers to the north and south of the Poles'ye.

28. Duffour, Lectures read in the French higher military school.

Chapter Three

1. Editor's note. This refers to the brief (October, 1935-May, 1936) Italian-Abyssinian (Ethiopian) War. The Italians invaded and occupied Ethiopia until they were driven out by British and Ethiopian forces in 1941.

2. Editor's note. This refers to the ongoing Spanish Civil War (1936–39). This refers to the Second Sino-Japanese War (1937–45), which began with the invasion of China proper by Japanese forces. The war dragged on until Japan's surrender to the Allies in 1945.

3. Editor's note. This is a reference to an important episode during the Spanish Civil War (1936–39). Following the revolt of Nationalist forces in July 1936, they sought to capture the capital, which was held by supporters of the Spanish Republic. The latter, supported by the Soviet Union and communist sympathizers, managed to hold off the Nationalists' initial attempts to take Madrid during the fall of 1936. The struggle soon developed into a siege, which was only ended when the Nationalist occupied the capital in March, 1939.

4. Editor's note. Gen. Erich von Falkenhayn (1861–1922) joined the German army in 1880 and rose through the ranks at home and abroad. In 1913 he was appointed to head the Prussian war ministry and the following year he was made chief of the General Staff, following the German army's reverses along the Marne River. He favored an offensive strategy in the West and was the principal author of the Battle of Verdun (February-December, 1916), which caused the German and French armies hundreds of thousands of casualties. Falkenhayn was relieved in the summer of 1916, although he later served with distinction in Romania and in Palestine.

5. Editor's note. Nikolai Nikolaevich Yudenich (1862–1933) was commissioned in the imperial Russian army in 1881. He served in the Russo-Japanese War (1904–05) and afterwards in various military districts. During World War I he served as an army chief of staff and later commanded the Caucasus Front. During the civil war he commanded the While forces in northwestern Russia, but failed to capture Petrograd (St. Petersburg/Leningrad) in 1919. Following his defeat, Yudenich spent the remainder of his life in exile.

6. Editor's note. Anton Ivanovich Denikin (1872–1947) was commissioned in the imperial Russian army in 1892 and fought in the Russo-Japanese War. During World War I he commanded a brigade, a division and a corps. He commanded the White forces in southern Russia during the civil war, although he was defeated by the Red forces at Orel in 1919. Following the war, he spent the remainder of his life in exile.

7. Editor's note. Marshal Marie Emile Fayolle (1852–1928) served in the French army from 1873 until 1914, when he retired. He was recalled to service upon the outbreak of World War I, where he commanded a division and several armies and an army group. Fayolle also commanded French forces in Italy and occupation troops in Germany following the war.

8. Editor's note. Heinz Wilhelm Guderian (1888–1954) joined the imperial German army in 1907 and served in World War I on the Western Front. During the interwar period he was instrumental in the theoretical and organizational development of the nascent German armored forces. During World War II he commanded a panzer corps in Poland and France and a panzer group/army during the initial invasion of the Soviet Union. He was relieved following the failure of the final offensive on Moscow, but in 1943 was appointed inspector general of armored troops. He was appointed chief of staff of the army the following year, but was relieved in early 1945 as the result of serious disagreements with Hitler. Guderian was a prisoner of war until his release in 1948.

9. Editor's note. This is a reference to any number of truck models produced by the Stalin Auto Plant (Zavod imeni Stalina) during these years.

10. Editor's note. The T-28 was a multi-turret medium tank that first entered production in 1932. One model weighed 28 tons and carried a crew of six. This model was armed with a 76.2-mm gun and four to five 7.62-mm machine guns. Although adequate for its time, the T-28 was obsolete by the start of World War II.

11. Editor's note. This is the army's operational section.

12. Editor's note. Isserson is referring here to the concluding stages of the Battle of Bilbao, which took place in June 1937. Here the Nationalist forces under Gen. Mola broke through the Republican defenses and drove them out of the Basque country.

13. Editor's note. Bourchant was a French military writer active during the interwar period.

14. {The defense of Leningrad in the Great Patriotic War was also conducted in such conditions.}

15. {This was the case, for example, in the operation along the Kursk salient in the Great Patriotic War.}

Chapter Four

1. Editor's note. The removal of the reference to Stalin indicates that Isserson probably did this after the start of the de-Stalinization campaign in 1956.
2. Editor's note. Isserson left a handwritten note that reveals the extent of the loss. The note does not list the subject of the first installment, which probably dealt with general questions of war. The installments are as follows: III—the meeting battle; IV—the breakthrough and the development of the breakthrough; V—the defensive operation and the launching of a counterblow, and; VI—the organization of the operational rear and the control of the operation (the work of the army and front staffs). Isserson closed this list with the notation that "the manuscripts of these installments have been lost."
3. Editor's note. The cavalry-mechanized group (*konno-mekhanizirovannaya gruppa*, or KMG) first appeared during the Soviet occupation of eastern Poland in 1939. Cavalry-mechanized groups were widely employed in the Red Army's major offensive operations during World War II.

Chapter Five

1. Stalin, "K Voprosu o Strategii i Taktiki Russkikh Kommunistov." *Kommunisticheskaya Revolyutsiya*, No. 7 (46) (1923), p. 14.
2. Lenin, vol. XIX, p. 202.
3. Editor's note. The reference to World War II as the "second imperialist war" was standard phraseology in the Soviet Union during this period.
4. Editor's note. This passage implies that it was written after the German breakthrough of the French front in May 1940.
5. A small war, from the point of view of its scale, and not of its character.
6. Editor's note. Generalissimo Francisco Franco (1892–1975) joined the Spanish army in 1907 and fought for several years against tribesmen in Spanish Morocco and thereafter quickly rose through the ranks. In 1936, as commander of the Army of Africa, he joined the Nationalist revolt against the Spanish Republic and gradually exerted his control over the state apparatus. Franco ruled Spain from as an absolute dictator from 1939 to his death.
7. Editor's note. Isserson here is referring to the Battle of Verdun (February–December, 1916), during which the German army attempted to bleed the French army white around the fortress city of Verdun. The battle has since become the apotheosis of attrition warfare.
8. The Bilbao area was defended along a 70-kilometer front by a total of 50,000 soldiers. The fortifications consisted only of 2 to 3 trench lines and had no depth whatsoever.
9. Editor's note. Here Isserson is referring to German and Italian military support for the Nationalist forces.
10. Getting ahead of ourselves, we will only say that such an operation was carried out in 1940 in France.
11. Editor's note. Lt. Gen. Charles Nicolas Victor Oudinot (1791–1863) was the son of Nicolas Oudinot, one of Napoleon's marshals. He served in the Napoleonic Wars and afterwards during the Bourbon restoration. He commanded the French troops that took Rome and restored the pope in 1849. He retired following Louis Napoleon's overthrow of the Second Republic in 1851.
12. Editor's note. Gen. Louis-Eugene Cavaignac (1802–57) joined the army in 1824 and played a distinguished role in the French conquest of Algeria. Called home in 1848, he was appointed minister of war by the government that followed the overthrow of King Louis Phillipe. As head of the National Guard, he decisively put down the June uprising of 1848.
13. Editor's note. This refers to the uprising by French workers, angered at the closing of the National Workshops, against the French government following the deposition of King Louis Phillipe. Gen. Cavaignac put down the uprising in June 1848, which resulted in 10,000 killed or wounded.
14. Marx and Engels, vol. VIII, p. 456.
15. *Kratkii Kurs Istorii VKP(b)*, chap. IV, section 2, p. 101.
16. From a *Pravda* editorial of September 14, 1939.
17. Ibid.
18. Editor's note. The Battle of Jena was fought on October 16 1806, following the Prussian declaration of war on Napoleonic France. Here, and near the neighboring town of Auerstedt, the Prussians were decisively defeated. The French quickly followed up on their victory and subjected Prussia to French domination for several years.
19. Editor's note. Here Isserson is referring to the Polish government's refusal to allow the passage of Soviet troops through the country in the event of German aggression. This effectively scuttled the prospect of a united front between the Western powers and the USSR against Germany.
20. Editor's note. Franz Mehring (1846–1919) was a German newspaperman and historian. He later joined the German Social Democratic Party and gravitated to its left wing during World War I.
21. Mehring, *Ocherki po Istorii Voin i Voennogo Iskusstva*. 3rd expanded and revised edition. Voengiz, p. 254.
22. Ibid.
23. Clausewitz, *O Voine*, vol. III, p. 67.
24. Editor's note. Maxime Weygand (1867–1965) joined the French army in 1886 and served in World War I in a number of higher staff positions. During the interwar period he served as military adviser to the Polish forces fighting the Red Army in 1920. He also served in the Middle East and as chief of the general staff, before retiring in 1938. He was recalled to duty following the French disaster in May 1940 and made commander-in-chief, although he could do nothing to stave off defeat. Weygand later played a prominent role in the collaborationist Vichy regime, before being arrested by the Germans.
25. Editor's note. George Fielding Eliot (1894–1971) was born in the United States but immigrated to Australia as a child. He served in the Australian army at Gallipoli and in France. Following his retirement from the US Army reserve, he wrote extensively on military subjects.
26. From now until the release of the official German history of the war, this figure cannot be said to be precisely established. In contemporary descrip-

tions of the German-Polish war, the number of German divisions varies from 45 to 70. One should necessarily consider the latter figure to be exaggerated.

27. Editor's note. Karl Rudolf Gerd von Rundstedt (1875–1953) joined the imperial army in 1892 and served primarily in staff positions during World War I, on both the western and eastern fronts. During World War II he commanded army groups in Poland, the West and the Soviet Union, before being dismissed in late 1941. He was appointed commander-in-chief West in 1942 and commanded German troops through the Normandy invasion. He was later dismissed, only to be reappointed again and oversaw preparations for the Ardennes offensive, before being relieved again. Following the war, Rundstedt was charged with war crimes and held for a number of years, before being released.

28. Editor's note. Sigmund Wilhelm Walther List (1880–1971) joined the imperial army in 1898 and served as a staff officer during World War I. During World War II he commanded armies in Poland, France and Greece. In 1942 he was appointed commander of an army group on the Eastern Front, but was soon afterward dismissed by Hitler. Following the war, List was convicted of war crimes, but was released in 1952.

29. Editor's note. Walter von Reichenau (1884–1942) joined the imperial army in 1903 and served as a staff officer during World War I. He was an enthusiastic supporter of Hitler and commanded armies in Poland and France. Reichenau died of a heart attack.

30. Editor's note. Hermann Hoth (1885–1971) joined the imperial army in 1903 and served in World War I. During World War II he commanded a motorized corps in Poland and France. He commanded panzer and infantry armies on the Eastern Front, before being relieved in 1943. Following the war, Hoth was found guilty of war crimes, but was released in 1954.

31. Editor's note. Johannes Albrecht Blaskowitz (1883–1948) joined the imperial army in 1901 and fought on both fronts during World War I. During World War II he commanded an army in Poland and later an army and army group in France, Germany and the Netherlands. Following the war, Blaskowitz was charged with war crimes, but committed suicide.

32. Editor's note. Moritz Albrecht Franz Friedrich Fedor von Bock (1880–1945) joined the imperial army in 1898 and served in World War I. During World War II he commanded army groups in Poland, France and the Soviet Union, before being relieved in 1942. Von Bock was killed by Allied aircraft.

33. Editor's note. Georg Karl Friedrich Wilhelm von Kuchler (1881–1968) joined the imperial army in 1900 and fought in World War I. During World War II he commanded armies in Poland, France and the Soviet Union. He commanded an army group on the Eastern Front from 1941 to his relief in 1944. Kuchler was convicted of war crimes, but released in 1953.

34. Editor's note. Gunther Adolf Ferdinand von Kluge (1882–1944) served in the imperial army during World War I in staff positions. During World War II he commanded armies in Poland, France and the Soviet Union. He commanded an army group on the Eastern Front during 1941–43. He was later appointed commander-in-chief West following the Normandy landings. Von Kluge was implicated in the anti–Hitler plot and committed suicide.

35. Editor's note. The *Landwehr* was an armed militia consisting of older-age cohorts for carrying out secondary military tasks in support of the regular army.

36. In a number of descriptions this battle is assigned to the class of preliminary activities, which is fundamentally incorrect.

37. Editor's note. Janusz Tadeusz Gasiorowski (1889–1949) fought in the Russian army during World War I, before deserting and joining the Polish independence movement. He fought against the Soviets in 1920 and served as chief of the General Staff during 1931–35. He was captured by the Germans in 1939 and spent the entire war in a prison camp. Gasiorowski died in France.

38. Editor's note. Georg-Hans Reinhardt (1887–1963) served in the imperial army during World War I. During World War II he commanded a panzer division in Poland and a panzer corps in France and the Soviet Union. He later commanded a panzer army and an army group on the Eastern Front. Reinhardt was convicted of war crimes, but released in 1952.

39. However, people could say, of course, that, essentially, nothing was left of the remainder. This only proves that the essence of the Polish army's situation by September 10 was precisely that its rear had been suppressed, its transportation paralyzed and its command centers put out of action, while the German highly-mobile formations had broken through into the depth.

40. Editor's note. The Fieseler Fi 156 Storch was a small liaison aircraft that appeared in the 1930s and was employed throughout the war and afterwards. The plane carried a crew of two, with a maximum speed of 175 kilometers per hour and a maximum range of 380 kilometers.

41. The German motor-mechanized forces were represented by three organizational types of a definite operational designation. These were armored divisions, light divisions and motorized divisions.

The **armored division** is the expressed carrier of strike power and is designated for the attack. Its core is the tank brigade, consisting of two regiments of two battalions each (in all, 180 tanks of the line, 70 headquarters tanks and 20 reserve tanks).

The division's second brigade, on three-axle all-terrain vehicles, is an infantry one. It consists of two regiments of transported infantry and automatic weapons, with two battalions in a regiment. This brigade offers the tanks security and support, clears and occupies territory and thus imparts independence to the armored division.

Besides this, the armored division includes:

An artillery regiment consisting of two battalions, with three batteries of 105-mm howitzers in each, for a total of 24 guns; an anti-tank battalion of three companies of 37-mm guns (36 guns); a reconnaissance group on armored vehicles and motorcycles; an engineer battalion, which includes pontoon sections, and a communications battalion consisting of a telegraph company and a radio company.

Thus the armored division is a completely independent formation capable of independently waging all types of the engagement.

The **light division** is designated mainly for carrying out operational intelligence tasks and should replace the former cavalry division. It has more intelligence organs and rifle sections than the armored division, but, on the other hand, significantly fewer tanks. Its offensive shock force is not very great. On the other hand, it is very mobile and may be successfully employed for rapidly seizing important lines and installations and for long-range operational intelligence. By their designation, we could compare the light division with a light cruiser, and the armored division with a battleship.

The **motorized division** is the maneuver reserve in the hands of the higher command and has, as a rule, the same organization as the infantry division. It is basically designated for the rapid support of the armored and light divisions that have been pushed forward.

42. Editor's note. This is probably a reference to the Ju-52, which served as the German army's main transport aircraft during the war.

43. Editor's note. Wladyslaw Bortnowski (1891–1966) fought with the Polish Legions during World War I, under Austro-Hungarian auspices. He later joined the new Polish army and took part in the Soviet-Polish War of 1920. He commanded the Pomorze Army against the Germans, was taken prisoner and held until the end of the war. Following the war, Bortnowski spent the rest of his life in exile.

44. Rydz-Smigly had already renounced the post of commander-in-chief.

45. Editor's note. Field Marshal August von Mackensen (1849–1945) joined the Prussian army in 1869 and took part in the Franco-Prussian War. During World War I he commanded a corps, army and army group against the Russians, Serbs and Romanians.

46. Editor's note. Edward Rydz-Smigly (1886–1941) was drafted into the Austro-Hungarian army in 1914 and served in the Polish Legions. He joined the new Polish army in 1918 and fought against the Soviets in 1920. He later served as head of the Polish armed forces and was the *de facto* leader of the country in the last years before World War II. He fled to Romania and Hungary following the Polish defeat. Rydz-Smigly later made his way back into Poland to join the underground movement and died of a heart attack there.

47. Editor's note. This refers to the Crimean peninsula and contiguous areas.

48. Author's note. Gen. Petr Nikolaevich Wrangel (1878–1928) joined the Russian imperial army in 1902 and fought in the Russo-Japanese War. During World War I he commanded a cavalry squadron and an infantry regiment and brigade. In 1918 he joined Gen. Denikin's White movement and was placed in command of the Caucasian Army. Wrangel took command of all White forces in the Crimea, but was defeated later that year and forced to leave the country. He died in exile.

49. Editor's note. This refers to the Red Army's victory over the Japanese forces in August-September 1939 along the Khalkhin-Gol River. The Soviet forces were commanded by the future deputy supreme commander-in-chief, Georgii Konstantinovich Zhukov.

Chapter Six

1. During the 1930s G.S. Isserson occupied the positions of chief of the operational department of the M.V. Frunze Military Academy, and then chief of the department of operational art in the General Staff Academy. In this article, the author, employing his personal reminiscences, lays out his views on the development of the theory of Soviet operational art during the 1930s. *The editors.*

2. J.F. Fuller, *Operatsii Mekhanizirovannykh Sil.* Translated from the English. Moscow, Voennoe Izdatel'stvo, 1933, p. 13.

3. Editor's note. Mikhail Nikolaevich Tukhachevskii (1893–1937) joined the Russian army in 1914, but was captured the following year. He escaped and returned to Russia in 1917 and then joined the Red Army in 1918. During the civil war he commanded armies and fronts and later put down internal rebellions. Following the war, he served as chief of staff, commanded a military district and was a deputy defense commissar. He also oversaw the Red Army's rearmament program. Tukhachevskii was falsely accused of an anti–Soviet plot and executed.

4. This memorandum was known to a narrow circle of workers in the RKKA Staff. This is discussed in greater detail in "Zapiski Sovremennika o M.N. Tukhachevskom," *Voenno-Istoricheskii Zhurnal*, no. 4 (1963).

5. At first Triandafillov referred to the latter two groups as TIP (machine gun destroyer tanks) and TIA (artillery destruction tanks).

6. Editor's note. The acronyms translate as NPP, or direct infantry-support tanks (*neposredstvennaya podderzhka pekhoty*) tanks, DPP, or long-range infantry-support tanks (*dal'nyaya podderzhka pekhoty*) and DD, or long-range (*dal'nego deistviya*) tanks.

7. Editor's note. Soviet manuals during these years were referred to as PU (*Polevoi Ustav*), followed by the year of publication.

8. Editor's note. Konstantin Bronislavovich Kalinovskii (1897–1931) joined the Russian army and served in World War I and joined the Red Army in 1918 and commanded an armored train. He graduated from the Frunze Military Academy in 1925 and was later a military advisor in China. From 1929 he was inspector of the army's armored forces and commander of a mechanized regiment. In 1931 he was appointed chief of the RKKA Mechanization and Motorization Directorate. Kalinovskii died in the same plane crash that killed Triandafillov.

9. Editor's note. The Frunze Military Academy was the successor to the RKKA Military Academy. It was renamed in honor of the recently-deceased defense commissar Mikhail Vasil'evich Frunze (1885–1925).

10. Editor's note. Robert Petrovich Eideman (1895–1937) joined the Russian army in 1916 and the Red Army two years later. During the civil war he commanded divisions and an army and also occupied political posts. Following the war, he served as a military district commander and chief of the Frunze Military Academy. Eideman was executed along with Tukhachevskii and several others.

11. Editor's note. Nikolai Yakovlevich Kotov (1893–1938) joined the Russian army and fought in World War I, at the same time he was engaged in underground revolutionary activity as a member of the Socialist Revolutionary Party. He joined the Red Army in 1918 and commanded a regiment and brigades during the civil war. Following the war, he served in the central administrative apparatus and was Eideman's deputy at the Frunze Military Academy, before transferring to the air force. Kotov was arrested in 1937 and later executed.

12. Editor's note. Kas'yan Aleksandrovich Chaikovskii (1893–1938) joined the Russian army in 1914 and fought in World War I. He joined the Red Army in 1918 and served primarily as a political commissar. Following the war, he commanded a brigade and a di-

vision. He completed the Frunze Military Academy in 1924 and later served as deputy chief under Eideman. In 1936 he was appointed head of the army's Combat Training Directorate. Chaikovskii was arrested in 1937 and died in prison.

13. Editor's note. Pavel Ivanovich Vakulich (1890–1937) joined the Russian army in 1908. He graduated from the RKKA Military Academy in 1924 and later served as chief of the academy's operational department. He also served as chief of the department of the tactics of higher formations in the General Staff Academy. Vakulich was arrested in 1937 and executed.

14. Editor's note. Sergei Nikolaevich Krasil'nikov (1893–1971) joined the Russian army in 1913 and the Red Army in 1918. Following the civil war, he held a variety of staff and teaching assignments. During the Great Patriotic War he served in the central administrative apparatus. Following the war, Krasil'nikov resumed his teaching duties in the General Staff Academy.

15. Editor's note. Pavel Grigor'evich Ponedelin (1893–1950) joined the Russian army in 1914 and the Red Army in 1918. During the civil war he commanded units on various fronts. Following the war, he served in a variety of command and staff positions and taught at the Frunze Military Academy. During the Great Patriotic War he commanded an army, but was captured in 1941. Ponedelin was arrested in 1945 and executed in 1950.

16. Editor's note. Il'ya Pavlovich Kit-Viitenko (1898–1977) was drafted into the Austro-Hungarian army in 1914 and was later captured by the Russians. During the civil war, he served in the secret police and joined the Red Army in 1922. He graduated from the Frunze Military Academy in 1928 and taught there as well. Kit-Viitenko was arrested in 1937 and released from prison in 1956.

17. Editor's note. Richard Stanislavovich Tsifer (1898–1937) was a senior instructor at the General Staff Academy. He was arrested in 1937 and executed shortly thereafter.

18. Editor's note. The Revolutionary Military Council of the USSR (RVS USSR) was the successor to the Revolutionary Military Council of the Republic (RVSR), which was established by the Bolsheviks in 1918 as a measure to more effectively prosecute the civil war. From its establishment in 1923, the RVS USSR exercised control over the USSR's armed forces.

19. In November 1933 Tukhachevskii once again approached Voroshilov on this question and in a memorandum wrote "...following your address at the RVS plenum, the impression was created among many people that in spite of the new weapons in the army, tactics must remain as before.... Following the plenum, a complete intellectual ferment has begun in the minds of the commanders. There is talk of renouncing the new tactical forms and their development..." (see M.N. Tukhachevskii, *Izbrannye Proizvedeniya*, Moscow, Voennoe Izdatel'stvo, 1964, vol. I, p. 18).

20. Editor's note. Iona Emmanuilovich Yakir (1896–1937) joined the Red Army in 1918 and commanded a number of small units and a rifle division. Following the war, he served in a number of command and administrative posts and commanded the Kiev Military District. Yakir was arrested and executed along with Tukhachevskii and others.

21. Editor's note. Ieronim Petrovich Uborevich (1896–1937) joined the Russian army in 1916 and the Red Army two years later. During the civil war he commanded a division and armies in European Russia and the Far East. Following the civil war, he served in a variety of command and administrative posts and commanded the Belorussian Military District. Uborevich was arrested and executed along with Tukhachevskii and others.

22. Editor's note. Sergei Sergeevich Kamenev (1881–1936) joined the Russian army in 1900 and graduated from the General Staff Academy in 1907. During World War I he commanded an infantry regiment and served as a corps chief of staff. He joined the Red Army in 1918 and commanded an army and a front. From 1919 to 1924 he was the commander-in-chief of Armed Forces of the Republic. Kamenev later served as chief of the RKKA Staff and later served in a variety of administrative positions.

23. In 1936 Tukhachevskii significantly reworked the first part of his book, *The New Problems of War*, with the rebirth of a large aggressive army in fascist Germany in mind. Unfortunately, the revised manuscript was lost.

24. M.N. Tukhachevskii, *Izbrannye Proizvedeniya*, vol. I, p. 12.

25. Editor's note. This refers to those czarist-era officers who threw in their lot with the Soviet regime and elected to serve in the Red Army.

26. Editor's note. Anatolii Vasil'evich Fedotov (1892–1938) joined the Russian army in 1909 and commanded an artillery brigade in World War I. He joined the Red Army in 1919 and served in various capacities as an artillery officer. He continued to serve in the artillery following the war and was deputy chief of the Frunze Military Academy's operational department under Isserson. He was later appointed chief of staff of the Leningrad Military District. Fedotov was arrested in 1937 and later executed.

27. Editor's note. Yevgenii Nikolaevich Sergeev (1887–1937) joined the Russian army in 1908 and served in staff assignments during World War I. He joined the Red Army in 1918 and served as a staff officer. He continued his staff service following the war and also taught tactics and operational art at the Frunze Military Academy and in the General Staff Academy. Sergeev was arrested in 1937 and executed.

28. Editor's note. Aleksei Makarovich Peremytov (1888–1938) joined the Russian army in 1908 and served in World War I in the field and on staffs. He joined the Red Army in 1918 and served as a staff officer in various postings. Following the war, he continued his staff work and also taught at the Frunze Military Academy and the General Staff Academy. Peremytov was arrested in 1938 and executed.

29. Editor's note. Aleksandr Nikolaevich Lapchinskii (1882–1938) joined the Russian army in 1909 and the Red Army in 1918. He served as a pilot during World War I and commanded air units during the civil war. Following the war, he occupied a number of administrative and teaching positions and wrote a number of works on the employment of air power. Lapchinskii was later arrested and executed.

30. Editor's note. Dmitrii Mikhailovich Karbyshev (1880–1945) joined the Russian army in 1898 and fought in the Russo-Japanese War. During World War I he served as chief engineer in a number of formations. During the civil war he occupied the same positions in the Red Army, which he joined in 1918. Following the war, he served in a variety of administrative

and teaching posts. Karbyshev was captured during the early days of the Great Patriotic War and executed by the Germans.

31. Editor's note. Ivan Ivanovich Trutko (1888–1941) later served in the General Staff's operational department. During the Great Patriotic War he served as assistant commander of the 26th Army for rear questions. Trutko was killed in trying to break out of the encirclement around Kiev.

32. Editor's note. Boris Konstantinovich Leonardov (1892–1939) joined the Russian army in 1915 and the Red Army in 1919, serving as a doctor in both capacities. Following the civil war, he continued to study and teach. Leonardov taught in the Frunze Military Academy and was deputy chief of the Main Military-Medical Directorate.

33. Editor's note. Aleksandr Ignat'evich Sedyakin (1893–1938) joined the Russian army in 1914 and fought in World War I. He joined the Red Army in 1918 and commanded divisions and other units. Following the civil war, he commanded armies and served in the central administrative apparatus. Sedyakin was arrested in 1937 and executed the following year.

34. Editor's note. Marshal Aleksandr Il'ich Yegorov (1883–1939) joined the Russian army in 1901 and commanded a company, battalion and regiment in World War I. He joined the Red Army in 1918 and commanded armies and fronts. Following the civil war, he commanded an army, a front and military districts and later served as chief of the General Staff. Yegorov was arrested in 1938 and later executed.

35. Editor's note. Vladimir Iosifovich Mikulin (1892–1961) joined the Russian army in 1914, where he trained as a pilot, and the Red Army in 1918, where he switched to the cavalry. Following the civil war, he taught at the Frunze Military Academy and was appointed chief of the higher cavalry school. Mikulin was arrested in 1937 and released in 1946, without the right to live in Moscow.

36. Editor's note. Nikolai Yefimovich Varfolomeev (1890–1939) fought in World War I and joined the Red Army in 1918, where he served primarily in staff positions. Following the civil war, he continued his staff work and was active in the army's system of military education. He taught in the Frunze Military Academy and was chief of staff of the Volga Military District. Varfolomeev was arrested in 1938 and later executed.

37. Editor's note. Yevgenii Aleksandrovich Shilovskii (1889–1952) joined the Russian army in 1907 and served in World War I. He joined the Red Army in 1918 and served staff and command posts during the civil war. Following the war, he served in staff and teaching positions. During the Great Patriotic War Shilovskii taught in the General Staff Academy and wrote a number of works on the army's war experience.

38. Editor's note. Nikolai Nikolaevich Shvarts (1882–1944) joined the Russian army in 1902 and fought in the Russo-Japanese War. He also took part in World War I as a staff officer. He joined the Red Army in 1918 and took part in the civil war as a staff officer. Following the war, he taught in the RKKA Military Academy and later in the General Staff Academy.

39. Editor's note. Fedor Platonovich Shafalovich (1884–1952) joined the Russian army in 1903 and served in staff positions during World War I. He joined the Red Army in 1918 and continued as a staff officer in Siberia and Central Asia. Following the civil war, he served in staff and administrative pots. In 1928 he began teaching at the Frunze Military Academy, and from 1936 in the General Staff Academy. Shafalovich continued his teaching activities during the Great Patriotic War.

40. Editor's note. Aleksei Ivanovich Gotovtsev (1883–1969) joined the Russian army in 1902. He served in World War I in various staff posts. He joined the Red Army in 1921 and taught in the RKKA Military Academy. From 1936 Gotovtsev taught in the General Staff Academy.

41. Editor's note. Aleksandr Andreevich Svechin (1878–1938) joined the Russian army in 1895, completed the General Staff Academy in 1903 and served in the Russo-Japanese War. During World War I he commanded an infantry regiment, a division and was chief of staff of an army. He joined the Red Army in 1918 and served in staff and academic assignments. Following the civil war, he taught in the Frunze and General Staff academies and was briefly under arrest in 1932. Svechin was again arrested in 1937 and executed.

42. Editor's note. Sergei Aleksandrovich Mezheninov (1890–1937) joined the Russian army in 1910 and fought in World War I. He was conscripted into the Red Army in 1918 and served as chief of staff of various armies and commanded several armies. Following the civil war, he served as chief of the army's air force directorate and deputy chief of the RKKA Staff. Mezheninov was arrested in 1937 and executed shortly afterward.

43. Editor's note. Sidor Pavlovich Obysov (1896–1937) graduated from the staff school and RKKA Military Academy following the civil war. He later served as chief of the RKKA Staff's operational section and deputy chief of the General Staff Academy. Obysov was arrested in 1937 and later executed.

44. Unfortunately, not one copy of this work exists today, as they were all destroyed during the period of Stalin's personality cult. Editor's note. Isserson here is mistaken. The work survived and is featured in this collection.

45. Editor's note. This is a reference to the N.Ye. Zhukovskii Air Academy, in honor of Nikolai Yegorovich Zhukovskii. It is now known as the N.Ne. Zhukovskii and Yu.A. Gagarin Military Air Academy.

46. Editor's note. The OKDVA, or Special Red Banner Far Eastern Army was organized in 1929 as the Special Far Eastern Army and, under the command of V.K. Blyukher, took part in the brief conflict with China over the Chinese Eastern Railway that same year and its name was changed to OKDVA the following year. In 1938 the OKDVA was reorganized into the Red Banner Far Eastern Front.

47. Editor's note. Boris Iosifovich Bobrov (1896–1937) joined the Russian army in 1915 and took part in World War I. He joined the Red Army in 1918 and primarily served on various staffs. Following the war, he continued his staff work and was appointed chief of staff of the Belorussian Military District in 1935. Bobrov was arrested in 1937 and executed.

48. Editor's note. Dmitrii Aleksandrovich Kuchinskii (1898–1938) joined the Russian army in 1917 and the Red Army a year later. During the civil war, he mainly served in staff positions, which he continued to do after the war. He was chief of staff of the Kiev Military District before being appointed chief of the General Staff Academy in 1936. Kuchinskii was arrested in 1937 and later executed.

49. Editor's note. Ivan Kensoforovich Gryaznov (1897–1938) joined the Russian army in 1916 and

fought in World War I. He joined the Red Army in 1918 and commanded a regiment, brigade and division. Following the war, he commanded a division, corps and several military districts. Gryaznov was arrested in 1937 and later executed.

50. Editor's note. Markian Yakovlevich Germanovich (1895–1937) joined the Russian army in 1915 and fought in World War I. He joined the Red Army in 1918 and commanded a company, battalion, regiment, and brigade. Following the war, he continued to serve in various military districts. In 1933 he was appointed chief of the RKKA Mechanization and Motorization Academy and was later deputy commander of the Leningrad Military District. He was arrested in 1937 and later executed.

51. Editor's note. Yakov Moiseevich Fishman (1887–1961) joined the revolutionary movement early on and spent considerable time abroad before the revolution. He joined the Red Army in 1921 and worked abroad as an intelligence officer. In 1925 he was appointed chief of the army's chemical directorate and in 1928 chief of Chemical Defense Institute. He was arrested in 1937 and freed in 1947. He was arrested again in 1949 and freed in 1954.

52. Editor's note. Vasilii Konstantinovich Blyukher 1890–1938) joined the Russian army in 1914 and the Red Army in 1918. During the civil war he commanded troops in Siberia, the Far East and in the Crimea. Following the civil war he served as a military advisor in China and in 1929 was appointed commander of the Special Far Eastern Army. In 1938 he commanded the Red Banner Far Eastern Front against the Japanese at Lake Khasan. Blyukher was arrested and executed later the same year.

53. Editor's note. Ivan Fedorovich Fed'ko (1897–1939) joined the Russian army in 1916 and the Red Army two years later. During the civil war, he commanded divisions and an army. Following the war, he commanded a division and a corps against anti–Soviet forces in Central Asia. Afterward, he commanded an army and military districts and was a deputy defense commissar. He was arrested in 1938 and executed.

54. Editor's note. Mikhail Vladimirovich Sangurskii (1894–1938) joined the Russian army in 1914 and the Red Army in 1918. During the civil war, he commanded a brigade and a division. Following the war, he commanded a division and a corps and was a military adviser in China. He later served as chief of staff of the OKDVA. Sangurskii was arrested in 1937 and later executed.

55. Editor's note. Stepan Nikolaevich Bogomyagkov (1890–1966) joined the Russian army in 1914 and fought in World War I. He joined the Red Army in 1918 and served in various staffs. Following the war, he continued his staff work and was also the chief of the air force's combat training directorate. He was arrested several times in the 1930s and spent ten years in the camps, before being freed in 1948. Bogomyagkov was arrested again in 1949 and freed in 1954.

56. Editor's note. Vladimir Davydovich Grendal' (1884–1940) joined the Russian army in 1902 and completed the artillery academy in 1911. During World War I he served as commander of an artillery battalion. He joined the Red Army in 1918 and served as artillery inspector with a number of fronts. Following the war, he served as artillery chief in a number of military districts and was head of the artillery academy. He later taught at the Frunze Military Academy and was deputy chief of the army's Main Artillery Directorate. Grendal' later commanded an army in the 1939–40 war with Finland.

57. Editor's note. Aleksei Vladimirovich Kirpichnikov (1889–1974) took part in World War I. He joined the Red Army in 1918 and served in staff positions. Following the civil war, he continued to serve in staff and command positions. Kirpichnikov taught at the Frunze Military Academy and the General Staff Academy.

58. Editor's note. Vasilii Konstantinovich Mordvinov (1892–1971) joined the Russian army in 1911 and fought in World War I. He was conscripted into the Red Army in 1918 and served in an army staff and commanded a division. Following the civil war, he taught at the Frunze Military Academy and then at the General Staff Academy. During World War II Mordvinov served in staff positions with several fronts and was acting chief of the General Staff Academy.

59. Editor's note. Petr Dmitrievich Korkodinov (1894–1968) briefly served as chief of staff of the 39th Army during the Great Patriotic War.

60. Editor's note. Boris L'vovich Teplinskii (1899–1972) joined the Red Army in 1918 and fought in the civil war. Following the war, he became a pilot and taught at the General Staff Academy during 1938–1941. During the Great Patriotic War he served in staff assignments until his arrest in 1943. Teplinskii was sentenced to ten years in 1952, but released the following year after Stalin's death.

61. Among the scientific works published during this time were: N.Ye. Varfolomeev's *Udarnaya Armiya*, M.R. Galaktionov's *Tempy Operatsii*, V.A. Melikov's *Problema Strategicheskogo Razvertyvaniya*, Ya.M. Zhigur's *Sovremennye Operatsii* (this work was not published and the manuscript has been lost), and works by A.N. Lapchinskii and a number of others. In 1932 the first edition of G.S. Isserson's short book, *Evolyutsiya Operativnogo Iskusstva*, appeared (a second and expanded edition appeared in 1937), in which the development of the forms and methods of armed struggle in modern war was studied and the theory of the deep operation was grounded. The book emphasized that we are at the dawn of a new era in military art and must change from the linear strategy to the strategy of depth. In December 1932 a critique of this book took place in the Central House of the Soviet Army. A.I. Sedyakin, the chief of the military training directorate, made the main report and stated that "this short book is quite instructive and offers a useful and creative fillip and the correct path for operational thinking." According to Sedyakin's conclusion, "on the whole, this work, with some amendments, quite correctly illuminates the problem of operational art" and "is perhaps the first and quite valuable contribution to this important matter." *Voina i Revolyutsiya* (1933), nos. 1–2, pp. 113–118. The editors.

62. M.N. Tukhachevskii, *Izbrannye Proizvedeniya*, vol. I, p. 261.

63. Editor's note. Georgii Pavlovich Sofronov (1893–1970) was drafted into the Russian army in 1914 and joined the Red Army four years later. He fought in the civil war and took part in the execution of Emperor Nicholas II and his family. Following the war, he commanded a division, corps and a military district and also served in the army's academic apparatus. During the Great Patriotic War, he commanded an army, but poor health prevented him from rising further. Following the war, Sofronov taught in the General Staff Academy.

64. This conversation is set forth in greater detail

in "Zapiski Sovremennika s M.N. Tukhachevskim." *Voenno-Istoricheskii Zhurnal* (1963), no. 4.

65. M.N. Tukhachevskii, *Izbrannye Proizvedeniya*, vol. I, p. 24.

66. Editor's note. Matvei Vasil'evich Zakharov (1898–1972) first served in the Red Guard and joined the Red Army in 1918, serving on a number of fronts in the civil war. Following the war, he served in staff and administrative positions and studied at the General Staff Academy, after which he served in a number of high-ranking staff positions. During the Great Patriotic War he served in a number of staff positions, mostly at the front level. Following the war, Zakharov commanded troops and was chief of the General Staff (1960–63, 1964–71).

67. A.V. Kirpichnikov (now a retired Lt. Gen.) drew up this task.

68. Editor's note. Isserson is referring to the Red Army's occupation of western Belorussia and western Ukraine in September 1939, as a result of the German-Polish War and the German-Soviet Non-Aggression Pact, which divided Eastern Europe into German and Soviet spheres of influence.

69. Editor's note. This refers to the period of World War II from the outbreak of war in September 1939 to the German attack through the Low Countries and France in May 1940. This period saw almost no action on land, and very little in the air and sea.

70. Editor's note. The Maginot Line was the name commonly applied to the French system of fortifications built along the German-French border during 1929–36, at the behest of defense minister Andre Maginot. The fortifications stretched some 400 kilometers from the upper Rhine to the border with Belgium. In May-June 1940 the Germans outflanked the Maginot Line.

71. There were a number of specific conditions in the German-Polish war of 1939, which favored the conduct of the deep operation. From the first, the German deployment occupied a flanking position *vis a vis* Poland. The Polish theater had not been fortified and offered complete freedom of maneuver. The front was not continuous and the Germans enjoyed a great superiority in men and equipment and complete air superiority.

72. Editor's note. Ivan Khristoforovich Bagramyan (1897–1982) joined the Russian army in 1915 and the Red Army in 1920. Following the civil war, he served for several years in the cavalry. He completed the Frunze Military Academy in 1934 and the General Staff Academy in 1938, and then took up teaching duties in the latter. During the Great Patriotic War he served primarily in high-ranking staff positions, before switching to command duties, which included an army and a front. Following the war, Bagramyan occupied a number of high-rank command, administrative and academic posts.

73. Editor's note. This is probably Fedor Mikhailovich Isaev (1894–1967), who during the Great Patriotic War served as deputy head of the General Staff's operational directorate and head of the operational section of the 18th Army.

74. Editor's note. Vladimir Yefimovich Klimovskikh (1895–1941) joined the Russian army in 1912 and the Red Army in 1918. During the civil war and for several years afterward he occupied staff positions, before switching to academic work, teaching in the Frunze Military Academy and General Staff Academy. At the beginning of the Great Patriotic War he was serving as chief of staff of the Western Front and was among those held responsible for the German breakthrough. Klimovskikh was recalled from the front and executed.

75. Editor's note. Nikolai Vasil'evich Korneev (1900–76) joined the Red Army in 1919. Following the civil war, he served predominantly in the intelligence branch and was an instructor in the General Staff Academy. During the Great Patriotic War he commanded an army and was chief of staff of several other armies, as well as head of the Soviet military mission to the Yugoslav partisans. Following the war, Korneev resumed teaching in the General Staff Academy.

76. Editor's note. Aleksandr Vasil'evich Sukhomlin (1900–70) joined the Red Army in 1918 and fought as a partisan. Following the civil war, he held a number of command and staff posts. He taught at the Frunze Military Academy and General Staff Academy. During the Great Patriotic War he served in a number of high-ranking staff positions and commanded an army. Following the war, Sukhomlinov continued his academic work.

77. Editor's note. Nikolai Iustinovich Trubetskoi (1890–1942) joined the Russian army in 1915 and the Red Army in 1918. Following the civil war, he chiefly served in the military supply apparatus and headed the department of military supply in the General Staff Academy. He was arrested at the beginning of the war and executed the following year.

78. Editor's note. This is probably Aleksei Ivanovich Shimonaev (1896–1959), who served as head of the General Staff's Rear Supply Directorate during the Great Patriotic War.

79. Editor's note. Petr Grigor'evich Yarchevskii (1895–1950) joined the army in 1913 and fought in World War I. He joined the Red Army in 1918 and served in a variety of staff positions. Following the war, he continued in staff postings. During the Great Patriotic War and afterwards, Yarchevskii taught in the General Staff Academy.

80. In the first half of the article, published in *Voenno-Istoricheskii Zhurnal* (no. 1), in tallying up the military men who made a contribution to the elaboration of the theory of the deep operation, the author committed an oversight in not naming A.Ya. Lapin, who in the beginning of the 1930s worked in the RKKA Military Training Directorate. A.Ya. Lapin actively participated in elaborating the theory of the deep engagement and introduced many substantial suggestions to it. In the same issue it was mistakenly stated that I.I. Trutko became a victim of lawlessness during the years of Stalin's personality cult. Actually, Maj. Gen. I.I. Trutko perished in September 1941, while breaking out of an encirclement in the area of Bedanovka village, Lokhvitsa district, Poltava Oblast' (Southwestern Front).

81. Editor's note. Isserson is mistaken, as the Higher Military Council (*Vysshii Voennyi Sovet*) existed as the armed forces' highest organ of strategic control only from March to September, 1918, when it was disbanded and replaced by the RVSR. The session Isserson refers to was a gathering of the country's highest military leaders, convened to review the state of the army and the state of its military theory. Defense Commissar Marshal Semyon Konstantinovich Timoshenko delivered the closing address on the nature of modern operations, which Isserson may have helped draw up.

82. M.N. Tukhachevskii, *Izbrannye Proizvedeniya*, vol. I, p. 261.

Index

airborne 51, 63, 77, 105, 142, 205, 217, 219–20, 222–23, 228–30, 233, 287, 289, 292–95, 298, 300–1
airborne detachment (ADO) 85–87, 91, 136, 142–43, 205
airborne infantry 280
Aisne River 161
Algeria 248
Amiens 161, 166
ANT-14 87
anti-aircraft 50–51, 95, 111, 169, 178, 200, 205–6
anti-tank 51, 111, 119–20, 134, 142, 170–71, 177, 187–88, 190, 192, 194, 202–3, 205–7, 303
Aragon 245–46
Arctic Ocean 45
army aviation group (AGA) 99, 115, 118, 120, 122–23, 126, 132, 136–37, 141–42, 151, 199, 220, 232
army defensive zone 173, 193, 197, 201, 303
Army Group North 258–59, 264
Army Group South 258–60, 262, 265
army operation 98, 109, 115–16, 128, 136, 214, 227, 231–33, 292
army rear line 192
army rear zone 173
army vanguard 58, 62–63, 115
artillery 26, 30, 37, 50, 52, 57–58, 61, 63, 81, 85–88, 90–94, 105, 126–27, 141, 146, 153, 163–65, 167–69, 171, 176–77, 181, 183, 190–94, 199–202, 204, 206–7, 219–20, 278, 241, 245, 268, 273, 278, 287–88, 292, 294, 296
assault air brigade 85–87, 92
attack echelon (EA) 62–65, 91, 126, 129–30, 133–38, 140–41, 146, 148, 223, 292
Austria 25
Austro-Prussian War 18, 25–26, 28, 259, 278
aviation 45–46, 49, 57, 59, 61, 63, 79–80, 83, 93, 98–99, 101, 103–4, 125–26, 131, 137, 141, 147, 153, 167, 169, 177, 183, 198–99, 201–2, 209, 213–14, 219–22, 224, 228–30, 233, 247, 256, 262, 268–69, 271, 278–79, 287, 289–90, 292–93, 295, 300

Bagramyan, Ivan Khristoforovich 302
Baltic Sea 254
Baranovichi 84
Barcelona 247
battle 23–24, 28, 33, 54, 60, 65–67, 80, 215, 225
Battle of the Frontiers 34, 55, 114, 267
Battle of the Marne 30
Bazaine, Francois Achille 25
Belgium 79
Belorussia 301
Belorussian Military District 294
Benedek, Ludwig August Ritter 25, 255, 278
Berlin 250, 256, 281
Bernhardi, Friedrich Adolf Julius 33, 74
Beskids 254
Bialystok 255, 261, 274, 277, 281–82
Bilbao 169, 190
Bismarck, Otto Eduard Leopold 54
Black Sea 45
Blaskowitz, Johannes Albrecht 259–60, 262, 265
Blyukher, Vasilii Konstantinovich 295
Bobrov, Boris Iosifovich 294
Bock, Moritz Albrecht Franz Friedrich Fedor 259
Bogomyakov, Stepan Nikolaevich 295
Bohemia 255, 257, 259, 278
Borodino 23
Bortnowski, Wladyslaw 273
Bouchacourt 196
breakthrough 21, 29, 36, 38–40, 61–62, 68, 75–77, 80, 90–91, 113–14, 121, 124–26, 129–30, 134–39, 141–43, 147–49, 153, 169–71, 186, 193, 199, 202–3, 205, 207, 209, 216–18, 221–24, 227, 229–30, 233, 245, 247, 289–90, 292–93, 298, 301, 301, 304, 306
breakthrough development echelon (ERP) 62–64, 68, 82, 91–92, 100, 121, 126, 129–31, 129–43, 146, 148–49, 153, 202, 223, 230, 232, 244, 292–93, 296, 300, 304
breakthrough operation 62, 81, 126, 138, 142–43, 146–48
Breslau 259
Brest-Litovsk 264, 274, 277, 281–82
Bromberg 260
Brunete 244
BT tank 85
Bzura River 268

Cannae 27, 36, 64, 279, 282
Cantabrian Mountains 242
Carpathian Mountains 254, 266, 279
Catalonia 245–47
Cavaignac, Louis-Eugene 249
cavalry 58–59, 62–64, 66–67, 84, 97–98, 100–1, 103–6, 117, 120, 122, 153, 175–76, 181, 194, 218, 222–23, 232, 291–92, 299
cavalry corps 85–87, 91–92, 107, 116–17, 119, 129, 133, 136, 138–39, 157, 180–81, 219–20, 301
cavalry division 93
cavalry-mechanized army 300–1
cavalry-mechanized group (KMG) 218–20, 223, 227–30
Chaikovskii, Kas'yan Aleksandrovich 287
Chalons 17
Chassepot rifle 24
chemical 47, 51, 80, 85, 87, 92, 94, 126, 178, 181, 183, 194, 204, 219–20, 222, 225–26, 294
China 156, 166, 239, 250
Ciechanow 264

322   Index

Clausewitz, Carl 17, 20,22, 24, 35, 41, 53, 55, 64, 66, 69–71, 159–60, 195, 238, 251
Colombey 27
combined operations 214
Comintern 20
communications 25, 68, 95–96, 128, 181, 184, 207, 213–14, 227–28, 255, 264–66, 268–69, 280–81
Communits Party 307
consecutive operation 31, 35, 48, 215, 247
counteroffensive 226
Cracow 256, 258, 260, 265, 273
Cullman, Frederic 18, 40
Czestochowa 256, 259–60, 262

D-8 87
Danzig 254–59, 262–64, 266, 268, 273
DD tank 287
Debeney, Marie Eugene 39
Deblin 276
deep breakthrough operation 64, 68, 136–37, 146
deep engagement 52, 74, 80, 82, 91, 129, 132–34, 137, 141, 153, 286–89, 295–96, 304
deep offensive 52–53
deep operation 15, 22, 48, 51, 55, 58, 66–68, 74–76, 80, 82–84, 87, 90, 96–102, 108–10, 114–17, 120, 122–24, 126–27, 133–34, 136, 147–53, 167, 237, 268–69, 283, 286, 288–97, 301, 303–5, 307
deep strategy 41, 56, 58–61, 64–66, 69–71, 236
deep tactics 288, 295
defensive operation 156, 174, 179, 183, 194–209, 216–17, 303–4
Denikin, Anton Ivanovich 161
Dirschau 262
Dnestr River 78–79
DPP tank 287
Dreyse rifle 24
Duffour, Gaston 39, 66
Dvina River 35, 78–79

East Prussia 31, 124, 160, 254–57, 259–60, 263–64, 268, 273, 279, 307
Eastern Front 41, 160
Ebro River 246
Eideman, Robert Petrovich 287, 290
Eighteenth Army (German) 63
Eighth Army (German) 259, 273, 281
Eleventh Army (German) 279
Eliot, George Fielding 256
encirclement 64, 136, 139, 217–8, 227, 229–30, 259–60, 266, 268, 273–76, 279–82, 300, 306
engagement 21, 23, 26–28, 57, 61, 67, 74, 76, 82, 88, 133, 137,
200, 215, 249, 271, 287–89, 294–96, 304
Engels, Friedrich 22, 24–25, 30, 32, 42, 54, 236, 248
engineer 51, 79–80, 95–97, 105, 128, 173–74, 177–79, 181, 183, 188, 194, 204, 219–20, 225–26, 230, 262, 264, 303
England 253–54, 266
Entente 40, 47, 286
Europe 237, 239, 247, 249–50, 296
The Evolution of Operational Art 74, 237

Falkenhayn, Erich 160, 279
Far East 45, 295
Fayolle, Marie Emile 165
Fed'ko, Ivan Fedorovich 295
Fedotov, Anatolii Vasil'evich 290, 304
field army 218–19, 221
field manual: (1925) 18, 304; (1929) 287, 295; (1936) 160, 168, 170–71, 209, 295, 304; (1939 draft) 304
Fieseler Storch 269
Fifth Army (French) 34
Fifth Army (German) 105
Finland 302
First Army (Bulgarian) 279
First Army (French) 165
First Army (German) 105
first attack echelon 176
Fishman, Yakov Moiseevich 294
Five-Year Plan 51, 74, 293
flexible defense 165
Foch, Ferdinand 40, 54
forward army 299
forward defensive echelon 174, 180–81
forward defensive zone 170, 174, 181–82, 187, 192, 200
forward detachment 181
forward echelon 222–23, 229, 231–32
forward theater 105
Fourteenth Army (German) 258, 260
Fourth Army (German) 259–60, 262, 264–65, 270, 281
4th Army (Soviet) 35
France 18–19, 25, 30, 51, 57, 77, 89, 242–44, 247–48, 253, 302, 306
Franco, Francisco 241, 246–47
Franco-Prussian War 15, 18–19, 25–29, 36, 54–55
French Revolution 17–18, 23, 43
front 88, 92, 96, 103–5, 108–9, 128, 131, 136, 142, 147, 151, 153, 187, 209, 218–21, 227–32, 291, 298, 300, 306
front aviation 199, 219, 224, 228–29, 293
front operation 89, 147, 159, 186, 209, 214, 219–20, 227–28, 230–32

Frunze, Mikhail Vasil'evich 43, 279
Frunze Military Academy (RKKA Military Academy) 287, 290, 293, 296, 303
Fuller, John Frederick Charles 19, 42, 50, 248
The Fundamentals of Conducting Operations 305
The Fundamentals of the Deep Operation 293, 303
The Fundamentals of the Defensive Operation 304

Galicia 269, 279
Gamelin, Maurice Gustave 76
Gasiorowski, Janusz Tadeusz 262
Gdynia 273
General Staff Academy 296–99, 301–3, 305, 307
German-Polish War 235, 237–40, 250–53, 256–57, 263, 266–67, 269, 271–72, 276–81, 283, 301, 307
German Wars of Unification 100
Germanovich, Markian Yakovlevich 294
Germany 18, 57, 77, 252–54, 257, 297
Gitschin 26
Gotovtsev, Aleksei Ivanovich 291
Graudenz 256
Gravelotte 27, 29, 36
Great Patriotic War 286, 294–95, 300, 305–7
Grendal, Vladimir Davydovich 295
Grodno 255, 275
Groener, Karl Edmund Wilhelm 18
ground operations 213–14
Gryaznov, Ivan Kensoforovich 294
Guadalajara 241–42
Guderian, Heinz Wilhelm 167, 169, 259–60, 262, 264, 268, 270, 274, 296
Gulf of Finland 79

heavy bomber corps 85–87
Hel peninsula 273, 276
Hentsch, Richard 35
Hermann-Brunhilda position 161
High Command Artillery Reserve 85–86, 91, 177
High Command Reserve 95, 109, 177, 191, 219
High Command Tank Reserve 85, 91
Higher Military Council 304
Hitler, Adolf 307
Holy Alliance 248
Hoth, Hermann 259–60, 262, 265–66, 270, 274
Hungary 275

Iberian peninsula 242
Immanuel, Friedrich 19
independent army 218
Isaev, Fedor Mikhailovich 302
Italian-Abyssinian War 157, 239

Jarama River 241
Jena 250–51

Kalinovskii, Konstantin Bronislavovich 287–88
Kamenev, Sergei Sergeevich 288
Karbyshev, Dmitrii Mikhailovich 290
Kirpichnikov, Aleksei Vladimirovich 295
Kit-Viitenko, Il'ya Pavlovich 287
Klimovskikh, Vladimir Yefimovich 302
Kluck, Alexander Heinrich Rudolph 105
Kluge, Gunther Adolf Ferdinand 259
Koniggratz (Sadowa) 27–28
Korkodinov, Petr Dmitrievich 295
Korneev, Nikolai Vasil'evich 302
Kotov, Nikolai Yakovlevich 287
Kovel' 128
Krasil'nikov, Sergei Nikolaevich 287
Krasne 128
Kreuzberg 258
Kuchinskii, Dmitrii Aleksandrovich 294, 297
Kuchler, Georg Karl Friedrich Wilhelm 259
Kulm 256, 262, 264
Kutno 265–66, 273–74, 292

Lake Ladoga 79
Lake Pskov 79
Landwehr 259, 282
Langlois, Hippolyte 58
Lapchinskii, Aleksandr Nikolaevich 290
Latvia 275
Leer, Genrikh Antonovich 28, 89
Leipzig 24
Lenin, Vladimir Il'ich 20, 25, 41–43, 158, 238
Leonardov, Boris Konstantinovich 290
Leonidas 29, 36
Lewal, Jules Louis 59, 67
Lida 128
Liege 105, 299
light bomber brigade 85–87, 92
linear operation 82, 149, 151
linear strategy 25, 29, 33–37, 40, 45, 52–54, 56, 60–61, 64–65, 70, 77, 151, 245, 279
List, Sigismund Wilhelm Walther 258, 260, 265
Lithuania 275
Lodz 256, 259–60
long-range aviation 82

Lowicz 273
Lublin 265, 274–75, 282
Ludendorff, Erich Friedrich Wilhelm 17
L'vov 256, 260, 266, 27375, 277, 281–82

Mackensen, August 279
MacMahon, Marie Esme Patrice Maurice 25
Madrid 159, 169, 209, 241, 246–47, 267
Maginot Line 302
main defensive echelon 174–76, 189
main defensive zone 170–71, 173–76, 178, 182–83, 187–89, 192–94, 199–206, 208–9
main echelon 109–11, 113–17, 119–23, 123, 135, 148, 201–2, 232–33, 292
Malaga 242
Manzanares River 241, 263
March (1918) offensive 17, 40, 47, 63, 165–66
Marengo 23
Marne River 31, 34–35, 37, 46, 52–53, 152, 160–61, 241, 263
Mars-La-Tour 27
Marxism 16, 288
Marxism-Leninism 20–21, 286, 289
mechanized brigade 86, 93
mechanized cavalry 83
mechanized corps 85–87, 91–92, 100, 106–7, 116–19, 129–30, 133, 136, 138–39, 141, 151, 153, 291, 301, 304
mechanized formation 83, 99–102, 106, 122
meeting battle 76, 112–15, 117–25, 134–35, 147, 216, 222, 226
meeting engagement 23, 26, 58, 118, 233
meeting operation 58
Mehring, Franz 251
Metz 27
Metzsch, Horst 19
Meuse River 34, 105
Mezheninov, Sergei Aleksandrovich 291
Mikulin, Vladimir Iosifovich 291
military art 16–19, 21–23, 25, 29, 48, 50–52, 64–65, 70, 75, 156–57, 167, 236–38, 240, 249–50, 277, 281, 283, 286, 307
Military-Medical Academy 290
military specialists 290–91
Minsk 84
Mlawa 262
Mnyuta river 55
mobile defense 161, 181, 216
Modlin 255–56
Moltke, Helmuth Carl Bernard 19–20, 25–33, 54, 56–57, 70, 110, 148
Mongolia 279

Moravia 257
Mordvinov, Vasilii Konstantinovich 295
Moscow 17
motorized division 85, 87, 93–94, 102, 106, 116–17, 119, 129, 136, 141, 151, 153, 252, 258–59, 269, 291, 301
motorized infantry 87, 91, 101, 105, 107, 109, 121, 232, 300
Mukden (Shenyang) 30

Nachod 26
Napoleon I (Bonaparte) 24–25, 28, 30, 53, 64, 115
Napoleon III (Louis) 25
Narew River 35, 259, 262–64, 266–68, 273
The Nature of Operations of Modern Armies 19
naval operations 214
Neman River 35, 52
Neuilly 27
The New Problems of War 288
Ninth Army (German) 279
NPP tank 287–88

obstacle zone 170
Obysov, Sidor Pavlovich 291
offensive operation 17, 34–35, 44, 46, 50, 102, 147, 212, 216, 221–22, 224–28
OKDVA (Special Red Banner Far Eastern Army) 294
operational art 16–17, 19–22, 28, 31–41, 44, 46, 53, 61–62, 65–68, 71, 74–75, 88–89, 97–98, 108, 110, 151–52, 212–13, 289–91, 294–98, 302, 305–6
operational defensive zone 128–32, 134, 136, 139, 141, 143, 171, 173, 175, 178, 187–88, 193–94, 202–3, 205–8
operational direction 89–91, 95, 103, 216, 218–19, 224, 228
operational echelon 59–60, 68
Orel 161
Ostrow Mazowiecka 266
Oudinot, Charles Nicolas Victor 249

Paris 27, 34, 46, 55
Peremytov, Aleksei Makarovich 290
Phony War 302
Picardy 47, 185–86
Pilica River 268
pinning army 228
poison 38
Poland 250–55, 257, 259–60, 267, 273, 275, 277–78, 280–82, 302, 306–7
Poles'ye 78–79, 162
Polish Corridor 252
Polish-Soviet War 43
Pomerania 255–57, 259–60, 262, 280
Pomorze 266, 273, 276, 282

Ponedelin, Pavel Grigor'evich 287
Poznan 254–56, 259, 262–63, 265–66, 268, 273, 281–82
Poznan Army 260, 265
Pratzen Heights 68
Prussia 18, 25, 250, 255
Przemysl 256
pursuit 216, 227
Pyrenees Mountains 243

Radom 259–60, 265–66, 273
Rawa River 268
rear army line 173, 183
rear defensive zone 127–31, 134, 139, 141, 173, 175, 187, 199, 205, 207–8
Red Army (RKKA) 42–44, 74, 104, 212, 221, 235, 275–76, 286–87, 296, 301
Regensburg 24
Reichenau, Walter 258, 260, 262, 265–66
reinforced rifle corps 85–86, 90–93, 95–96, 110–11, 129, 135, 232
Reinhardt, Georg-Hans 265
Republicans (Spain) 166, 169, 241–47, 249, 267
Requin, Edouard-Jean 19
reserve defensive echelon 175–76, 193–94, 205, 207
reserve echelon 109–10, 113–14, 122–23, 135, 148, 176, 193–94, 201, 205, 207, 232–33, 292
Revolutionary Military Council (RVS) 288
Rhine River 34, 57, 79, 251
RKKA Military Academy 75
RKKA Staff/General Staff 290–91, 293–95, 297, 304, 306–7
Romania 274–75, 279
Rundstedt, Karl Rudolf Gerd 258
Russian Civil War 17–18, 42–44, 55, 156–57, 279, 286
Russo-Japanese War 30
Rydz-Smigly, Edward 275, 279
Rzeszow 266

St. Privat 27, 29, 36
St. Quentin 185
Sambor 260
Sambre River 34
Samsonov, Aleksandr Vasil'evich 17
San River 45, 259, 261, 263, 265–66, 268
Sandomierz 266–67
Sangurskii, Mikhail Vladimirovich 295
Saxony 259
Schlichting, Sigismund 22, 26, 29, 56, 59, 74
Schlieffen, Alfred 18, 29, 33, 36–37, 55, 62, 74, 100, 279
Schlieffen Plan 79
second defensive echelon 174–75, 194

second defensive zone 126, 171, 173, 175–76, 183, 188, 192–93, 201–3, 205–6
Second Empire 27, 54
second obstacle defensive zone 171
Sedan 27–28, 54
Sedyakin, Aleksandr Ignat'evich 290–91, 295
Seeckt, Johannes Friedrich 42, 253
Segre River 246
Serbia 279
Sere der Revieres, Raymond Adolph 79
Sergeev, Yevgenii Nikolaevich 290
Seventh Army (German) 63
Shafalovich, Fedor Platonovich 291
Shanghai 80
Shilovskii, Yevgenii Aleksandrovich 291
Shimonaev, Aleksei Ivanovich 302
shock army 89–93, 95–98, 102, 108, 110, 113–14, 119, 124–25, 129, 131, 135–36, 150, 153, 218–20, 228–30, 232, 292–93
Shvarts, Nikolai Nikolaevich 291
Siedlce 275
Siegfried position 161, 174
Sieradz 273
Silesia 255, 257, 259, 266, 280
Sino-Japanese War 80
Slovakia 255, 257
Sofronov, Georgii Pavlovich 298
Soldan, George 42
Southwestern Front (Russian) 185
Soviet-Finnish War 302
Soviet Union 254, 275, 286, 305
Spain 156, 166–67, 169, 190, 235, 237–50, 267, 277–78, 301
spatial operation 290–91
Spicheren 26–27
Stalin, Iosif Vissarionovich 35, 44, 236, 290, 296, 298, 301, 306–7
Stanislavov 128
strategic aviation 104, 142
strategy 17–18, 20–25, 27–28, 38, 45, 53–54, 89, 152, 166, 213, 225, 239, 274, 291, 297–98, 302, 304, 306
Strategy 18
strategy of a single point 23
strategy of destruction 44, 64, 281
success development echelon (ERU) 121–25, 193
Sukhomlin, Aleksandr Vasil'evich 302
Svechin, Aleksandr Andreevich 291
Switzerland 36

T-26 85
T-27 87
T-28 177
tactical defensive zone 127–32, 134, 137, 139, 162, 165–66
tactics 17–21, 23–26, 28, 38, 50, 53, 58, 68, 132, 134, 166, 276, 289, 295, 298
tank 38, 49–51, 61, 63, 80–81, 86, 91, 93, 95, 100, 132–33, 141, 153, 167–68, 170, 177, 181, 188, 192, 194, 200, 202–5, 207, 218–19, 222–23, 232, 248, 252, 254–55, 257, 278, 286–88, 292, 294, 296, 303
Tavria 279
TB-3 104
Tenth Army (German) 258, 270, 273
Teplinskii, Boris L'vovich 295
theater of military activities (TVD) 25, 28, 45, 48, 51, 70, 76, 78–80, 83–84, 88–89, 102–5, 107–9, 111, 124, 127–29, 135, 147, 149, 162, 166, 185–87, 208, 214–15, 218–21, 227–31, 240, 255, 257, 279, 297, 299–300, 302–3
theater of war 26, 70, 212, 299
Third Army (French) 165
Third Army (German) 259–60, 262, 264, 273, 281
Thirty Years' War 32
Toledo 241
Trautenau 26
Treaty of Versailles 18, 43, 57, 77, 252, 254, 258
Triandafillov, Vladimir Kiriakovich 19, 78, 287–88, 293
Trubetskoi, Nikolai Iustinovich 302
Trutko, Ivan Ivanovich 290
Tsifer, Richard Stanislavovich 287
Tukhachevskii, Mikhail Nikolaevich 286–91, 293, 295, 297–99, 306

Uborevich, Ieronim Petrovich 288, 294, 298
Ukraine 301
Ukrainian Military District 294
Ulm 24
Upper Silesia 255–56, 258

Vakulich, Pavel Ivanovich 287
Valdemorillo 241
Valencia 246–47
vanguard echelon (AVE) 62, 66, 105–10, 113–25, 135, 149, 151, 292
Varfolomeev, Nikolai Yefimovich 291
Verdun 241
Vesle River 161
Vilnius 128

Vistula River 34–35, 45, 52–53, 66–67, 78, 149, 151–52, 255–56, 259–68, 273–75
Vladimir-Volynskii 274
VLKSM (Komsomol) 44
Volkovysk 128
Voroshilov, Kliment Yefremovich 44, 158, 212, 288

Warsaw 255–56, 260–61, 264–68, 273–76, 281–82
Warta River 262, 265–66, 273
Western Bug River 256, 260, 274–76, 280
Western Europe 238–40, 255, 266

Western Front 36, 160, 191, 200, 243, 247
Western Front (Soviet) 52
Weygand, Maxime 255
Wkra River 35
Wlodawa 275–76
World War I 17–18, 21, 33, 36–38, 40–41, 44–50, 52, 54–55, 66, 77, 80–81, 89, 92–93, 138, 151, 156–57, 160, 162–63, 165–68, 185, 190–91, 196, 200, 202, 236, 238, 242, 244, 247, 249, 278–79, 288, 291, 299, 303
World War II 302
Worth 26–27

Wrangel, Petr Nikolaevich 279
Wusong 80

Yakir, Iona Emmanuilovich 288, 294, 298
Yarchevskii, Petr Grigor'evich 302
Yegorov, Aleksandr Il'ich 290–91, 293–94, 297, 299, 304–5
Yudenich, Nikolai Nikolaevich 161

Zakharov, Matvei Vasil'evich 300
Zamosc 274
Zhukovskii Air Academy 294

www.ingramcontent.com/pod-product-compliance
Ingram Content Group UK Ltd.
Pitfield, Milton Keynes, MK11 3LW, UK
UKHW050703160426
5217IPUK00038B/2044